The Extended Specimen: Emerging Frontiers in Collections-based Ornithological Research

STUDIES IN AVIAN BIOLOGY

A Publication of The American Ornithological Society

www.crcpress.com/browse/series/crcstdavibio

Studies in Avian Biology is a series of works founded and published by the Cooper Ornithological Society in 1978, and published by The American Ornithological Society since 2017. Volumes in the series address current topics in ornithology and can be organized as monographs or multi-authored collections of chapters. Authors are invited to contact the Series Editor to discuss project proposals and guidelines for preparation of manuscripts.

Volume 50
Studies in Avian Biology
American Ornithological Society

AMERICAN ORNITHOLOGICAL
SOCIETY

The Extended Specimen: Emerging Frontiers in Collections-based Ornithological Research

EDITED BY

Michael S. Webster

Cornell Lab of Ornithology
Cornell University
Ithaca, NY

CRC Press
Taylor & Francis Group
Boca Raton London New York

CRC Press is an imprint of the
Taylor & Francis Group, an **informa** business

Cover photo by Mike McDowell: Magnolia Warbler (*Setophaga magnolia*). Pheasant Branch, Wisconsin, May 17, 2014 (spring migration).

PERMISSION TO COPY

CRC Press
Taylor & Francis Group
6000 Broken Sound Parkway NW, Suite 300
Boca Raton, FL 33487-2742

First issued in paperback 2020

© 2018 by American Ornithological Society
CRC Press is an imprint of Taylor & Francis Group, an Informa business

No claim to original U.S. Government works

ISBN-13: 978-1-4987-2915-4 (hbk)
ISBN-13: 978-0-367-65782-6 (pbk)

Library of Congress Cataloging-in-Publication Data

Names: Webster, Michael S., (Ornithologist) editor.
Title: The extended specimen : emerging frontiers in collections-based ornithological research / editor, Michael S. Webster.
Description: Boca Raton : Taylor & Francis, 2017. | Series: Studies in avian biology | Includes bibliographical references.
Identifiers: LCCN 2016059788 | ISBN 9781498729154 (hardback : alk. paper)
Subjects: LCSH: Birds--Collection and preservation. | Birds--Research.
Classification: LCC QL677.7 .R65 2017 | DDC 598.072/3--dc23
LC record available at https://lccn.loc.gov/2016059788

Visit the Taylor & Francis Web site at
http://www.taylorandfrancis.com

and the CRC Press Web site at
http://www.crcpress.com

CONTENTS

CONTRIBUTORS

JOHN BATES
Life Sciences
Field Museum of Natural History
1400 S. Lake Shore Drive
Chicago, IL 60605
jbates@fieldmuseum.org

DAVID A. BLOOM
Museum of Vertebrate Zoology
3101 Valley Life Sciences Building
University of California
Berkeley, CA 94720-3160
dbloom@vertnet.org

KIMBERLY S. BOSTWICK
Cornell University Museum of Vertebrates
Department of Ecology and Evolutionary Biology
Cornell University
Sapsucker Woods Road
Ithaca, NY 14850
ksbostwick@icloud.com

KEVIN J. BURNS
Department of Biology
San Diego State University
5500 Campanile Drive
San Diego, CA 92182-4614
kburns@mail.sdsu.edu

JOSEPH I. BYINGTON
Cornell University Museum of Vertebrates
Department of Ecology and Evolutionary Biology
Cornell University
Ithaca, NY 14853
jib56@cornell.edu

CARLA CICERO
Museum of Vertebrate Zoology
3101 Valley Life Sciences Building
University of California
Berkeley, CA 94720-3160
ccicero@berkeley.edu

SANTIAGO CLARAMUNT
Department of Natural History
Royal Ontario Museum
Toronto, Canada

ELIZABETH P. DERRYBERRY
Department of Ecology and Evolutionary Biology
Tulane University
400 Lindy Bogs
New Orleans, LA 70118
ederrybe@tulane.edu

BRANT C. FAIRCLOTH
Department of Biological Sciences
Museum of Natural Science
202 Life Sciences Building
Louisiana State University
Baton Rouge, LA 70808
brant@faircloth-lab.org

EMMA I. GREIG
Cornell Laboratory of Ornithology
Cornell University
Ithaca, NY 14853
eig9@cornell.edu

ERIC R. GULSON-CASTILLO
Cornell University Museum of Vertebrates
Department of Ecology and Evolutionary Biology
Cornell University
Ithaca, NY 14853
erg57@cornell.edu

ROBERT P. GURALNICK
Department of Natural History
Florida Museum of Natural History
358 Dickinson Hall
University of Florida
Gainesville, FL 32611
rguralnick@flmnh.ufl.edu

SHANNON HACKETT
Life Sciences
Field Museum of Natural History
1400 S. Lake Shore Drive
Chicago, IL 60605
shackett@fieldmuseum.org

TODD ALAN HARVEY
Peabody Museum
Yale University
New Haven, CT
todd.harvey@yale.edu

JACK P. HRUSKA
Department of Ecology and Evolutionary Biology
University of Kansas
Lawrence, KS 66045
jph239@cornell.edu

HELEN F. JAMES
Department of Vertebrate Zoology
National Museum of Natural History
Smithsonian Institution
P.O. Box 37012, MRC 116
Washington, DC 20013-7012
jamesh@si.edu

LEO JOSEPH
Australian National Wildlife Collection
CSIRO National Facilities and Collections
GPO Box 1700
Canberra, ACT 2601 Australia
Leo.Joseph@csiro.au

MICHELLE S. KOO
Museum of Vertebrate Zoology
3101 Valley Life Sciences Building
University of California
Berkeley, CA 94720-3160
mkoo@berkeley.edu

HOLLY L. LUTZ
Department of Surgery
University of Chicago
5841 S. Maryland Avenue
Chicago, IL 60637
hlutz@surgery.bsd.uchicago.edu

and

Department of Population Medicine and
    Diagnostic Sciences College of Veterinary
    Medicine
Cornell University
Ithaca, NY 14853

and

Lab of Ornithology
Cornell University
159 Sapsucker Woods Road
Ithaca, NY 14850

NICHOLAS A. MASON
Department of Ecology and Evolutionary Biology
Cornell University
Corson Hall, 215 Tower Road
Ithaca, NY 14853
nicholas.albert.mason@gmail.com

JOHN E. MCCORMACK
Moore Laboratory of Zoology
Occidental College
1600 Campus Rd
Los Angeles, CA 90041
mccormack@oxy.edu

KEVIN J. MCGRAW
School of Life Sciences
Arizona State University
Tempe, AZ 85287-4501
kevin.mcgraw@asu.edu

ADOLFO G. NAVARRO-SIGÜENZA
Museo de Zoología
Facultad de Ciencias
Universidad Nacional Autónoma de México
Circuito Exterior
Ciudad Universitaria
Mexico City 04510, Mexico
adolfon@ciencias.unam.mx

SOPHIA C. ORZECHOWSKI
Department of Wildlife Ecology and Conservation
University of Florida
Gainesville, FL 32611
sco24@cornell.edu

PEGGY H. OSTROM
Department of Integrative Biology
Ecology, Evolutionary Biology, and Behavior
    Program
Michigan State University Museum
203 Natural Science Building
Michigan State University
East Lansing, MI 48824
ostrom@msu.edu

JAVIER OTEGUI
Department of Natural History
Florida Museum of Natural History
358 Dickinson Hall
University of Florida
Gainesville, FL 32611
javier.otegui@gmail.com

BRET PASCH
Department of Biological Sciences
Northern Arizona University
617 S. Beaver Street
Flagstaff, AZ 86011
bret.pasch@nau.edu

TERESA M. PEGAN
Cornell University Museum of Vertebrates
Department of Ecology and Evolutionary Biology
Cornell University
Ithaca, NY 14853
tmp49@cornell.edu

A. TOWNSEND PETERSON
Biodiversity Institute
University of Kansas
Lawrence, Kansas 66045
town@ku.edu

FLOR RODRÍGUEZ-GÓMEZ
Moore Laboratory of Zoology
Occidental College
1600 Campus Rd
Los Angeles, CA 90041
rodriguezgomez@oxy.edu

LAURA A. RUSSELL
Biodiversity Institute
and Natural History Museum
1345 Jayhawk Boulevard
University of Kansas
Lawrence, KS 66045
larussell@vertnet.org

EDWIN SCHOLES III
Macaulay Library
Johnson Center for Birds and Biodiversity
Sapsucker Woods Road
Ithaca, NY 14850
es269@cornell.edu

ALLISON J. SHULTZ
Department of Organismic and Evolutionary
    Biology
Museum of Comparative Zoology
Harvard University
26 Oxford Street
Cambridge, MA 02138
ashultz@fas.harvard.edu

CAROL L. SPENCER
Museum of Vertebrate Zoology
3101 Valley Life Sciences Building
University of California
Berkeley, CA 94720-3160
atrox@berkeley.edu

MARY C. STODDARD
Department of Ecology and Evolutionary Biology
Princeton University
Princeton, NJ, 08544
mstoddard@princeton.edu

DANIEL B. THOMAS
Institute of Natural and Mathematical Sciences
Massey University
Auckland 0632, New Zealand
d.b.thomas@massey.ac.nz

VASYL V. TKACH
Department of Biology
University of North Dakota
10 Cornell Street STOP
Grand Forks, ND 58202
vasyl.tkach@email.und.edu

WHITNEY L. E. TSAI
Moore Laboratory of Zoology
Occidental College
1600 Campus Rd
Los Angeles, CA 90041
wtsai@oxy.edu

BENJAMIN M. VAN DOREN
Edward Grey Institute of Field Ornithology
Department of Zoology
University of Oxford
South Parks Road
Oxford OX1 3PS UK
bmv25@cornell.edu

MICHAEL S. WEBSTER
Cornell Lab of Ornithology
Cornell University
159 Sapsucker Woods Road
Ithaca, NY 14850
msw244@cornell.edu

JASON D. WECKSTEIN
Academy of Natural Sciences of Drexel University
Ornithology Department
Department of Biodiversity, Earth, and
    Environmental Sciences
Drexel University
1900 Benjamin Franklin Parkway
Philadelphia, PA 19103
jdw342@drexel.edu

JOHN R. WIECZOREK
Museum of Vertebrate Zoology
3101 Valley Life Sciences Building
University of California
Berkeley, CA 94720-3160
tuco@berkeley.edu

ANNE E. WILEY
Department of Biology
University of Akron
185 East Mill Street
Akron, OH 44325-3908
awiley@uakron.edu

DAVID W. WINKLER
Cornell University Museum of Vertebrates
Department of Ecology and Evolutionary Biology
Cornell University
Ithaca, NY 14853
dww4@cornell.edu

ERIC M. WOOD
Department of Biological Sciences
California State University
Los Angeles, CA 90032
ericmwood@calstatela.edu

NATALIE A. WRIGHT
Division of Biological Sciences
University of Montana
32 Campus Dr., HS104
Missoula, MT 59812
nataliestudiesbirds@gmail.com

PREFACE

The dawn of the 21st century has seen a broadening gap between views of ornithological and other biological collections. On the one hand, these collections and the specimens they contain are recognized by many as central to modern-day research ranging from ecology through functional anatomy to genomics. Indeed, dramatic new technical advancements have diversified the types of specimens that we can collect and preserve, and have opened new doors to the types of questions that can be answered with both traditional and new specimen types. Accordingly, to many, this century heralds expanding potential for collections-based ornithological research. On the other hand, the broader public, and even many in the scientific research community, appear to view biological collections as something of a holdover from a bygone age of discovery. Many picture museum collections as arcane repositories of dusty old specimens, and this view has fueled opposition from some to modern specimen collecting. This opposition, in turn, has led to increasing challenges for modern collections, and a seismic shift away from research for many.

The growing disconnect between the perceived and actual research value of ornithological collections led to a symposium, held at the Joint Meeting of the American Ornithologists' Union (AOU) and Cooper Ornithological Society (COS) in August 2013, which was hosted—appropriately enough—by the Field Museum of Natural History in Chicago, Illinois. The goals of that symposium

were to survey the many uses of traditional ornithological specimens, to illustrate new technologies for using them in modern-day research, and to introduce the concept of the "extended specimen"—a constellation of specimen and data types that jointly add research value to each other. At the conclusion of that symposium, I was approached by Brett Sandercock, who at that time was Series Editor for the Studies in Avian Biology series. Brett suggested that the symposium might make a useful contribution to that series, and so the concept for this volume was hatched.

The intent of this volume is very much in line with the original AOU/COS symposium: to introduce both professional and amateur ornithologists to the many modern-day uses of ornithological research collections, and thereby bridge the perception gap regarding their scientific value. The authors of the various chapters hope to inspire creative young minds to use specimens and associated data in exciting new ways to address ever more challenging research questions. At the same time, we hope to encourage and support those in charge of ornithological collections to embrace the concept of the extended specimen, thereby further broadening the research impact of those collections. Happily, many in the collections community have been thinking in this way for years, and have been growing their own collections and/or building strategic partnerships to include new specimen and data types that have historically fallen somewhat outside of most ornithological collections. This volume celebrates those efforts and encourages more. Finally, we aim to help the broader public understand that, rather than being dusty old drawers filled with specimens of questionable research value, ornithological collections are dynamic centers of modern-day research, utilizing cutting-edge methods drawn from across the scientific spectrum to help us understand and preserve the diversity of life on our planet.

This volume would not have been possible without the help and input of many. I wish to thank my co-organizers of the AOU/COS symposium—Kim Bostwick, Edwin Scholes III, and David Winkler—for pulling together that original collection of presentations. Thanks also, of course, to the participants in that original symposium, many of whom are authors of chapters in this volume. The Local Committee, particularly John Bates, was instrumental in making that symposium a reality. Brett Sandercock encouraged us to publish the symposium as a volume in the Studies in Avian Biology series, and Kate Huyvaert picked up the baton to help make that happen when she took over as Series Editor. Thanks to Chuck Crumly at CRC Press for pushing this project along with appropriate carrots and sticks. Most of all, thanks to the many researchers and staff working hard to expand our ornithological collections to meet the opportunities and challenges of research in the 21st century.

MICHAEL S. WEBSTER
Ithaca, New York

CHAPTER ONE

The Extended Specimen*

Michael S. Webster

Abstract. The purpose of biological research collections is to preserve and make accessible specimens that capture the individual phenotype for research on the patterns of biodiversity across taxa, time, and space. In ornithology, the most common specimen type is the "study skin," which primarily captures the outward appearance of the individual. That outward appearance, though, is just one aspect of a complex multidimensional phenotype. The research value of ornithological collections is enhanced when they include specimen types and data that capture other aspects of the phenotype and its underlying genotype. Fortunately, recent advances that make new use of traditional specimens (study skins), as well as new specimen and data types, now make this possible. The "extended specimen" is a constellation of specimen preparations and data types that, together, capture the broader multidimensional phenotype of an individual, as well as the underlying genotype and biological community context from which they were sampled. Ornithological collections are now growing to embrace these specimen and data types, which creates enormous potential for specimen-based research in the 21st century.

Key Words: biodiversity media, ornithology, research collections, specimens.

Ornithological collections preserve and make accessible specimens that capture the phenotypic variation of birds: variation across taxonomic boundaries (species and subspecies), across space (geographic locations and populations), and across time (collecting events). As such, these specimens have been used to explore a host of questions central to understanding the ecology, evolution, and conservation of birds (Chapter 13, this volume). The most typical type of specimen in these ornithological collections is, by far, the "study skin" (Chapter 2, this volume), which is essentially a cotton-stuffed skin that preserves the outward appearance of the bird while discarding most other "skin-in" parts of the bird's anatomy. Numerous studies in the past have used study skins to examine and explore variation in features such as plumage coloration, molt patterns, and size/shape of external anatomical features (e.g., tarsus, wing, or bill). Such studies are sometimes used to define species/subspecies limits (i.e., alpha taxonomy), to explore clinal variation associated with ecology or other factors, to understand the selective factors that have shaped the external anatomy of birds, and to address a broad range of other questions. Accordingly,

---

* Webster, M. S. 2017. The extended specimen. Pp. 1–9 in M. S. Webster (editor), The Extended Specimen: Emerging Frontiers in Collections-based Ornithological Research. Studies in Avian Biology (no. 50), CRC Press, Boca Raton, FL.

study skins have been—and continue to be—an invaluable source of data for ornithological studies (Bates et al. 2004, Joseph 2011, Clemann et al. 2014, Rocha et al. 2014).

Yet these traditional ornithological specimens also are somewhat limited because they do not capture all aspects of the phenotype, which is complex and multidimensional. Over three decades ago, Dawkins (1982) introduced the concept of the *extended phenotype*, his main idea being that genes have phenotypic effects that stretch beyond the body of the individual organism. As Dawkins argued, defining the phenotype as merely the anatomical features of the individual (external or otherwise) is arbitrary and limiting, as that definition excludes other important features of the phenotype. Take, for example, bird nests. These structures all serve essentially the same purpose across birds—to protect the eggs from environmental conditions such as weather and predators—yet the diversity of these structures is stunning (Collias and Collias 1984). Moreover, the nest is a physical manifestation of the parent bird's behavior (Clutton-Brock 1991), is shaped by the interaction of genes and environment (Muth and Healy 2011; Hoi et al. 2012; Hall et al. 2013, 2014), is adapted to local environmental conditions (Crossman et al. 2011, Mainwaring et al. 2014), and evolves over time in response to selective pressures (e.g., Collias 1997, Winkler and Sheldon 1993, Irestedt et al. 2006, Hall et al. 2015). Yet the nest itself lies well outside of the organism's actual body, and hence would be excluded from overly restrictive definitions of the phenotype.

Dawkins's (1982) concept proved highly influential, but also highly controversial, particularly with respect to its strong advocacy for a gene-centric view of evolution that ignores polygenic and gene-by-environment interactions (e.g., see Dawkins 2004, Jablonka 2004, Turner 2004). But the starting premise of Dawkins's assertion remains intact and relevant: the phenotype of an organism, whether it be the product of gene(s) or environment or a combination of the two, is more than just that organism's outward appearance. For example, behavioral attributes such as nest structure, male courtship displays, foraging strategies, and parental care are all manifestations of the individual phenotype, produced by the interaction of genotype and environment, and yet are not anatomical. Similarly,

tissue-specific gene expression, molecular microstructure of feathers, internal anatomy, and physiological processes are all aspects of the phenotype that cannot be easily discerned with the naked eye. Moreover, these phenotypic attributes of the individual organism can also affect the phenotype of other organisms, as when Sociable Weaver (*Philetairus socius*) nesting colonies become the nesting substrate for other birds (Maclean 1973), or when young exhibit traits that increase feeding rates from the parents (e.g., Lyon et al. 1994, Wright and Leonard 2002), even when those parents are a different species (e.g., Kilner et al. 1999).

If the central mission of ornithological collections is to preserve specimens that capture the phenotypic diversity of birds across space and time, then a broader concept is needed with regard to what those specimens are. We need to think more broadly about the multidimensional individual phenotype, and to preserve in our ornithological collections those specimen types, associated assets, and data that capture that phenotype. We need specimen types that preserve phenotypic traits that are not easily discernable from the outward appearance of a bird. We need an *extended specimen* that captures the *extended phenotype* (Figure 1.1).

Fortunately, recent technological and analytical developments now make it possible to collect, catalog, and use the information that comprises the extended specimen. We currently are witnessing a renaissance in collections-based ornithological research that embraces a broader concept for the types of specimens and data that can be preserved in our collections (Chapter 13, this volume). These new technological and analytical approaches have opened the gates to a broad diversity of questions that lay somewhat outside traditional concepts of collections-based research. In some cases, the "new" developments began decades ago and are now accumulating a track record of research successes. In other cases, the new developments are just now getting their start, but hold enormous potential for future research. The various chapters in this volume detail several different areas where ornithological specimens—both traditional and new—are being used in exciting new ways to broaden and deepen our understanding of birds and biodiversity, and to conserve that biodiversity in the face of ever-increasing anthropogenic pressure. This chapter provides a brief introduction to

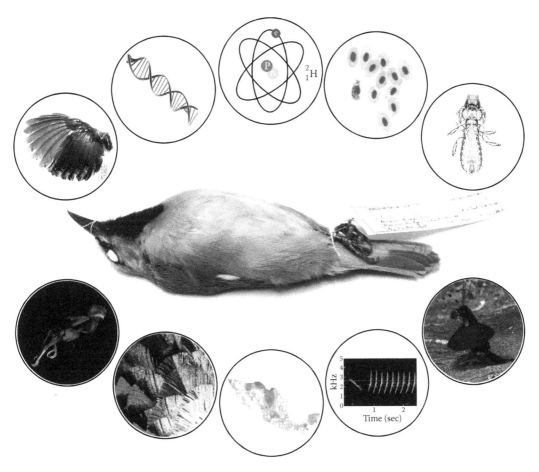

Figure 1.1. The "extended specimen" is a constellation of specimen and data types that, in combination, capture the multidimensional phenotype (and genotype) of an individual. At the center is a traditional "study skin" that captures the external morphology. Across the top are images depicting specimens and extracts that capture other dimensions of the phenotype (left to right): a spread wing, DNA extracted from the study skin or a separately preserved subsample, chemical isotopes (e.g., deuterium) also extracted from the specimen tissue, a malarial parasite among healthy red blood cells, and an ectoparasitic louse collected from a specimen during preparation. Across the bottom are other specimen and data types that capture still other dimensions of the phenotype (left to right): a CT scan of the internal morphology (made from a fluid-preserved specimen), iridescent feathers that can be used for photospecrometric analyses of reflectance, a range map developed from specimen metadata, an acoustic spectrogram of a vocalization, and a frame from a video depicting a courtship display. In principle, most or all of these various specimen and data types could be collected from a single individual (e.g., audio and video recordings could be taken before the bird is collected, and parasites can be collected during specimen preparation), and so would be tied together by their association with the physical study skin and with each other. It is also possible that some specimen/data types would be collected from individuals that are not prepared as physical specimens; for example, audio and video recordings or DNA samples might be collected from individuals from the same population/location but that were not sacrificed. This figure illustrates the conceptual connections between specimen types, rather than actual connections, as in this case the specimens/data depicted were not collected from the same individual. Actual images shown are as follows: the center study skin is a Spectacled Weaver (*Ploceus ocularis suahelicus* FMNH441159; photo by Holly Lutz); the spread wing is from a Striped Wren-Babbler (*Kenopia striata*; photo by Reid Rumelt); the endoparasite is a malarial parasite (*Haemoproteus zosteropis* ex. African Yellow White-eye, *Zosterops senegalensis*; photo by Holly Lutz); the ectoparasite is an avian louse (*Cotingacola lutzae* ex. Cinereous Mourner, *Laniocera hypopyrra*; photo by Michel Valim); the iridescent feathers are from a specimen of a Wahnes's Parotia (*Parotia wahnesi*; photo by Edwin Scholes); the CT scan is from a Black-and-yellow Broadbill (*Eurylaimus ochromalus*; CUMV44530; image courtesy of Eric Gulson); the map depicts avifaunal species turnover in Mexico based on analyses of museum specimen records (reprinted with permission from Peterson et al. 2015); the audio spectrogram is from a recording of a Northern Cardinal (*Cardinalis cardinalis*; Macaulay catalogue #191165, recorded by Wilbur Hershberger in 2012); and the video frame is from a video of a Wahnes's Parotia (Macaulay catalogue #469255; filmed by Edwin Scholes and Tim Laman).

the central concept of this volume: the extended specimen (Figure 1.1).

EXTENDING THE TRADITIONAL SPECIMEN

Recent developments are broadening the utility of ornithological specimens in two key ways. The first is by creating new uses for traditional physical specimens (i.e., study skins). Consider, for example, studies of plumage coloration. Although museum skins have long been used for studies of external morphology and coloration (e.g., West 1962, Rising 1970, Greenberg and Droege 1990), new technological advances have created unprecedented opportunities for addressing exciting new questions in this area (Chapter 3, this volume). These techniques include sophisticated and objective approaches to measuring coloration (reflectance) of feathers that are pigmented or have non-iridescent structural coloration (e.g., Eaton and Lanyon 2003; Stoddard and Prum 2008, 2011) as well as those that are iridescent (Meadows et al. 2011), methods for measuring and decomposing the patterns on birds' eggs (Stoddard and Stevens 2010), and methods for directly examining the pigments found in feathers (e.g., McGraw et al. 2005). A complementary but somewhat different approach is the use of scanning and transmission electron microscopy (SEM and TEM, respectively) to examine melanosomes and color-producing nanostructures in feathers, the morphology and density of which affect the color produced (Prum 2006, Vinther et al. 2008). Remarkably, some of these methods can be applied to fossil specimens as well as to specimens of extant species, leading to discoveries about the coloration of long extinct animals, such as dinosaurs and early birds (e.g., Li et al. 2010, Vinther et al. 2016). Moreover, studies of integument of both extant and extinct species from museum specimens have shown that the physiological mechanisms producing diverse melanin-based coloration in modern birds arose abruptly with the evolution of feathers themselves (Li et al. 2014). In addition to shedding light on the mechanisms and evolution of bird coloration, these methods have allowed researchers to address broader interdisciplinary questions. For example, integrative studies of study skins from museum collections have recently shed light on the processes of coevolution, the degree to which underlying mechanisms affect evolutionary patterns, and the general rules governing the design of signals of individuality (see Maia et al. 2013, Stoddard et al. 2014, Eliason et al. 2015).

Advanced methods for studying plumage coloration are only the tip of the iceberg, as other developing technologies are similarly creating new research uses for traditional study skins. A prime example is extraction of chemicals from small subsamples of a physical specimen, which can be used for isotopic studies that reveal hidden dimensions of the extended phenotype, such as movement patterns and diet (Chapter 6, this volume). Similarly, traditional physical specimens such as study skins can also yield "ancient DNA" for sophisticated population genetic and genomic studies, allowing us to explore phylogenetic relationships and the genotypes that underlie phenotypic diversity at unprecedented levels of detail (Chapter 9, this volume). Finally, advances in the analysis of distributional data, as can be gleaned from the metadata associated with specimens (e.g., on the specimen tag), have transformed our ability to explore animal distributions and how these change over time (Chapter 7, this volume).

The second way that we are now broadening the utility and scope of specimen-based research is through expansion of the types of specimen preparations and other materials that are collected and preserved in ornithological collections. Although specimen types such as skeletons and anatomical (fluid-preserved) specimens have been collected for many decades, the number of these specimens in our collections is dwarfed by the number of traditional study skins (Chapter 2, this volume). New techniques that allow for sophisticated 3D imaging and analysis of anatomy from micro- to macroscales promise to revolutionize the analysis of internal anatomy and portend a dramatic increase in the utility and collection of such specimens (Chapter 2, this volume). Similarly, spread wing preparations are currently rare in research collections, yet these preparation types are extremely useful for analyses that can reveal, among other things, important phenotypic traits such as flight behavior and dispersal ability (Chapter 8, this volume). These various specimen types are rare compared to traditional study skins, yet their value to exploring key patterns of avian biodiversity, now being unlocked by developing new methodologies, cannot be overemphasized.

EXTENDING THE SPECIMEN CONCEPT

The preceding discussion centers on new uses for physical specimen types that have long had a home in ornithological collections, albeit only in small numbers for some specimen types. Just as impactful are other assets and data types that are mostly just now finding their way into ornithological collections as those collections expand the scope of the materials they curate and preserve. Although not "specimens" by the traditional meaning of the word, these assets capture key aspects of the extended phenotype that physical specimens simply cannot. Accordingly, many ornithological collections are now becoming repositories for these materials or are partnering with institutions that have the expertise to handle them.

One prime example is "biodiversity media": audio recordings, video recordings, and still photographs of living animals. Audiovisual recordings of this sort can capture the behavioral phenotype of an individual, including its voice, courtship displays, foraging techniques, and other behaviors that cannot be discerned easily, if at all, from examination of a physical specimen (Figure 1.2). Accordingly, these media can be used for a multitude of studies, such as those on the evolution of acoustic (Chapter 4, this volume) and physical (Chapter 5, this volume) courtship displays. Yet, with a few key exceptions (e.g., the Museum of Vertebrate Zoology, University of California, Berkeley), few traditional ornithological collections currently preserve or curate biodiversity media of this sort. In large part this is because

handling such media requires specialized infrastructure and expertise that differs substantially from that needed to handle physical specimens (Gaunt et al. 2005, Webster and Budney 2016). As a result, most biodiversity media tend to be preserved and curated in collections that specialize on such material, such as the Macaulay Library at the Cornell Lab of Ornithology and the Borror Laboratory at Ohio State University (see Alström and Ranft 2003, Ranft 2004, Webster and Budney 2016). Nonetheless, such media are valuable to research on the extended phenotype, and the institutions that house them are part of a global network of ornithological collections that is expanding in the 21st century.

In addition to standard physical specimens, some ornithological collections also preserve assets that are associated with, but not really a part of, an individual bird. Excellent examples of this are nest and egg collections, such as those found at the Museum of Vertebrate Zoology, the Field Museum in Chicago, and the Peabody Museum of Natural History at Yale University. Nests and eggs are not a part of a bird taken as a specimen, but rather are things that birds produce, and as such they serve as prime examples of the extended phenotype. Eggs and nests also show staggering diversity across space and time, and the collections that house them allow for detailed analyses of that diversity (e.g., Chapter 3,

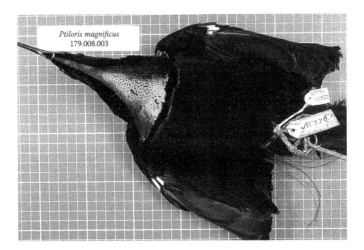

Figure 1.2. Physical specimen (study skin) of a Magnificent Riflebird (*Ptiloris magnificus*) from the Naturalis Biodiversity Center, Leiden, Netherlands. A specimen such as this can yield data on phenotypic traits such as external morphology, coloration, and molt, as well as DNA and chemicals for genetic and isotopic analysis. For "media specimens" that can be used to study the behavioral phenotype of this same species, see http://macaulaylibrary.org/video/455444 and http://macaulaylibrary.org/audio/173711.

this volume). Similarly, some researchers are now beginning to collect endo- and ectoparasites from birds, including from birds collected as specimens, and these invertebrate specimens tell us a good deal about the birds from which they were collected (Chapter 10, this volume). Such samples are typically deposited in appropriate invertebrate or entomological collections, and can be used to study epidemiology, population health, coevolution, and a host of other questions focused on the biology of the avian hosts (as well as the parasites themselves) if linked to the physical specimen with which they are associated.

Another category of "nontraditional" ornithological materials is data and other assets associated with a specific collecting event. These materials can include information and photographs describing the collecting locale, field notes about the collecting event itself, or data about organisms in the area that were not actually collected (e.g., count data and lists of bird species present in the area but not collected). Such data can tell us a good deal about the populations and communities of birds sampled. Although some larger collections have routinely preserved and maintained this information for decades (e.g., collections of field notebooks), the practice is now becoming more routine as scanning and databasing technologies have made the process simpler and more robust. Importantly, the collection and archiving of collecting event data is being facilitated by the recent explosion in observational databases such as eBird (Wood et al. 2011, Sullivan et al. 2014), which allows bird census data to be easily uploaded and archived online, including observational data gathered during a collecting event (Chapter 11, this volume).

Mechanisms like eBird are also lowering barriers to participation in biodiversity data collection, and have spurred an explosion in "citizen science" projects (Bonney et al. 2014). Just as ornithological specimens can provide important insights into bird distributions (Chapter 7, this volume) and how they change over short periods of time (e.g., Peterson et al. 2015), eBird observations can be used to explore the dynamics of bird distributions at contemporary timescales (Sullivan et al. 2009, Wood et al. 2011). Moreover, traditional specimen-based approaches to such studies are by their very nature coarse-grained, because collecting expeditions are complicated and expensive, and so sampling occurs only at broad spatial and temporal scales. Citizen science approaches sidestep this constraint, and can be used for powerful and fine-grained analyses of distributions, including detailed analyses of contemporary change as populations adjust to anthropogenic pressures (Pimm et al. 2014). Ornithological and other biological collections should embrace and work with these "specimen-free" data, as these collections already have expertise in organizing and curating large amounts of data. Indeed, incorporation of specimen-free data into traditional collections is already beginning to happen. Many citizen science projects have originated from and/or are housed at research collections that handle physical specimens: eBird, along with the Macaulay Library and the Cornell Museum of Vertebrates, is housed at the Cornell Lab of Ornithology; and iNaturalist finds its home at the California Academy of Sciences in San Francisco. This trend is likely to expand in the future, as ornithological and other natural history collections expand their own citizen science efforts, harnessing the power of the crowd and blurring the distinction between specimens and data (Chapter 13, this volume).

CONCLUSION: JUST WHAT IS THE "EXTENDED SPECIMEN?"

These are exciting times for specimen-based ornithological research. In recent years, other authors have written about the value of ornithological specimens and the important research opportunities that they bring (e.g., Suarez and Tsutsui 2004, Bates et al. 2004, Joseph 2011, Clemann et al. 2014, Rocha et al. 2014). But the purpose of this brief essay—and the rest of this book—is to illustrate the value of new approaches using specimens to examine the broader extended phenotype of birds and other organisms. New specimen and data types, as well as new uses for traditional specimens, are now being harnessed to explore variation in previously hidden aspects of the avian phenotype. These approaches allow for studies examining the causes and consequences of phenotypic variation in both space and time, including dynamic responses to anthropogenic pressures. The *extended specimen* is the sum of these various specimen and data types, which, together, reveal key aspects of the extended phenotype of the individual (Figure 1.1).

An important goal for the collections community, then, is to embrace these diverse specimen and data types to make our research collections maximally useful for a broad array of future analyses, all aimed at a better understanding and conservation of birds. To do so, though, will require not just application of new methodologies and collection of new specimen types, but also maintaining strong connections across associated specimens and data assets. Case studies given by Mason et al. (Chapter 4, this volume) and Bostwick et al. (Chapter 5, this volume) illustrate well the power that comes from simultaneous analysis of physical specimens and other assets collected from the same populations or individuals. Similarly, coevolutionary and epidemiological studies will require simultaneous examination of bird and symbiont or parasite specimens (Chapter 10, this volume). Given the infrastructure and expertise needed to handle various specimen and data types, it is likely that different samples and assets from an individual bird often will end up being housed in different institutions (Chapter 13, this volume). Accordingly, it will be important to link these different specimen types across the various institutions that house them. Fortunately, recent databasing advances such as VertNet are already facilitating the ability of researchers to locate key specimens from across diverse research collections (Chapter 12, this volume) and can be used to forge such specimen connections across institutional databases.

Ornithological collections are now expanding the types of specimens and data that they preserve, and these specimens and data are being used in powerful new ways. This constellation of specimens and data is the extended specimen that can be used to explore the multidimensional extended phenotypes of birds, to understand the factors that have shaped the diversity of those phenotypes, and to develop strategies to preserve that diversity.

ACKNOWLEDGMENTS

This book and chapter grew out of a symposium held at the Joint Meeting of the American Ornithologists' Union and Cooper Ornithological Society (Field Museum of Natural History, Chicago, Illinois, August 2013). I thank the co-organizers and participants in that symposium for discussions that have shaped the ideas expressed here. In particular, I thank J. Bates, K. Bostwick, G. Budney, C. Cicero, L. Joseph, E. Scholes, M. Shawkey, and D. Winkler for stimulating discussions and comments (although they may not agree with all of the opinions expressed here). Brett Sandercock originally encouraged us to publish the symposium as a volume in the Studies in Avian Biology series, and Kate Huyvaert picked up the baton to help make that happen. Holly Lutz kindly created Figure 1.1. Thanks most of all to the many researchers and staff working to expand our ornithological collections to meet the opportunities and challenges of research in the 21st century.

LITERATURE CITED

Alström, P., and R. Ranft. 2003. The use of sounds in bird systematics, and the importance of bird sound archives. Bulletin of the British Ornithological Club (Suppl.) 123A:114–135.

Bates, J. M., R. C. K. Bowie, D. E. Willard, G. Voelker, and C. Kahindo. 2004. A need for continued collecting of avian voucher specimens in Africa: why blood is not enough. Ostrich 75:187–191.

Bonney, R., J. L. Shirk, T. B. Phillips, A. Wiggins, H. L. Ballard, A. J. Miller-Rushing, and J. K. Parrish. 2014. Next steps for citizen science. Science 343:1436–1437.

Clemann, N., K. M. C. Rowe, K. C. Rowe, T. Raadik, M. Gomon, P. Menkhorst, J. Sumner, D. Bray, M. Norman, and J. Melville. 2014. Value and impacts of collecting vertebrate voucher specimens, with guidelines for ethical collection. Memoirs of Museum Victoria 72:141–151.

Clutton-Brock, T. H. 1991. The evolution of parental care. Princeton University Press, Princeton, NJ.

Collias, N. E. 1997. On the origin and evolution of nest building by passerine birds. Condor 99:253–270.

Collias, N. E., and E. C. Collias. 1984. Nest Building and Bird Behavior. Princeton University Press, Princeton, NJ.

Crossman, C. A., V. G. Rohwer, and P. R. Martin. 2011. Variation in the structure of bird nests between northern Manitoba and southeastern Ontario. PLoS One 6(4):e19086.

Dawkins, R. 1982. The extended phenotype: the long reach of the gene. Oxford University Press, Oxford, UK.

Dawkins, R. 2004. Extended phenotype—but not too extended. A reply to Laland, Turner and Jablonka. Biology and Philosophy 19:377–396.

Eaton, M. D., and S. M. Lanyon. 2003. The ubiquity of avian ultraviolet plumage reflectance. Proceedings of the Royal Society B 270:1721–1726.

Eliason, C. M., R. Maia, and M. D. Shawkey. 2015. Modular color evolution facilitated by a complex nanostructure in birds. Evolution 69:357–367.

Gaunt, S. L. L., D. A. Nelson, M. S. Dantzker, G. F. Budney, and J. W. Bradbury. 2005. New directions for bioacoustics collections. Auk 122:984–987.

Greenberg, R., and S. Droege. 1990. Adaptations to tidal marshes in breeding populations of the swamp sparrow. Condor 92:393–404.

Hall, Z. J., M. Bertin, I. E. Bailey, S. L. Meddle, and S. D. Healy. 2014. Neural correlates of nesting behavior in Zebra Finches (*Taeniopygia guttata*). Behavioural Brain Research 264:26–33.

Hall, Z. J., S. E. Street, S. Auty, and S. D. Healy. 2015. The coevolution of building nests on the ground and domed nests in Timaliidae. Auk 132:584–593.

Hall, Z. J., S. E. Street, and S. D. Healy. 2013. The evolution of cerebellum structure correlates with nest complexity. Biology Letters 9:20130687.

Hoi, H., A. Krištín, F. Valera, and C. Hoi. 2012. Traditional versus non-traditional nest-site choice: alternative decision strategies for nest-site selection. Oecologia 169:117–124.

Irestedt, M., J. Fjeldså, and P. G. P. Ericson. 2006. Evolution of the ovenbird–woodcreeper assemblage (Aves: Furnariidae)—major shifts in nest architecture and adaptive radiation. Journal of Avian Biology 37:260–272.

Jablonka, E. 2004. From replicators to heritably varying phenotypic traits: the extended phenotype revisited. Biology and Philosophy 19:353–375.

Joseph, L. 2011. Museum collections in ornithology: today's record of avian biodiversity for tomorrow's world. Emu 111:i–vii.

Kilner, R. M., D. G. Noble, and N. B. Davies. 1999. Signals of need in parent-offspring communication and their exploitation by the common cuckoo. Nature 397:667–672.

Li, Q., J. A. Clarke, K.-Q. Gao, C.-F. Zhou, Q. Meng, D. Li, L. D'Alba, and M. D. Shawkey. 2014. Melanosome evolution indicates a key physiological shift within feathered dinosaurs. Nature 507:350–353.

Li, Q., K.-Q. Gao, J. Vinther, M. D. Shawkey, J. A. Clarke, L. D'Alba, Q. Meng, D. E. G. Briggs, and R. O. Prum. 2010. Plumage color patterns of an extinct dinosaur. Science 327:1369–1372.

Lyon, B. E., J. M. Eadie, and L. D. Hamilton. 1994. Parental preference selects for ornamental plumage in American coot chicks. Nature 371:240–243.

Maclean, G. L. 1973. The sociable weaver, part 4: predators, parasites and symbionts. Ostrich 44:241–253.

Maia, R., D. R. Rubenstein, and M. D. Shawkey. 2013. Key ornamental innovations facilitate diversification in an avian radiation. Proceedings of the National Academy of Sciences USA 110:10687–10692.

Mainwaring, M. C., D. C. Deeming, C. I. Jones, and I. R. Hartley. 2014. Adaptive latitudinal variation in Common Blackbird *Turdus merula* nest characteristics. Ecology and Evolution 4:841–851.

McGraw, K. J., J. Hudon, G. E. Hill, and R. S. Parker. 2005. A simple and inexpensive chemical test for behavioral ecologists to determine the presence of carotenoid pigments in animal tissues. Behavioral Ecology and Sociobiology 57:391–397.

Meadows, M. G., N. I. Morehouse, R. L. Rutowski, J. M. Douglas, and K. J. McGraw. 2011. Quantifying iridescent coloration in animals: a method for improving repeatability. Behavioral Ecology and Sociobiology 65:1317–1327.

Muth, F., and S. D. Healy. 2011. The role of adult experience in nest building in the Zebra Finch, *Taeniopygia guttata*. Animal Behaviour 82:185–189.

Peterson, A. T., A. G. Navarro-Sigüenza, E. Martínez-Meyer, A. P. Cuervo-Robayo, H. Berlanga, and J. Soberón. 2015. Twentieth century turnover of Mexican endemic avifaunas: landscape change versus climate drivers. Science Advances 1:e1400071.

Pimm, S. L., C. N. Jenkins, R. Abell, T. M. Brooks, J. L. Gittleman, L. N. Joppa, P. H. Raven, C. M. Roberts, and J. O. Sexton. 2014. The biodiversity of species and their rates of extinction, distribution, and protection. Science 344:1246752.

Prum, R. O. 2006. Anatomy, physics, and evolution of avian structural colors. Pp. 295–353 in G. E. Hill and K. J. McGraw (editors), Bird coloration, volume 1. Harvard University Press, Cambridge, MA.

Ranft, R. 2004. Natural sound archives: past, present and future. Anais da Academia Brasileira de Ciências 76:455–465.

Rising, J. D. 1970. Morphological variation and evolution in some North American orioles. Systematic Zoology 19:315–351.

Rocha, L. A., A. Aleixo, G. Allen, F. Almeda, C. C. Baldwin, M. V. Barclay, J. M. Bates, A. M. Bauer, F. Benzoni, C. M. Berns, M. L. Berumen, D. C. Blackburn, S. Blum, F. Bolaños, R. C. Bowie, R. Britz, R. M. Brown, C. D. Cadena, K. Carpenter, L. M. Ceríaco, P. Chakrabarty, G. Chaves, J. H. Choat, K. D. Clements, B. B. Collette, A. Collins, J. Coyne, J. Cracraft, T. Daniel, M. R. de Carvalho, K. de Queiroz, F. Di Dario, R. Drewes, J. P. Dumbacher, A. Engilis Jr., M. V. Erdmann, W. Eschmeyer, C. R. Feldman, B. L. Fisher, J. Fjeldså, P. W. Fritsch, J. Fuchs,

A. Getahun, A. Gill, M. Gomon, T. Gosliner, G. R. Graves, C. E. Griswold, R. Guralnick, K. Hartel, K. M. Helgen, H. Ho, D. T. Iskandar, T. Iwamoto, Z. Jaafar, H. F. James, D. Johnson, D. Kavanaugh, N. Knowlton, E. Lacey, H. K. Larson, P. Last, J. M. Leis, H. Lessios, J. Liebherr, M. Lowman, D. L. Mahler, V. Mamonekene, K. Matsuura, G. C. Mayer, H. Mays Jr., J. McCosker, R. W. McDiarmid, J. McGuire, M. J. Miller, R. Mooi, R. D. Mooi, C. Moritz, P. Myers, M. W. Nachman, R. A. Nussbaum, D. Ó. Foighil, L. R. Parenti, J. F. Parham, E. Paul, G. Paulay, J. Pérez-Emán, A. Pérez-Matus, S. Poe, J. Pogonoski, D. L. Rabosky, J. E. Randall, J. D. Reimer, D. R. Robertson, M. O. Rödel, M. T. Rodrigues, P. Roopnarine, L. Rüber, M. J. Ryan, F. Sheldon, G. Shinohara, A. Short, W. B. Simison, W. F. Smith-Vaniz, V. G. Springer, M. Stiassny, J. G. Tello, C. W. Thompson, T. Trnski, P. Tucker, T. Valqui, M. Vecchione, E. Verheyen, P. C. Wainwright, T. A. Wheeler, W. T. White, K. Will, J. T. Williams, G. Williams, E. O. Wilson, K. Winker, R. Winterbottom, and C. C. Witt. 2014. Specimen collection: an essential tool. Science 344:814–815.

Stoddard, M. C., R. M. Kilner, and C. Town. 2014. Pattern recognition algorithm reveals how birds evolve individual egg pattern signatures. Nature Communications 5:4117.

Stoddard, M. C., and R. O. Prum. 2008. Evolution of avian plumage color in a tetrahedral color space: a phylogenetic analysis of New World buntings. American Naturalist 171:755–776.

Stoddard, M. C., and R. O. Prum. 2011. How colorful are birds? Evolution of the avian plumage color gamut. Behavioral Ecology 22:1042–1052.

Stoddard, M. C., and M. Stevens. 2010. Pattern mimicry of host eggs by the common cuckoo, as seen through a bird's eye. Proceedings of the Royal Society B 277:1387–1393.

Suarez, A. V., and N. D. Tsutsui. 2004. The value of museum collections for research and society. BioScience 54:66–74.

Sullivan, B. L., J. L. Aycrigg, J. H. Barry, R. E. Bonney, N. Bruns, C. B. Cooper, T. Damoulas, A. A. Dhondt, T. Dietterich, A. Farnsworth, D. Fink, J. W. Fitzpatrick, T. Fredericks, J. Gerbracht, C. Gomes, W. M. Hochachka, M. J. Iliff, C. Lagoze, F. A. La Sorte, M. Merrifield, W. Morris, T. B. Phillips, M. Reynolds, A. D. Rodewald, K. V. Rosenberg, N. M. Trautmann, A. Wiggins, D. W. Winkler, W-K. Wong, C. L. Wood, J. Yu, and S. Kelling. 2014. The eBird enterprise: an integrated approach to development and application of citizen science. Biological Conservation 169:31–40.

Sullivan, B. L., C. L. Wood, M. J. Iliff, R. E. Bonney, D. Fink, and S. Kelling. 2009. eBird: a citizen-based bird observation network in the biological sciences. Biological Conservation 142:2282–2292.

Turner, J. S. 2004. Extended phenotypes and extended organisms. Biology and Philosophy 19:327–352.

Vinther, J., D. E. G. Briggs, R. O. Prum, and V. Saranathan. 2008. The color of fossil feathers. Biology Letters 4:522–525.

Vinther, J., R. Nicholls, S. Lautenschlager, M. Pittman, T. G. Kaye, E. Rayfield, G. Mayr, and I. C. Cuthill. 2016. 3D camouflage in an ornithischian dinosaur. Current Biology 26:1–7.

Webster, M. S., and G. F. Budney. 2016. Sound archives and media specimens in the 21st century. Pp. 462–485 in C. H. Brown and T. Riede (editors), Comparative bioacoustic methods eBook. Bentham Science Publishers, Oak Park, IL.

West, W. A. 1962. Hybridization in grosbeaks (Pheucticus) of the Great Plains. Auk 79:399–424.

Winkler, D. W., and F. H. Sheldon. 1993. Evolution of nest construction in swallows (Hirundinidae): a molecular phylogenetic perspective. Proceedings of the National Academy of Sciences USA 90:5705–5707.

Wood, C., B. Sullivan, M. Iliff, D. Fink, and S. Kelling. 2011. eBird: engaging birders in science and conservation. PLoS Biology 9:e1001220.

Wright, J., and M. L. Leonard. 2002. The evolution of begging: competition, cooperation and communication. Springer, New York, NY.

CHAPTER TWO

Getting under the Skin

A CALL FOR SPECIMEN-BASED RESEARCH
ON THE INTERNAL ANATOMY OF BIRDS*

Helen F. James

Abstract. Study of the comparative internal anatomy of birds is undergoing a renaissance, spurred by technological and methodological advances. Our ability to image the soft anatomy and bones in 3D using x-ray computed tomography (CT), magnetic resonance imaging (MRI), and optical surface imaging has opened the door to a wide range of analyses using avian skeletal and anatomical specimens. For anatomical specimens, simple staining techniques that enhance the contrast between different soft tissues, and at the same time raise the opacity of soft tissues to x-rays, enable the simultaneous 3D visualization of skeletal and soft tissue anatomy. Image processing software further allows anatomical features such as individual muscles to be segregated and measured on a computer monitor, without necessitating dissection of the anatomical specimen. Perfusion techniques can allow the vascular or respiratory system to be similarly imaged. For the skeleton, CT and optical scans enable the production of detailed computer models of the bones for biometric and biomechanical studies. Online repositories of morphological image files can make internal anatomy widely accessible. Avian skeletal and anatomical collections are far less comprehensive than traditional study skin collections, yet they represent a wealth of relatively unexplored phenotypic variation in birds. The purpose of this chapter is to review and encourage the use of these techniques in the study of avian phenotypes, emphasizing the various specimen types that can be used as well as the deeper understanding of the ecological and behavioral context of the phenotype that emerges from such studies.

Key Words: anatomical specimens, avian paleontology, computed tomography, functional anatomy, geometric morphometrics, skeletons, spirit collections, 3D imaging.

* James, H. F. 2017. Getting under the skin: a call for specimen-based research on the internal anatomy of birds. Pp. 11–22 in M. S. Webster (editor), The Extended Specimen: Emerging Frontiers in Collections-based Ornithological Research. Studies in Avian Biology (no. 50), CRC Press, Boca Raton, FL.

dvances in instrumentation and methods for studying vertebrate morphology offer exciting new ways to reveal and analyze internal phenotypic variation in birds. Emblematic of these advances are the captivating 3D visualizations of internal anatomy that have recently appeared in a diversity of journals (e.g., Figure 2.1). The primary tools employed to create those images are computed tomography (CT), magnetic resonance imaging (MRI), and laser or other optical surface scanning technologies, combined with software programs for creating 3D images from the scans (Chatham and Blackband 2001, Rosset et al. 2004, Goldman 2007, Marshall and Stutz 2012). These approaches, together with a suite of other techniques, some old and some new, have opened a frontier in our ability to see and study the insides of birds.

My objective in this chapter is to spur ornithologists to adopt these techniques, to incorporate internal anatomical traits more frequently in their study designs, and to collect the anatomical specimens that make these studies possible. The chapter touches on topics in avian biology to which studies of skeletal and anatomical specimens principally contribute, and highlights emerging techniques in imaging, data gathering, and data analysis that can facilitate research using those collections. I also offer brief comments on the readiness of avian skeletal and anatomical collections to fulfill this research mission.

In consonance with the theme of this volume, I omit discussion of several important research areas that admittedly make good use of skeletal and anatomical collections, but that do not aim specifically to understand the phenotype in an evolutionary or ecological context. Thus, I discuss avian paleontology but not zooarchaeology, and I omit ancient biomolecules such as the gene fragments and proteins that are often preserved in skeletal and sometimes anatomical specimens. Good reviews are available on these topics, such as Wiley et al. (Chapter 6, this volume) for stable isotope analysis of avian museum specimens, Wood and De Pietri (2015) for emerging paleo-ornithological techniques including the study of ancient biomolecules, and McCormack et al. (Chapter 9, this volume) for genomic approaches that utilize specimens.

AVIAN ANATOMICAL AND SKELETAL COLLECTIONS

Traditional museum study skin collections lie at the heart of our knowledge about the species and taxonomy of birds, their geographic distributions, plumages, and other external traits. The major scientific collections of birds were built up primarily from the 1880s through the 1960s (Winker 1996, Livezey 2003), a period when much ornithological research effort was devoted to establishing basic information about the systematics and biogeography of the world's birds. The avian study skin was

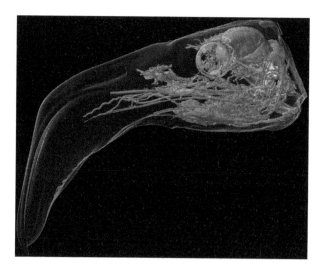

Figure 2.1. Visualization of the cranial vasculature of an American Flamingo (*Phoenicopterus ruber*), created by injecting a latex/barium medium into the vasculature and producing a CT scan. (From Holliday et al. 2006.)

adopted early on as a common unit of comparison for this research, a unit that was also conveniently fungible for curators engaged in trades with other institutions. When, in the 1960s and 1970s, the attention of ornithologists turned increasingly to ecology, the study skin was still the specimen of choice because it preserved external traits of birds that could be compared with those of birds studied in the field.

A consequence of this love for the beautiful study skin is that collectors of birds have preserved far fewer skeletal and anatomical specimens than skins. As an example, at the National Museum of Natural History (NMNH; Washington, DC), the avian holdings fall out as roughly 84% study skins representing 75% of the world's species, but only 11% skeletons representing 54% of the species and 5% anatomical specimens representing 44% of the species; eggs and nests are excluded from these calculations. The skeletal and anatomical holdings at NMNH are the most comprehensive anywhere, yet they continue to lag far behind the study skin collection, even though, in recent decades, the curators at NMNH have placed a priority on closing those gaps (Figure 2.2). The emphasis on study skins for alpha taxonomy is further highlighted by the composition of avian type collections. Again using the NMNH as an example, 3,971 type specimens are in the bird collection, only five of

which are anything other than a study skin. It almost goes without saying that the phenotypic variation that lies beneath the skin of birds has not been studied to nearly the same extent as has the variation observable in study skins, particularly when it comes to intrageneric and intraspecific patterns. Thus, specimens that capture the internal anatomy of birds can be considered as part of the "extended specimen" that is the focus of this book (see Chapter 1, this volume).

COMPARATIVE ANATOMY

The comparative study of avian anatomy, traditionally based on dissection and histology, has fallen into a relatively quiet period in recent decades (Livezey 2003), although certainly some notable work has continued to appear (e.g., Moreno and Carrascal 1993, Patak and Baldwin 1998, Maxwell and Larsson 2007). The rise in interest in avian ecology and behavior, and the supplanting of comparative anatomy with molecular genetics in the field of systematics, must partly account for this. The result has been that fluid-preserved avian specimens have by and large been languishing in their jars, and curators have received very few requests for their use. Yet this situation is beginning to reverse itself, and anatomical specimens are coming back in demand thanks to exciting

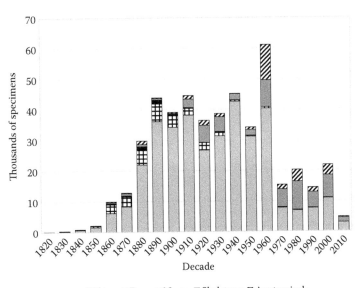

Figure 2.2. Bird specimens added to the scientific collection of the National Museum of Natural History, Smithsonian Institution, by decade and specimen type. The 2010 time bin represents the first half of the decade.

developments in imaging and computer-based analytical methods.

Some recent examples of this trend include studies of the vascular system of the head of flamingoes that revealed a previously unknown paralingual sinus that may be part of the filter feeding mechanism (Holliday et al. 2006), a study of the anatomy of the knee joint of ostriches that detailed the skeletomuscular system associated with their unusual second patella (Chadwick et al. 2014), and a study of the tongue apparatus of waterfowl that characterized anatomical differences between grazers and filter-feeders and used them to interpret key fossil anatids (Li and Clarke 2015). Each of these studies used CT scanning to create 3D anatomical models. In the flamingo study, a barium/latex medium was injected into the carotid artery and jugular vein to produce spectacular scans of the vascular system (Figure 2.1). In the ostrich study, MRI was used to visualize the muscular system and CT to image the bones of the knee joint. Finally, the study of the tongue apparatus in waterfowl imaged both bones and muscles simultaneously, using CT scanning enhanced with iodine-based perfusion staining of the whole specimen. It was then possible to segment the individual muscles using computer models and create a 3D, in situ, virtual dissection (Figure 2.3). This technique is part of a revolution sparked by Metscher (2009), who highlighted the advantages of staining whole fluid-preserved anatomical specimens of small animals with iodine or iodine potassium iodide (I_2KI) to enhance the contrast between soft tissues, such as individual muscles and organs, in CT scans. A mineral stain like iodine also raises the opacity of soft tissues to x-rays, making them similar enough to bone in opacity that bones and soft anatomy can be visualized in a single CT scan. This simple but exciting insight about the advantages of iodine-based staining facilitates simultaneous visualization and precise measurements of the soft anatomy and skeleton in 3D. It opens the door to detailed anatomical and biomechanical investigations of fluid-preserved specimens, with traditional dissection and histology needed only to confirm the computer-based interpretations. Further experiments on iodine-based staining of anatomical specimens for CT scanning have refined the methods and adapted them for use with larger animals (e.g., Gignac and Kley 2014, Li et al. 2015).

SYSTEMATICS OF EXTANT SPECIES

Skeletons and anatomical specimens have long played a strong role in avian systematics. In classic taxonomies, higher categories of birds were arranged in part based on a small number of important anatomical traits, drawn, for example, from the hind-limb myology and configuration of palatal bones (e.g., Fürbringer 1888, Gadow 1892).

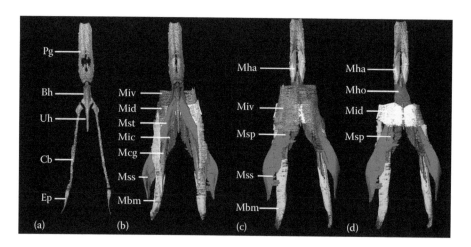

Figure 2.3. Virtual dissection of the tongue apparatus in a Canada Goose (*Branta canadensis*), created by segmenting individual muscles in enhanced-contrast CT scans. Dorsal (a, b) and ventral (c, d) views. Abbreviations: Bh, basihyal; Cb, ceratobranchial; Ep, epibranchial Pg, paraglossal; Uh, urohyal; Mbm, M. branchiomandibularis; Mcg, M. ceratoglossus; Mha, M. hyoglossus anterior; Mho, M. hyoglossus obliquus; Mic, M. interceratobranchialis; Mid, M. intermandibularis dorsalis; Miv, M. intermandibularis ventralis; Msp, M. serpihyoideus; Mss, M. stylohyoideus; Mst, M. stylohyoideus. (From Li and Clarke 2015.)

Other authors took a fine-scale approach, using comparative myology and osteology to inform the classification of all taxa within a group or clade, such as in Raikow's (1993) study of woodcreepers and Bertelli and Giannini's (2005) study of penguins. Despite this deep history of anatomical investigation, the internal anatomy of many groups of birds, for example, in the Passeriformes, has yet to receive very much attention at all. As one example, the notarium—a series of fused thoracic vertebrae independent of the synsacrum—was thought to be absent in Passeriformes until a relatively recent osteological survey revealed these structures to be widespread in the order, including in such familiar taxa as the European Starling (*Sturnus vulgaris*) and the crossbills (*Loxia* spp.; Storer 1982, James 2009). Establishing the systematic distribution of a trait that has evolved repeatedly, like the notarium, can open the door to studies of the trait's function and evolutionary development.

Beginning in the 1970s for anatomical specimens (e.g., Raikow 1977) and somewhat later for skeletons (e.g., Livezey 1986), internal anatomy has been used to build character state matrices for phylogenetic analysis, primarily using the parsimony criterion. The ambitious morphological analysis of higher-level classification of birds by Livezey and Zusi (2006, 2007) is a prime example. However, by now, molecular systematics has understandably supplanted morphology as the main source of data for phylogenetics of modern birds (see Jarvis et al. 2014, Prum et al. 2015). Yet anatomical characters remain relevant in systematics because they are a valuable data partition often used in tandem with molecular data (Fleischer et al. 2001, James et al. 2003), because they enable interpretation of the fossil record (discussed later), and because they reveal evolutionary patterns in the phenotype.

MORPHOLOGY AND BEHAVIOR

Research that extends beyond the boundaries of traditional systematics to place morphology in an ecological or behavioral context can be particularly satisfying because it provides a more holistic explanation of morphological diversity. For example, Bostwick et al. (2012) used CT scanning to document the unique construction of forelimb bones in the Club-winged Manakin (*Machaeropterus deliciosus*), and then related the bony morphology to the evolution of unusual feathers via sexual

selection (see Chapter 5, this volume). Riede et al. (2015) examined the myology and histology of the syrinx in male Pectoral Sandpipers (*Calidris melanotos*), detailing the anatomical specializations in the inflatable esophagus that allow them to produce hooting calls during courtship display. Prince et al. (2011) showed that sexual dimorphism in morphology of the syrinx and upper vocal tract underlies the larger vocal repertoires and more frequent vocalizations of male versus female European Starlings. As a final example, Krilow and Iwaniuk (2015) identified seasonal changes in the volumes of two specific brain regions, the striatopallidal complex and acropallium, that occur in association with drumming behavior in male Ruffed Grouse (*Bonasa umbellus*).

PALEONTOLOGY

The portion of an avian phenotype that is most likely to fossilize is the skeleton, which, as a consequence, has not experienced the same slow rate of use in research as have anatomical specimens. Indeed, avian paleontology—including considerable work with skeleton specimens—has a burgeoning literature that is bringing forth new findings for every major geological period in the history of birds. Some recent highlights include a greatly improved and still expanding Mesozoic fossil record of birds (e.g., Carvalho et al. 2015, Hu et al. 2015), Paleogene fossils from several continents that are particularly informative about the biogeographic and temporal origins of modern bird orders (Mayr 2009), a diverse Neogene record that encompasses extinct avian giants as well as fossils that are clearly relevant to modern genera and species (e.g., Olson and Rasmussen 2001, Tambussi et al. 2012), and rich Quaternary island records that are informative about the history of interactions between birds and people (e.g., James 1995, Worthy and Holdaway 2002, Steadman 2006, Meijer et al. 2013).

Phylogenetic placement of fossils relies on comparative osteology, and consequently many osteological character state matrices have been developed primarily to analyze fossils (e.g., Mayr and Clarke 2003, James 2004, Worthy 2009, Smith 2010). There is also great interest in how best to make use of the fossil record to reliably calibrate molecular phylogenetic trees to a timescale (e.g., Ksepka and Clarke 2015, Smith and Ksepka 2015). Figure 2.4 shows the type specimen of *Limnofregata azygosternon*,

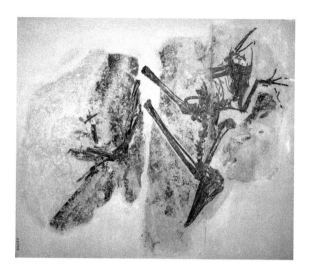

Figure 2.4. Holotype of *Limnofregata azygosternon*, USNM 22753, an Eocene relative of frigatebirds from Wyoming that constrains the minimum age of the split between Fregatidae and Suloidea (including Sulidae, Phalacrocoracidae, and Anhingidae; see Smith and Ksepka 2015).

a fossil relative of frigatebirds that is a minimum of 51.8 million years old, and illustrates the quality of preservation in older fossils that is particularly helpful for calibrating the molecular phylogeny of modern birds.

Once the systematic position and time frame of a fossil have been worked out, its morphology can be further studied to answer diverse questions about the history of avian phenotypes. CT and optical scanning are excellent tools to visualize and model fossils, and are good companions to more traditional techniques like caliper morphometrics and histology. To give just one of many possible examples, the ability to create virtual endocasts of the braincase from CT scans has facilitated comparisons of fossil and modern neuroanatomy (Ksepka et al. 2012, Smith and Clarke 2012, Tambussi et al. 2015).

MORPHOMETRICS

Morphometrics, or the measurement and analysis of linear dimensions, shapes, and volumes, is employed in research on all topics surrounding the internal anatomy of birds. For instance, caliper measurements of the skeleton have long been used to characterize the diversity of avian phenotypes in relation to ecology and evolutionary diversity. A classic example is Hertel's (1994) discriminant function analysis of cranial morphometrics in vultures, in relationship to feeding guilds. If the same study were performed today, we could characterize skull geometry using 3D landmark-based geometric morphometrics, as Kulemeyer et al. (2009) did in their analysis of morphological integration of the neurocranium and bony beak of corvids. Kulemeyer et al.'s study generated several hypotheses about the adaptation of beak shapes and binocular visual fields in corvids, as they relate to pecking compared to probing foraging modes, and raised a possible relationship among skull shape, head posture, and their propensity to take lengthy flights (as indicated by wing tip shape; see Chapter 8, this volume). This is just one example of how the ability to take precise measurements, including geometric measurements based on landmarks and measurements of surface areas and volumes, is leading to a deeper understanding of avian morphology and stimulating new hypotheses to explain the ecological, functional, and developmental context of morphologies.

Geometric morphometric analysis is an umbrella term that encompasses a variety of specific analytical approaches based on landmarks, but it has some general limitations when it comes to birds. Because birds lack teeth and have many fused bones with few open sutures in adulthood, they provide fewer good skeletal landmarks than is typical of other vertebrate groups. It may thus be necessary to resort to interpolated or "semi-landmarks" in birds, as Kulemeyer et al. (2009) did to characterize skull and bill shape. To examine whether pygostyle shape is related to locomotory mode in waterbirds, Felice and O'Connor (2014) compared two-dimensional outlines of the bones using elliptical Fourier analysis, another variant of geometric morphometrics. As a final example, Sievwright and MacLeod (2012) used an eigensurface analysis of the proximal humerus of diurnal raptors (Figure 2.5) to show that the fine-scale geometry of just one end of an important bone can be correlated with flight style and migratory behavior.

Another limitation of geometric morphometric approaches is that they are most appropriate for single bony elements or articulated structures like the skull, and less applicable when the study concerns multiple bony elements that have no fixed geometric relationship to each other, such as the bones in a limb. This helps to explain why basic caliper morphometrics are still in widespread use. Recent examples include a regression analysis of scaling relationships between skeletal measurements and avian body mass (Field et al. 2013), and

a study showing that skeletal limb proportions in bluebirds (*Sialia* spp.) are convergent on those of flycatchers (Corbin et al. 2013).

In the context of morphometrics, I would emphasize that most studies in the past have examined interspecific patterns. With notable exceptions (e.g., McKitrick 1986, Johnston 1990, Rising 2001, Murphy 2007, Krilow and Iwaniuk 2015), intraspecific variation is an aspect of avian biology that has barely been touched upon using internal phenotypic characters. Instead, the common way to characterize intraspecific variation using museum specimens is through analysis of external measurements and plumages of study skins. Arguably, the skeleton offers more and perhaps better landmarks, in the sense that skeletal measurements are likely to have higher repeatability and less variability within an individual. For example, studying skeletal museum specimens can avoid seasonal changes due to the growth and wear of feathers and rhamphotheca or to body mass changes that might add unwanted variance if that is not the topic of study. In addition, new

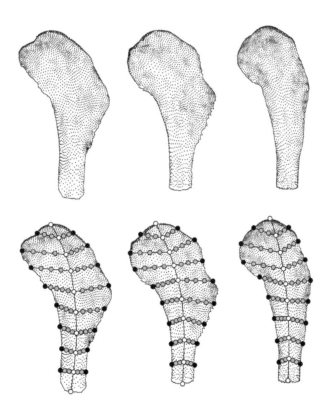

Figure 2.5. Point models of the proximal humeri of diurnal raptors, derived from laser scans (top row), showing the placement of semi-landmarks for eigensurface analysis (bottom row). (From Sievwright and MacLeod 2012.) The species are (left to right) *Falco sparverius*, *Accipiter cooperii*, and *Gyps africanus*. (Copyright: Trustees of the Natural History Museum.)

imaging techniques provide a means to visualize and precisely measure variation in internal traits, including soft and skeletal anatomy, and therefore begin to reveal their relationship to ecological influences at the intraspecific level.

PHYSIOLOGY AND FUNCTIONAL ANATOMY

One of the most spectacular uses of CT imaging of avian anatomy published to date is the study by Düring et al. (2013) on the detailed morphology of the syrinx in the Zebra Finch (*Taeniopygia guttata*). This paper exemplifies how the combination of CT and MRI scans coupled with traditional histology and dissection can reveal anatomical structures and stimulate novel hypotheses about the biomechanical processes that underlie complex behaviors as never before. Another eye-catching example of internal anatomy laid bare is O'Connor and Claessens' (2005) latex cast of the extensive pulmonary air sac system in a typical anatid. The impetus for that study was to extrapolate from avian anatomy to interpret the physiology of dinosaurs, but the image never fails to surprise and impress any ornithologist who has not seen it before.

X-ray videography, in some cases with landmarks in the form of small metal beads surgically implanted, can reveal the internal anatomy in motion in the living bird. For example, Dawson et al. (2011) used this technique to observe the motion of the quadrate during feeding in Mallards (*Anas platyrhynchos*), a motion that had been much written about but not previously studied *in vivo*. Although the action takes place in living birds, such studies rely on specimens to design the research and to interpret the results in an anatomical context. Another biomechanical technique, finite element analysis, is borrowed from engineering and, simply put, models the form under study as a mesh of tetrahedral elements and then uses the model to calculate stresses and strains under different conditions, for instance, to query the biomechanics of pecking in an ostrich skull (Rayfield 2011, Cuff et al. 2015). An intriguing aspect of finite element analysis is its ability to assess the mechanics of alternate morphologies, for instance, by removing a bony process from the skull and asking how the change would affect function (Bright 2014). This method has not been widely applied in ornithology but could prove quite useful in the future.

The aforementioned papers intensively studied one or a few specimens, usually representing a single species, in order to understand avian functional anatomy writ large. Good comparative studies of morphological diversity in the sensory system and organs of birds are also emerging, and these comparative studies tend to make much more extensive use of museum specimens. As examples, Corfield et al. (2015) studied diversity in olfactory bulb size in birds in relationship to potential ecological and behavioral correlates, and Smith and Clarke (2012) studied the association between brain endocast morphology and locomotory mode in Charadriiformes.

EVOLUTION AND DEVELOPMENT

Evolutionary developmental biology is an expanding research area thanks to a suite of improved technologies for studying the genes and mechanisms that influence an animal's development. Laboratory manipulation of model organisms like the chick and the duck continues to be very useful for understanding the embryology of birds (Ealba et al. 2015), but, more recently, a broader, comparative approach is taking hold. The research by Abzhanov and collaborators revealing the developmental mechanisms that determine beak shape in Darwin's Finches (Geospizinae) is a prime example (Abzhanov et al. 2004, 2006; Forster et al. 2008). By examining ontogenetic development in a group like Darwin's Finches, whose ecology has already been intensively studied, Abzhanov et al. were able to integrate knowledge from the two major disciplines of evolutionary developmental biology and ecology, a research interface sometimes termed "eco-evo-devo."

A good review of the modern comparative approach in the field of animal evo-devo is provided by Mallarino and Abzhanov (2012). What is notable for us is that the comparative approach relies strongly on museum specimens. It relies both on establishing the phylogenetic relationships among taxa and on a fine-scale analysis of the morphological differences between them to lay the groundwork for laboratory experiments.

CONCLUSIONS AND FUTURE DIRECTIONS

The instruments, techniques, and computational methods for studying internal vertebrate anatomy are improving in exciting ways. Yet, the internal anatomy of the bird has been underrepresented in research on avian systematics, evolution, and ecology. This underrepresentation is made clear by the state of avian anatomical and skeletal collections, which are very incomplete compared with traditional study skin collections. For many avian taxa, collections are not adequate for studying the internal anatomical traits of birds, and this is particularly true for studies at the specific or subspecific level. Now is an excellent time for ornithology, as a field, to make up for lost time and to more frequently incorporate internal anatomical variation in research designs that address patterns in ecology and evolution. Enabling further research of the sort discussed in this chapter will require that we build up our skeletal and anatomical collections of birds to meet this new level of demand (see also Chapter 11, this volume). High-quality anatomical specimens, in particular, are in short supply, in part because many older specimens have already been dissected. Emerging research areas, such as comparative evo-devo and neuroanatomy, will require special collections that barely exist now.

In reviewing recent literature on avian anatomy, some noteworthy themes and trends emerge. First, the broadly comparative anatomical papers that are being published using new techniques are often motivated by an interest in interpreting the fossil record. Second, but related to the first point, comparative neuroanatomy and sensory system anatomy are growing research themes, and when these studies broadly survey museum specimens, they too are often motivated by a desire to interpret fossils. Third, ornithological collections are supporting a shift in the field of evolutionary developmental biology toward study of nonmodel organisms, but, to date, only a handful of such studies have been published. We can expect growth in this field and will need to enhance collections to support it. Fourth, functional and biomechanical studies use museum specimens but tend to intensively examine only a few specimens rather than to compare a broad selection. And fifth, some impressive work has been recently published that relates internal morphology to behavior or ecology, but, considering that most practicing ornithologists are primarily interested in behavior and ecology, the amount of work being published in this area is relatively little. Hence, I call specifically upon this large group of ornithologists to think about the untapped potential that lies under the skin of birds.

I would also like to reemphasize that the most common units of comparison in studies of avian anatomy are the species, genus, or family. Ornithology has a deep legacy of research on adaptation and speciation that focuses on variation (often geographic variation) at lower taxonomic levels, such as within species and superspecies or across hybrid zones. I perceive an opportunity to capitalize on this legacy of research by extending it to encompass intraspecific variation in internal anatomical traits, which will lead to studies that better relate anatomy, physiology, and development to the processes of adaptation and speciation. In this sense, we could finally follow the lead of Charles Darwin, who analyzed both internal and external variation in domesticated pigeons, ducks, and chickens to build his case for the evolution of species by natural selection (Darwin 1868).

ACKNOWLEDGMENTS

I thank B. K. Schmidt for the collections data shown in Figure 2.2, L. M. Witmer for providing Figure 2.1, Z. Li for providing Figure 2.3, and H. Sievwright for providing Figure 2.5. Permission to reproduce Figures 2.1, 2.3, and 2.5 was granted (in order) by *Anatomical Record*, *Evolutionary Biology*, and *Zoological Journal of the Linnean Society*. I am indebted to M. S. Webster and K. P. Huyvaert for excellent editorial feedback and to an anonymous reviewer for approving the draft.

LITERATURE CITED

Abzhanov, A., W. P. Kuo, C. Hartmann, B. R. Grant, P. R. Grant, and C. J. Tabin. 2006. The calmodulin pathway and evolution of elongated beak morphology in Darwin's finches. Nature 442:563–567.

Abzhanov, A., M. Protas, B. R. Grant, P. R. Grant, and C. J. Tabin. 2004. Bmp4 and morphological variation of beaks in Darwin's Finches. Science 305:1462–1465.

Bertelli, S., and N. P. Giannini. 2005. A phylogeny of extant penguins (Aves: Sphenisciformes) combining morphology and mitochondrial sequences. Cladistics 21:209–239.

Bostwick, K. S., M. L. Riccio, and J. M. Humphries. 2012. Massive, solidified bone in the wing of a volant courting bird. Biology Letters 8:760–763.

Bright, J. A. 2014. A review of paleontological finite element models and their validity. Journal of Paleontology 88:760–769.

Carvalho, I. D., F. E. Novas, F. L. Agnolin, M. P. Isasi, F. I. Freitas, and J. A. Andrade. 2015. A Mesozoic bird from Gondwana preserving feathers. Nature Communications 6:7141.

Chadwick, K. P., S. Regnault, V. Allen, and J. R. Hutchinson. 2014. Three-dimensional anatomy of the ostrich (Struthio camelus) knee joint. PeerJ 2:e706.

Chatham, J. C., and S. J. Blackband. 2001. Nuclear magnetic resonance spectroscopy and imaging in animal research. ILAR Journal 42:189–208.

Corbin, C. E., L., K. Lowenberger, and R. P. Dorkoski. 2013. The skeleton flight apparatus of North American bluebirds (Sialia): phylogenetic thrushes or functional flycatchers? Journal of Morphology 274:909–917.

Corfield, J. R., K. Price, A. N. Iwaniuk, C. Gutierrez-Ibañez, T. Birkhead, and D. R. Wylie. 2015. Diversity in olfactory bulb size in birds reflects allometry, ecology, and phylogeny. Frontiers in Neuroanatomy 9:102.

Cuff, A. R., J. A. Bright, and E. J. Rayfield. 2015. Validation experiments on finite element models of an ostrich (Struthio camelus) cranium. PeerJ 3:e1294.

Darwin, C. 1868. The variation of animals and plants under domestication. Volume I. John Murray, London, UK.

Dawson, M. M., K. A. Metzger, D. B. Baier, and E. L. Brainerd. 2011. Kinematics of the quadrate bone during feeding in mallard ducks. Journal of Experimental Biology 214:2036–2046.

Düring, D. N., A. Ziegler, C. K. Thompson, A. Ziegler, C. Faber, J. Müller, C. Scharff, and C. P. H. Elemans. 2013. The songbird syrinx morphome: a three-dimensional, high-resolution, interactive morphological map of the zebra finch vocal organ. BMC Biology 11:1.

Ealba, E. L., A. H. Jheon, J. Hall, C. Curantz, K. D. Butcher, and R. A. Schneider. 2015. Neural crest-mediated bone resorption is a determinant of species-specific jaw length. Developmental Biology 408:151–163.

Felice, R. N., and P. M. O'Connor. 2014. Ecology and caudal skeletal morphology in birds: the convergent evolution of pygostyle shape in underwater foraging taxa. PLoS One 9:e89737.

Field, D. J., C. Lynner, C. Brown, and S. A. F. Darroch. 2013. Skeletal correlates for body mass estimation in modern and fossil flying birds. PLoS One 8:e82000.

Fleischer, R. C., C. L. Tarr, H. F. James, B. Slikas, and C. E. McIntosh. 2001. Phylogenetic placement of the Po'o-uli, Melamprosops phaeosoma, based on mitochondrial DNA sequence and osteological characters. Studies in Avian Biology 22:98–103.

Forster, D. J., J. Podos, and A. P. Hendry. 2008. A geometric morphometric appraisal of beak shape in Darwin's finches. Journal of Evolutionary Biology 21:263–275.

Fürbringer, M. 1888. Untersuchungen zur morphologie und systematic der vögel. Vols. 1, 2. Von Holkema, Amsterdam.

Gadow, H. 1892. On the classification of birds. Proceedings of the Zoological Society of London 1892:229–256.

Gignac, P. M., and N. J. Kley. 2014. Iodine-enhanced micro-CT imaging: methodological refinements for the study of the soft-tissue anatomy of post-embryonic vertebrates. Journal of Experimental Zoology, Part B–Molecular and Developmental Evolution 322:166–176.

Goldman, L. W. 2007. Principles of CT and CT technology. Journal of Nuclear Medicine Technology 35:115–128.

Hertel, F. 1994. Diversity in body size and feeding morphology within past and present vulture assemblages. Ecology 25:1074–1084.

Holliday, C. M., R. C. Ridgely, A. M. Balanoff, and L. M. Witmer. 2006. Cephalic vascular anatomy in flamingos (Phoenicopterus ruber) based on novel vascular injection and computed tomographic imaging analyses. Anatomical Record 288A:1031–1041.

Hu, D., Y. Liu, J. Li, X. Xu, and L. Hou. 2015. Yuanjiawaornis viriosus, gen. et sp. nov., a large enantiornithine bird from the Lower Cretaceous of western Liaoning, China. Cretaceous Research 55:210–219.

James, H. F. 1995. Prehistoric extinctions and ecological changes on oceanic islands. Ecological Studies 115:88–102.

James, H. F. 2004. The osteology and phylogeny of the Hawaiian finch radiation (Fringillidae: Drepanidini), including extinct taxa. Zoological Journal of the Linnean Society 141:207–255.

James, H. F. 2009. Repeated evolution of fused thoracic vertebrae in songbirds. Auk 126:862–872.

James, H. F., G. P. Ericson, B. Slikas, L. Fu-min, F. Gill, and S. L. Olson. 2003. Pseudopodoces, a misclassified terrestrial tit (Paridae) of the Tibetan Plateau: evolutionary consequences of shifting adaptive zones. Ibis 145:185–202.

Jarvis, E. D., S. Mirarab, A. J. Aberer, B. Li, P. Houde, C. Li, S. Y. W. Ho, B. C. Faircloth, B. Nabholz, J. T. Howard, A. Suh, C. C. Weber, R. R. da Fonseca, J. Li, F. Zhang, H. Li, L. Zhou, N. Narula, L. Liu, G. Ganapathy, B. Boussau, M. S. Bayzid, V. Zavidovych, S. Subramanian, T. Gabaldon, S. Capella-Gutierrez, J. Huerta-Cepas, B. Rekepalli, K. Munch, M. Schierup,

B. Lindow, W. C. Warren, D. Ray, R. E. Green, M. W. Bruford, X. Zhan, A. Dixon, S. Li, N. Li, Y. Huang, E. P. Derryberry, M. F. Bertelsen, F. H. Sheldon, R. T. Brumfield, C. V. Mello, P. V. Lovell, M. Wirthlin, M. P. Cruz Schneider, F. Prosdocimi, J. A. Samaniego, A. M. Vargas Velazquez, A. Alfaro-Nunez, P. F. Campos, B. Petersen, T. Sicheritz-Ponten, A. Pas, T. Bailey, P. Scofield, M. Bunce, D. M. Lambert, Q. Zhou, P. Perelman, A. C. Driskell, B. Shapiro, Z. Xiong, Y. Zeng, S. Liu, Z. Li, B. Liu, K. Wu, J. Xiao, X. Yinqi, Q. Zheng, Y. Zhang, H. Yang, J. Wang, L. Smeds, F. E. Rheindt, M. Braun, J. Fjeldsa, L. Orlando, F. K. Barker, K. A. Jonsson, W. Johnson, K.-P. Koepfli, S. O'Brien, D. Haussler, O. A. Ryder, C. Rahbek, E. Willerslev, G. R. Graves, T. C. Glenn, J. McCormack, D. Burt, H. Ellegren, P. Alstrom, S. V. Edwards, A. Stamatakis, D. P. Mindell, J. Cracraft, E. L. Braun, T. Warnow, W. Jun, M. T. P. Gilbert, and G. Zhang. 2014. Whole-genome analyses resolve early branches in the tree of life of modern birds. Science 346:1320–1331.

Johnston, R. F. 1990. Variation in size and shape in pigeons. Wilson Bulletin 102:213–225.

Krilow, J. M., and A. N. Iwaniuk. 2015. Seasonal variation in forebrain region sizes in male Ruffed Grouse (*Bonasa umbellus*). Brain, Behavior and Evolution 86:176–190.

Ksepka, D. T., A. M. Balanoff, S. Walsh, A. Revan, and A. Ho. 2012. Evolution of the brain and sensory organs in Sphenisciformes: new data from the stem penguin *Paraptenodytes antarcticus*. Zoological Journal of the Linnean Society 1666:202–219.

Ksepka, D. T., and J. A. Clarke. 2015. Phylogenetically vetted and stratigraphically constrained fossil calibrations within Aves. Palaeontologia Electronica 18.1.3FC:1–25.

Kulemeyer, C., K. Asbahr, P. Gunz, S. Frahnert, and F. Bairlein. 2009. Functional morphology and integration of corvid skulls—a 3D geometric morphometric approach. Frontiers in Zoology 6:2.

Li, Z., and J. A. Clarke. 2015. The craniolingual morphology of waterfowl (Aves, Anseriformes) and its relationship with feeding mode revealed through contrast-enhanced x-ray computed tomography and 2D morphometrics. Evolutionary Biology 43:12–25.

Li, Z., J. A. Clarke, R. A. Ketcham, M. W. Colbert, and Y. Fei. 2015. An investigation of the efficacy and mechanism of contrast-enhanced x-ray computed tomography utilizing iodine for large specimens through experimental and simulation approaches. BMC Physiology 15:5.

Livezey, B. C. 1986. A phylogenetic analysis of recent anseriform genera using morphological characters. Auk 103:737–754.

Livezey, B. C. 2003. Avian spirit collections: attitudes, importance, and prospects. Bulletin of the British Ornithologists' Club 123A:35–51.

Livezey, B. C., and R. L. Zusi. 2006. Higher-order phylogeny of modern birds (Theropoda, Aves: Neornithes) based on comparative anatomy: I. methods and characters. Bulletin of Carnegie Museum of Natural History 37:1–544.

Livezey, B. C., and R. L. Zusi. 2007. Higher-order phylogeny of modern birds (Theropoda, Aves: Neornithes) based on comparative anatomy. II. analysis and discussion. Zoological Journal of the Linnean Society 149:1–95.

Mallarino, R., and A. Abzhanov. 2012. Paths less traveled: evo-devo approaches to investigating animal morphological evolution. Annual Review of Cell and Developmental Biology 28:743–763.

Marshall, G. F., and G. E. Stutz (editors). 2012. Handbook of optical and laser scanning (2nd ed.). CRC Press/Taylor & Francis Group, Boca Raton, FL.

Maxwell, E. E., and H. C. E. Larsson. 2007. Osteology and myology of the wing of the Emu (*Dromaius novaehollandiae*), and its bearing on the evolution of vestigial structures. Journal of Morphology 268:423–441.

Mayr, G. 2009. Paleogene fossil birds. Springer-Verlag, Berlin, Germany.

Mayr, G., and J. A. Clarke. 2003. The deep divergences of neornithine birds: a phylogenetic analysis of morphological characters. Cladistics 19: 527–553.

McKitrick, M. C. 1986. Individual variation in the flexor cruris lateralis muscle of the Tyrannidae (Aves: Passeriformes) and its possible significance. Journal of Zoology 209:251–270.

Meijer, H. J. M., T. Sutikna, E. W. Saptomo, R. D. Awe, Jatmiko, S. Wasisto, H. F. James, M. J. Morwood, and M. W. Tocheri. 2013. Late Pleistocene-Holocene non-passerine avifauna of Liang Bua (Flores, Indonesia). Journal of Vertebrate Paleontology 33:877–894.

Metscher, B. D. 2009. MicroCT for developmental biology: a versatile tool for high-contrast 3D imaging at histological resolutions. Developmental Dynamics 238:632–640.

Moreno, E., and L. M. Carrascal. 1993. Leg morphology and feeding postures in four *Parus* species—an experimental approach. Ecology 74:2037–2044.

Murphy, M. T. 2007. A cautionary tale: cryptic sexual size dimorphism in a socially monogamous passerine. Auk 124:515–525.

O'Connor, P. M., and L. P. A. M. Claessens. 2005. Basic avian pulmonary design and flow-through ventilation in non-avian theropod dinosaurs. Nature 436:253–256.

Olson, S. L., and P. C. Rasmussen. 2001. Miocene and Pliocene birds from the Lee Creek Mine, North Carolina. Smithsonian Contributions to Paleobiology 90:233–365.

Patak, A. E., and J. Baldwin. 1998. Pelvic limb musculature in the emu Dromaius novaehollandiae (Aves: Struthioniformes: Dromaiidae): adaptations to high-speed running. Journal of Morphology 238: 23–37.

Prince, B., T. Riede, and F. Goller. 2011. Sexual dimorphism and bilateral asymmetry of syrinx and vocal tract in the European Starling (Sturnus vulgaris). Journal of Morphology 272:1527–1536.

Prum, R. O., S. B. Jacob, A. Dornburg, D. J. Field, J. P. Townsend, E. Moriarty Lemmon, and A. R. Lemmon. 2015. A comprehensive phylogeny of birds (Aves) using targeted next-generation DNA sequencing. Nature 526:569–573.

Raikow, R. J. 1993. Structure and variation in the hindlimb musculature of the woodcreepers (Aves, Passeriformes, Dendrocolaptinae). Zoological Journal of the Linnean Society 107:353–399.

Raikow, R. J. 1977. The origin and evolution of the Hawaiian honeycreepers (Drepanididae). Auk 94:331–342.

Rayfield, E. J. 2011. Strain in the ostrich mandible during simulated pecking and validation of specimen-specific finite element models. Journal of Anatomy 218:47–58.

Riede, T., W. Forstmeier, B. Kempenaers, and F. Goller. 2015. The functional morphology of male courtship displays in the Pectoral Sandpiper (Calidris melanotos). Auk 132:65–77.

Rising, J. D. 2001. Geographic variation in size and shape of Savannah Sparrows (Passerculus sandwichensis). Studies in Avian Biology 23:1–65.

Rosset, A., L. Spadola, and O. Ratib. 2004. OsiriX: an open-source software for navigating in multidimensional DICOM images. Journal of Digital Imaging 17:205–216.

Sievwright, H., and N. MacLeod. 2012. Eigensurface analysis, ecology, and modelling of morphological adaptation in the falconiform humerus (Falconiformes: Aves). Zoological Journal of the Linnean Society 165:390–419.

Smith, N. A., and J. A. Clarke. 2012. Endocranial anatomy of the Charadriiformes: sensory system variation and the evolution of wing-propelled diving. PLoS One 7:e49584.

Smith, N. D. 2010. Phylogenetic analysis of Pelecaniformes (Aves) based on osteological data: implications for waterbird phylogeny and fossil calibration studies. PloS One 5:e13354.

Smith, N. D., and D. T. Ksepka. 2015. Five well-supported fossil calibrations within the "Waterbird" assemblage (Tetrapoda, Aves). Palaeontologia Electronica 18.1.7FC:1–21.

Steadman, D. W. 2006. Extinction and biogeography of tropical Pacific birds. University of Chicago Press, Chicago, IL.

Storer, R. W. 1982. Fused thoracic vertebrae in birds: their occurrence and possible significance. Journal of the Yamashina Institute of Ornithology 14:86–95.

Tambussi, C. P., R. de Mendoza, F. J. Degrange, and M. B. Picasso. 2012. Flexibility along the neck of the Neogene terror bird Andalgalornis steulleti (Aves Phorusrhacidae). PLoS One 7:e37701.

Tambussi, C. P., F. J. Degrange, and D. T. Ksepka. 2015. Endocranial anatomy of Antarctic Eocene stem penguins: implications for sensory system evolution in Sphenisciformes (Aves). Journal of Vertebrate Paleontology 35:e981635.

Winker, K. 1996. The crumbling infrastructure of biodiversity: the avian example. Conservation Biology 10:703–707.

Wood, J. R., and V. L. De Pietri. 2015. Next-generation paleornithology: technological and methodological advances allow new insights into the evolutionary and ecological histories of living birds. Auk: Ornithological Advances 132:486–506.

Worthy, T. H. 2009. Descriptions and phylogenetic relationships of two new genera and four new species of Oligo-Miocene waterfowl (Aves: Anatidae) from Australia. Zoological Journal of the Linnean Society 156:411–454.

Worthy, T. H., and R. N. Holdaway. 2002. The lost world of the moa: prehistoric life of New Zealand. Indiana University Press, Bloomington, IN.

CHAPTER THREE

Advanced Methods for Studying Pigments and Coloration Using Avian Specimens[*,†]

*Kevin J. Burns, Kevin J. McGraw, Allison J. Shultz,
Mary C. Stoddard, and Daniel B. Thomas*

Abstract. Advanced analyses of feathers, eggs, and other colorful tissues in ornithology collections are revealing fresh insights into the life histories of birds. Here, we describe the methods used in these studies, including high-performance liquid chromatography, digital photography, Raman spectroscopy, and spectrophotometry. We use case studies from across the diversity of birds and from deep in the fossil record to illustrate method usage, limitations, and other considerations for analyzing museum specimens. Structural colors in feathers and the surface coloration of eggs are particularly emphasized.

Key Words: digital photography, egg pigmentation, high-performance liquid chromatography, hyperspectral imaging, Raman spectroscopy, spectrophotometry, structural coloration.

Colors are essential for the life history strategies of many animals, including birds (Hill and McGraw 2006a,b). Bright and vivid colors often function as visual signals, communicating the quality of an individual to potential mates, rivals, or predators (Hill 1991, Pryke and Griffith 2006, Maan and Cummings 2012). Muted and dark colors also can be important social signals (Møller 1987, Hoi and Griggio 2008, Karubian et al. 2011), or provide camouflage to animals at both lower and higher trophic levels (Götmark 1987, Montgomerie et al. 2001, Charter et al. 2014). These and other fitness benefits have driven the evolution of a remarkable diversity of pigments and color-producing structures among animals (McGraw 2006a,b; Prum 2006). For their intricate mechanisms of color generation, and the do-or-die importance of their colorful displays, birds are particular marvels of coloration. Birds have colorful eggs, eyes, plumage, skin, and scales, and can show substantial variation in coloration across species, populations, and individuals.

The significance and evolution of color variation among birds is often revealed through comparative analyses, and ornithology collections are ideal for large-scale comparisons of color and other phenotypes. The new layers of information added to ornithology specimens through analyses of plumage and eggshell color exemplify the "extended specimen" philosophy. Spectrophotometry, digital photography, chromatography, and Raman spectroscopy provide

* Burns, K. J., K. J. McGraw, A. J. Shultz, M. C. Stoddard, and D. B. Thomas. 2017. Advanced methods for studying pigments and coloration using avian specimens. Pp. 23–55 *in* M. S. Webster (editor), The Extended Specimen: Emerging Frontiers in Collections-based Ornithological Research. Studies in Avian Biology (no. 50), CRC Press, Boca Raton, FL.
† All authors contributed equally to the chapter.

detailed information about the light absorbance properties and pigment compositions of tissues. Accordingly, the development of these sophisticated new methodologies is opening new doors for the use of museum specimens in the analysis of pigmentation and coloration. And yet each technique also has characteristic advantages and disadvantages when studying the colors of feathers, eggshells, or other tissues.

In this chapter, we describe the modern toolkit for studying avian coloration from museum specimens. We begin with spectrophotometry, a fundamental technique for analyzing colorful tissues. From underlying principles to data analysis, we describe the use of spectrophotometry in ornithology collections and identify future areas of research. Spectrophotometry is showcased in special case studies on structural coloration and avian eggshell coloration. Next we discuss advances in digital photography and hyperspectral imaging, which are quickly becoming critical tools for the study of animal coloration. We next transition from surface coloration to the underlying chemistry of coloration: the chromatography section summarizes recent advances in analysis of avian pigmentation. High-performance liquid chromatography is highlighted as the gold standard technique for identifying carotenoids, porphyrins, and other pigments in avian tissues. The next section of the chapter focuses on Raman spectroscopy, a relatively unknown technique in ornithology. The key advantage of laser-based Raman spectroscopy is nondestructive analysis, which can be ideal for precious museum specimens. The final section of the chapter highlights the emergence of new methods, including computer vision algorithms to analyze egg patterns, for the study of eggshell coloration. Such methods can extend to the study of plumage. Collectively, these sections give an overview of the advanced techniques currently used to study avian coloration from museum specimens as well as living animals.

SPECTROPHOTOMETRY

The study of plumage coloration has a long history in ecology and evolutionary biology; until recently most studies have relied on human perception of avian color. Given the subjective nature of human assessments, researchers began arguing that the quantification of reflectance spectra provided a superior measure of avian color than is possible with human observers (e.g., Endler 1990, Johnson et al. 1998). The use of spectrophotometry has become even more important with the acknowledgment that human vision and avian vision differ in two major ways. First, unlike humans who only have three types of retinal cones, birds have four types with different spectral sensitivities (Vorobyev et al. 1998, Cuthill 2006). One of the avian cone types is most active in the ultraviolet (UV) portion of the spectrum, enabling birds to see into a portion of the spectrum that humans cannot perceive (Goldsmith 1980, Cuthill 2006). Second, birds also have oil droplets attached to each of these cone cells that act as long-pass cut-off filters that absorb all wavelengths of light below certain values, thus enhancing birds' discriminatory capabilities (Vorobyev and Osorio 1998, Vorobyev et al. 1998, Vorobyev 2003, Hart and Vorobyev 2005, Cuthill 2006).

Although scientists have appreciated these differences between birds and humans for many years, only recently have studies of plumage coloration incorporated this information through reflectance spectrophotometry (Cuthill 2006, Bennett and Théry 2007). Reflectance spectrophotometry studies have shown that quantification by human vision alone does not provide sufficient information for the study of avian plumage coloration (Bennett et al. 1994, Cuthill et al. 1999, Cuthill 2006, Håstad and Odeen 2008). For example, sexual dichromatism might be much more prevalent than can be appreciated by human perception (Andersson et al. 1998, Cuthill et al. 1999, Eaton 2005, Eaton and Johnson 2007, Burns and Shultz 2012). Humans might be able to accurately measure dichromatism, but only with some types of coloration (i.e., not UV), and it is difficult to predict under what conditions humans can discriminate dichromatic birds (Armenta et al. 2008b, Seddon et al. 2010). Similarly, several studies (e.g., Shultz and Burns 2013) have reported correlations between UV plumage signals and factors such as mating systems and habitat. Thus, studies that rely on human vision alone to investigate potential correlates to avian coloration are ignoring a potentially critical component of the signaling system of birds. Likewise, spectrophotometry provides a more objective and detailed approach to studies of geographic variation and taxonomy than human vision can provide (e.g., Schmitz-Ornés 2006).

Museum collections facilitate these studies by providing the large series of specimens needed—both within and across species—for detailed comparisons. It is often not practical nor possible for one researcher to collect such a series during a reasonable time span. For example, using museum specimens, Eaton (2005) assessed a broad array of birds to provide a general evaluation of the prevalence of sexual dichromatism across all birds, and Burns and Shultz (2012) surveyed nearly all species in one large group (Thraupidae, 372 species) to provide an assessment of the prevalence of UV plumage patches.

Limitations and Considerations for Using Specimens

Museum specimens can be a wonderful resource for filling in sampling gaps or surveying a wide range of species, but special considerations must be kept in mind when measuring coloration. Many studies of bird coloration focus on plumage coloration, but coloration in other body parts, such as bare skin, the bill, legs, or irises, may also play an important role in signaling and behavior (e.g., Murphy et al. 2009). Unfortunately, colors in these soft tissue parts are rarely preserved in museum skin specimens, as the integument dries out and eyes are rarely preserved. Plumage coloration is generally well preserved if specimens are stored in appropriate conditions including protection from insects, light, and damage from other abiotic factors. But even if stored in appropriate conditions, fading and color changes can still occur, particularly in older specimens (Hausmann et al. 2003, McNett and Marchetti 2005, Armenta et al. 2008b, Doucet and Hill 2009). Disagreement exists about the extent of change in coloration under differing storage conditions. For example, Hausmann et al. (2003) found no particular change in the UV part of the spectrum as specimens aged, whereas McNett and Marchetti (2005) found greater degradation of reflectance spectra at shorter wavelengths in older museum specimens. Dissimilarities in the types of changes can be observed in regions colored by different mechanisms (Doucet and Hill 2009). Potential storage and age-related effects are primarily relevant for intraspecific studies, as changes are typically small (Doucet and Hill 2009). By choosing specimens from similar time periods and museums, effects from changes in coloration due to storage can be

minimized. Finally, coloration changes are less severe in younger specimens, and Armenta et al. (2008a) demonstrated that specimens younger than 50 years old have spectral feather measurements similar to those of live birds.

The approach used to capture spectrophotometer measurements has remained largely unchanged for the last decade, and a detailed description of the equipment and setup is reviewed by Andersson and Prager (2006). However, note that iridescent colors, or colors with a hue that is angle-dependent, can be difficult to measure in an accurate and repeatable manner (Meadows et al. 2011). Thus, the conclusions of a study may depend on angle geometry if the angle of reflectance is not properly taken into account (Santos and Lumeij 2007). Nonetheless, by controlling for angle and quantifying maximum reflectance, it is possible to obtain highly repeatable measurements within an individual specimen, even for iridescent colors (Meadows et al. 2011).

Analytical Approaches

After obtaining raw reflectance spectra, a number of different ways exist to analyze the data and extract variables that can be used in studies of evolution, ecology, and behavior. Historically, a popular way to describe colors has been to calculate tristimulus color variables such as hue, saturation, and brightness (Montgomerie 2006), which are extracted directly from the reflectance spectrum. For example, "hue" is calculated as the wavelength of highest reflectance. In terms of color perception, hue depends on the relative stimulation of color cones, so the tristimulus "hue" value may not relate well to the "hue" actually perceived by the receiver, particularly when a reflectance curve has more than one peak (Montgomerie 2006). Depending on which tristimulus metrics are used, it can be very difficult to compare values across patches, much less across studies (Montgomerie 2006, Delhey et al. 2014). Principal component analysis (PCA) can also be used to describe variation across raw reflectance spectra (Montgomerie 2006) without any sensory system assumptions. Although PCA can be a useful way to objectively identify intra- or interspecific variation, it also has several drawbacks, including variation in brightness swamping out other signals, the inability to analyze spectra from different colors due to difficulty interpreting subsequent

PC axes, and lack of comparability across studies of other species or even other patches of the same species (Montgomerie 2006). The benefit to using methods based on raw spectral curves, such as tristimulus metrics or PCA, is that they make no assumptions about a particular visual system. This approach can be useful for studies of color production (e.g., Shawkey and Hill 2005) and systematics (e.g., McKay 2013), and can also be useful when little is known about the visual system of the intended signal receiver(s).

An alternative is to quantify reflectance spectra while incorporating details about the sensory perception of the receiver (Montgomerie 2006). Today it is common for researchers to incorporate models of avian perception when analyzing color. A number of approaches exist. We will discuss the two most common approaches later: chromaticity diagrams or tetrachromatic color space models and receptor-noise discrimination models. Both methods hinge on the calculation of the quantum or photon cone "catch," or total output, of each of the four color receptors (Cuthill 2006, Montgomerie 2006), which requires information about the spectral sensitivities of the four avian color cone types. These models can also include detailed information about oil droplets in the avian color cones, the irradiance spectrum of incident light, and the transmission properties of air and the bird's ocular media (Montgomerie 2006).

Often the spectral sensitivities of the species in question are not known, so researchers may use the most closely related species with this information, such as the Blue Tit (*Cyanistes caeruleus*) in the case of passerines (Hart et al. 2000). Spectral cone sensitivities are thought to be highly conserved, though there are two broad categories for the cone sensitive to the shortest wavelengths (Hart 2001). This cone can either be most sensitive in the UV wavelength range (UVS visual system) or shifted upward toward the violet wavelength range (VS visual system) (Cuthill 2006). Recent work sequencing the SWS1 opsin gene that includes the single nucleotide polymorphism correlated with this sensitivity shift has shown that the UVS/VS visual system has shifted across families in the bird phylogeny multiple times (Ödeen and Håstad 2013) and that the visual system can shift even within a relatively small family of birds (Maluridae; Ödeen et al. 2012). However, few species' visual systems have been physiologically or behaviorally characterized to quantify visual system variation (Kemp et al. 2015). Note that, in addition to the four single-cone types described earlier, birds also have double cones that are thought to be important in the detection of achromatic (luminance) signals, such as motion perception and the detection of pattern, texture, and form (Osorio and Vorobyev 2005, Hart and Hunt 2007). Because chromatic and achromatic cues are likely processed independently in birds (Vorobyev and Osorio 1998, Kelber et al. 2003, Endler and Mielke 2005), often luminance achromatic signals are modeled separately when considering signal contrasts (e.g., Doucet et al. 2007). Once calculated, the quantum cone catches are generally applied in one of two ways: they can be analyzed in avian tetrahedral color space (Goldsmith 1990, Endler and Mielke 2005, Montgomerie 2006, Stoddard and Prum 2008, Kemp et al. 2015, Renoult et al. 2015) or used in calculating discrimination thresholds, such as just noticeable differences (JNDs; Vorobyev and Osorio 1998). These different approaches are reviewed in Kemp et al. (2015) and are described in more detail later.

The avian tetrahedral color space (Endler and Mielke 2005, Endler et al. 2005, Stoddard and Prum 2008) is a kind of chromaticity diagram. For a given color patch, the quantum cone catches are calculated and transformed in a tetrahedron whose vertices represent each of the four photoreceptor classes (Figure 3.1), thus indicating the extent to which each cone class is stimulated. Within the avian tetrahedral color space, the vector drawn from the achromatic center to the color patch measurement can then be described with spherical coordinates. These coordinates include theta (or longitudinal measure) and phi (or latitudinal measure), both proxies for hue; and r (the length of the vector), a proxy for chroma (saturation; Stoddard and Prum 2008). In addition, by plotting all plumage regions from an individual or species together, it is possible to obtain measurements to describe the entire occupied color space using a series of measurements (Figure 3.1). These measures include color span, which is the distance between patches; hue disparity, which is the difference in vector angles; and color volume, which is the volume of the minimum convex polygon occupied by all plumage measurements (Figure 3.1; Stoddard and Prum 2008). One can also calculate the overlap of different

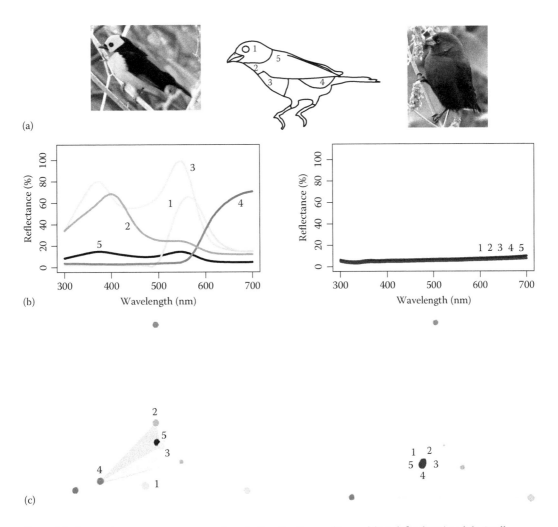

Figure 3.1. Spectrophotometer measurements from the Paradise Tanager (*Tangara chilensis*; left column) and the Small Ground Finch (*Geospiza fuliginosa*; right column). Measurements were taken from the crown (1), throat (2), breast (3), rump (4), and back (5), as depicted on the line drawing (top center). (a) Photographs of the birds show that, from a human perspective, the Paradise Tanager is quite colorful with many different color patches, whereas the Small Ground Finch is quite drab with uniform black plumage. (b) Spectrophotometer measurements show that the reflectance curves from the Paradise Tanager vary across body regions, and are likely produced by different coloration mechanisms that include feather structure (2 and 4), carotenoids (4), melanin (5), and a combination of structure and carotenoids (1). In contrast, the Small Ground Finch, consistent with the human view, has very similar reflectance curves from all regions and are all likely produced by melanin. (c) When plotted in the avian tetrahedral color space (plotted using pavo; Maia et al. 2013a), the Paradise Tanager plumage regions occupy a much greater proportion of the color space and therefore have a much greater color volume (depicted by the gray polygon), color span, and hue disparity. For the Small Ground Finch coloration, on the other hand, the plumage regions all cluster together in the avian tetrahedral color space, and have a small color volume, color span, and hue disparity. (Photos by K. J. Burns.)

color distributions (Stoddard and Stevens 2011) and quantify the full breadth of signaling in taxonomic groups (Stoddard and Prum 2011, Shultz and Burns 2013; see upcoming case study). Ultimately, tetrahedral color space analyses are powerful because they provide a way to make first approximations of color differences, make fewer assumptions than discrimination models (described in the next section), and provide a convenient way to visualize color variation across individuals, species, and broader taxonomic groups.

For determining whether two color stimuli can be discriminated, a second type of visual model is used: a receptor-noise limited model that models color differences in terms of JNDs (Vorobyev and Osorio 1998). The model is based on the idea that color discrimination is limited by noise arising in the photoreceptors, and early experiments showed that behavioral performance could be predicted, under some conditions, based on estimates of noise in each channel (Vorobyev and Osorio 1998, Vorobyev et al. 2001). The model produces a measure of the difference between two color stimuli, measured in JND, where a JND value less than one means that the colors are indistinguishable and a JND greater than one means that the colors can be discriminated. The model is very useful for asking questions about fine-scale color discrimination, which is highly relevant for questions about mimicry, crypsis, and mate choice. The Vorobyev-Osorio JND model is designed to describe color differences at or near threshold (i.e., two very similar colors); however, it is not known how well the model explains suprathreshold color variation. The question of how best to model colors that are quite different remains an open one (Kemp et al. 2015), given that we still know relatively little about neural color processing (Renoult et al. 2015) such as color constancy (Kelber and Osorio 2010). For estimating suprathreshold color differences in the absence of explicit behavioral data, JND measurements are unlikely to provide advantages relative to measurements derived from chromaticity diagrams (Kemp et al. 2015).

The R package, pavo, is useful for performing many of the analyses described in the preceding sections, and can be easily incorporated into scripts to automate analyses across many species (Maia et al. 2013a). The TetraColorSpace program (Stoddard and Prum 2008) in MATLAB® also provides users with a menu of options for analyses in tetrahedral color space.

After measurements are collected from raw reflectance data (tristimulus color variables, quantum cone catches, or avian tetrahedral color space measurements), PCA can be used on these calculated values to identify which components of coloration explain the most variation among individuals (Montgomerie 2006, Delhey et al. 2014) or among species (Mason et al. 2014). These principal component values can also be used in downstream analyses to simplify interpretation and to perform fewer tests on individual variables

to minimize the risk of false positive results (e.g., Mason et al. 2014).

Applications to Ecology and Evolution

Spectrophotometry has been applied to address a variety of questions of ecological and evolutionary importance. These include studies of sexual selection (e.g., Eaton and Johnson 2007, Mason et al. 2014), species delimitation (e.g., Schmitz-Ornés 2006, Maley and Winker 2007), seasonal changes in color (e.g., Tubaro et al. 2005, Barreira et al. 2007, Delhey et al. 2010), age-related social status (e.g., Bridge et al. 2007, Nicolaus et al. 2007), and to assess the relationship between plumage and ecological characteristics (e.g., Friedman et al. 2009, Shultz and Burns 2013, Dunn et al. 2015). Here we highlight a case study to illustrate the types of questions that can be addressed with spectrophotometry of museum specimens.

Plumage color may be shaped over evolutionary time by the particular light environment found in the habitat of a species. For example, selection may favor crypsis, with plumage color evolving to match that of the background (Endler and Théry 1996, Doucet et al. 2007), whereas conspicuousness might be favored in other situations (Marchetti 1993, Gomez and Théry 2007). Species can also experience a variety of light environments within broad habitat characterizations. For example, species found in the same forest habitat that spend more time in the canopy are subject to a different light environment than those in the understory (Gomez and Théry 2007). By objectively quantifying plumage color, spectrophotometry facilitates the study of these diverse selection pressures across species.

Shultz and Burns (2013) addressed the effect of the light environment on plumage color evolution in a group of 44 species of tanagers in the subfamily Poospizinae. They quantified plumage in these species from museum skins, and then mapped aspects of plumage color across a molecular phylogeny. They compared different models of trait evolution with varying degrees of complexity, including models where habitat had no effect, models that compared open versus closed habitats, and models that incorporated foraging strata as well as open versus closed habitats. Plumage was quantified by plotting reflectance spectra in the avian tetrahedral color space (Stoddard and Prum 2008) and extracting values for mean color span, color volume, mean hue disparity, mean chroma (saturation), and mean brilliance (brightness) for each sex of each species. The model-fitting results showed that habitat plays an important role in shaping the plumage of both males and females. Plumage measures of color diversity best fit a model that only included the selective regime of open versus closed habitat, but measures of plumage brightness best fit a model that included foraging strata as well as open versus closed habitat. These results suggest that species within this clade match background contrast and color diversity to increase crypsis, and that the way that this is achieved depends on environmental lighting variables.

Future Directions

As more spectral data are accumulated, we will need more complex methods and models to analyze these types of data in a meaningful way, so that they can be incorporated more broadly into studies of ecology and evolution. When collecting spectral data, it will be essential to archive and annotate the raw data in a manner that will be accessible to researchers in the future (e.g., the Dryad database; White et al. 2015). Finally, while reflectance spectra currently represent the most accurate way to quantify plumage coloration in many circumstances, digital photography (see following section) is becoming more popular and offers several advantages over spectrophotometry, such as the ability to consider not only color but also patterning and patch sizes (Stevens et al. 2007, McKay 2013).

DIGITAL PHOTOGRAPHY AND HYPERSPECTRAL IMAGING

Digital photography is quickly becoming a powerful tool for visual ecologists, due to the ease and speed with which it allows two-dimensional information to be captured and analyzed (Stevens et al. 2007, Pike 2011, Akkaynak et al. 2014). Unlike spectrophotometry, which allows for point-by-point color capture, digital images simultaneously capture color and spatial information. Once photographs are obtained, digital image analysis can be performed, often using custom-designed computer code for analyzing traits of interest. Digital photography has been employed in several recent museum-based studies involving egg

coloration and patterning (Cassey et al. 2010a, 2012a; Stoddard and Stevens 2010; Stoddard et al. 2014; also see the upcoming section "Advanced Methods for Studying Avian Egg Color"). Researchers have also used digital photography in skin collections to investigate the extent to which avian taxa differ in plumage coloration (McKay 2013, McKay et al. 2014), demonstrating the great potential of digital photography as a tool for systematics. Additionally, digital photography has proven useful for the quantification of more complex aspects of plumage appearance, including barring (Gluckman and Cardoso 2009).

To make a camera ready for use in studies of animal coloration, a series of custom calibrations and corrections must be performed (Stevens et al. 2007, 2009). In particular, calibrations typically involve linearizing and equalizing the RGB responses for each channel and determining the specific sensitivities of the camera's different sensors. To study avian colors using digital photography—as in spectrophotometry—it is important to capture light across the entire bird-visible range of wavelengths (approximately 300 to 700 nm). Note that many other animal taxa, including many reptiles, amphibians, fish, and insects, also have ultraviolet sensitivity. Capturing the full visible spectrum using a digital camera is usually achieved by taking one image with a visible pass filter to block the UV and infrared, and a second image through a UV-pass filter, which blocks non-UV wavelengths; these two images can then be combined to cover the full range of wavelengths required. Most cameras contain a UV-blocking filter, which must be removed, and care must be taken to use a lens that can transmit ultraviolet wavelengths. The main drawback to digital photography is that the modification and calibration process can be complex and time-consuming. However, once these steps are completed, a camera can be used to efficiently gather color data, including ultraviolet, in the field and in museum collections, and these data can then be analyzed objectively or with visual models.

New software packages designed for color analysis make digital photography increasingly attractive (Troscianko and Stevens 2015) for color studies. In addition, tools for the analysis of pattern and texture in digital images are becoming increasingly popular. For example, Stoddard et al. (2014) recently developed a pattern recognition and matching tool called NaturePatternMatch (www.naturepatternmatch.org) for the analysis of complex visual signals. The tool uses the scale-invariant feature transform (SIFT; Lowe 1999, 2000), a computer vision algorithm designed to detect informative local features in an image. These features are extracted and then matched across images. NaturePatternMatch is inspired by visual processes believed to be important in vertebrate recognition tasks, though more work needs to be done to establish which model of computer vision most accurately resembles true visual recognition in birds and other animals. Ultimately, NaturePatternMatch can be used to understand aspects of animal signaling, recognition, and camouflage, as well as to explore aspects of avian pattern formation and development.

As museums move to digitize their skin and egg collections, curators are advised to consider using carefully calibrated cameras. At the very least, color charts (such as those made by X-rite, Grand Rapids, MI) should be included as color standards in digital photographs of specimens. Note that although digital photography with proper calibrations can permit objective color measurements and, if combined with visual models, estimates of avian retinal cone stimulation values, it is not possible to reproduce the full reflectance spectrum of a given color, as is the case with a spectrophotometer. In this sense, spectrophotometry and digital photography both have an important and complementary place in the study of avian coloration. To achieve full spectral capture in two or three dimensions, a hyperspectral camera is required.

Hyperspectral cameras, which capture full spectrum information at each pixel in an image, have been developed (Chiao et al. 2011, Kim et al. 2012) and may soon become the gold standard for quantifying avian coloration. Already, hyperspectral imaging has been incorporated into field-based studies of animal coloration and camouflage (Chiao et al. 2011, Russell and Dierssen 2015). However, hyperspectral cameras are expensive and require sophisticated postcapture processing. In the future, it will be critical to develop advanced computational methods for analyzing hyperspectral data in a meaningful and efficient way.

As a final and important point, we strongly urge researchers not to rely on color plates, illustrations, or uncalibrated photographs when addressing questions about avian coloration in the context of signaling and communication. Not only do these media fail to convey information about ultraviolet coloration, they are highly variable, subjective, and sometimes misleading. Carefully controlled spectrophotometry, digital photography, and hyperspectral imaging—applied to specimens in the field or in museum collections—are critical for the correct and rigorous assessment of avian color signals.

CHROMATOGRAPHIC ANALYSES OF BIRD PIGMENTS

Some of the most striking colors displayed by birds are derived from chemical pigments deposited in integumentary tissue. Though we now know that many structural features of avian tissues also can play an integral role in color production (see later), much of the early work on the mechanisms of avian coloration centered on the types of pigments—some of which are endogenously produced and others of which are environmentally acquired—used to generate the array of colors seen in bird feathers and bare parts, including the beak, iris, eye ring, and legs. At least six major classes of avian integumentary colorants exist—carotenoids, melanins, psittacofulvins, porphyrins, pterins, and turacins (Figure 3.2)—and, due to their unique molecular characteristics, a host of biochemical procedures are available to analyze the colorants of birds. Recent technological improvements, including high-performance liquid chromatography (Stradi et al. 1995a) and Raman spectroscopy (see later; Stradi et al. 1995b), have aided in both identifying previously undiscovered compounds and in quantifying amounts of both major and minor forms of these pigment types, so that refined questions can be asked about the control, function, and evolution of avian pigmentation. Museum specimens have served as rich storage depots of material, especially feathers and eggshells (Thomas et al. 2014b, 2015), for extracting and analyzing pigments, and for testing hypotheses about the evolution of pigment-based color mechanisms and how this links, for example, to variation in coloration, ecology, phylogeny, and sexual dichromatism.

Analytical Approaches

Avian integumentary pigments were among the first pigments described in animals, just a few

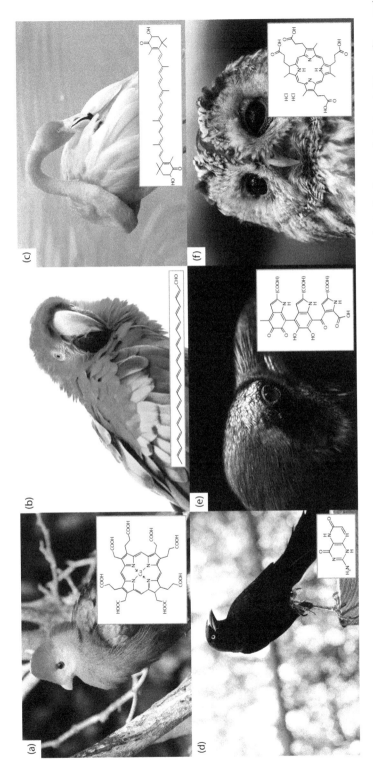

Figure 3.2. Diversity of pigments found in avian integument. (a) Red turacin (molecular structure depicted) and green turacoverdin in the plumage of a Red-crested Turaco (*Tauraco erythrolophus*); (b) red psittacofulvins in the plumage of Scarlet Macaws (*Ara macao*); (c) carotenoid pigment (molecular structure of astaxanthin depicted) in the pink plumage of American Flamingos (*Phoenicopterus ruber*); (d) pterin pigment (molecular structure of xanthopterin depicted) in the yellow iris of a male Great-tailed Grackle (*Quiscalus mexicanus*); (e) melanin pigment (incomplete molecular structure of eumelanin depicted; arrow at top denotes diverse functional groups that can be included) in the black feathers of Common Raven (*Corvus corax*); and (f) porphyrin pigment (molecular structure of coproporphyrin III depicted) in the brown plumage of a Tawny Owl (*Strix aluco*).

decades after carotenoids were described for carrots (Wackenroder 1831). The first biochemical investigations of avian pigments were done over a century ago on the unusual colorants in feathers of turacos (Musophagidae; Church 1869, Gamgee 1895) and parrots (Psittaciformes; Krukenberg 1882; Figure 3.2). These analyses, which predate chromatography techniques (Tswett 1906), were largely restricted to simple chemical testing, such as acid-base-heat reactivity and phosphate precipitation, and general spectral characterization, rather than molecular characterization *per se*. Interestingly, turacos and parrots have been found to be the only groups of animals that harbor these particular pigment classes (turacins and psittacofulvins, respectively); these rare, autapomorphic expressions of pigmentation merit further investigation, especially with respect to their distribution within each taxonomic group (sensu McGraw and Nogare 2005) as well as their molecular and genetic underpinnings.

Once adsorption chromatography became more accessible to chemists in the 20th century (e.g., Strain 1934), it facilitated the identification of particular types of pigments in birds. Different forms of carotenoids, for example, were separated from the plumages of bird groups ranging from woodpeckers (Völker 1934, Test 1969) to bishops (Kritzler 1943), and it was at this time that diet experiments coupled with feather analyses showed that birds could deposit both dietary forms, such as lutein and zeaxanthin, and metabolites, such as picofulvin, into colorful plumage. Adsorption chromatography also permitted the first elucidation of unique fluorescent porphyrins in the rufous-colored plumage of some birds, such as owls and bustards (Figure 3.2; Völker 1938).

Thin-layer chromatography served as the popular method for analyzing bird pigments throughout much of the mid-20th century, with work on carotenoids again grabbing the majority of attention by avian pigment chemists. Fox, Volker, and Brush were instrumental in using this technique to characterize additional feather carotenoids from new avian taxa (e.g., Ciconiiformes, Fox 1962; Cotingidae, Brush 1969) and in enabling biological inquiries such as the pigmentary origin of sexual dichromatism (Brush 1967), interspecific differences in coloration (Troy and Brush 1983, Hudon et al. 1989), intraspecific genetic plumage variation (Brush and Seifried 1968, Brush 1970), and the specific

precursor–product relationships for plumage carotenoids modified from dietary forms (Fox et al. 1969, Brush and Power 1976). Other than egg yolk and skin of chickens (Smith and Perdue 1966), and a few species of game birds (Czeczuga 1979) and wading birds (Fox 1962), soft tissue pigments of birds had been largely ignored. A major finding from this initial work, though, was that the fatty-acid esters of carotenoids, such as astaxanthin in pheasants, can be found in avian skin, indicating that birds may need to stabilize pigments in these bioactive tissues if they are to display them. This era also brought refined chemical analyses of feather porphyrins in owls and bustards, such as the identification of free and esterified forms of coproporphyrin, uroporphyrin, and protoporphyrin (With 1978); and the identification of new compounds including pterins and purines that create white, yellow, orange, and red color in the avian iris (e.g., Oliphant 1987), though hemoglobin and carotenoids can, instead, create these colors in some species (Oliphant 1988).

The invention of high-performance liquid chromatography (HPLC), coupled with improved extraction techniques (Hudon and Brush 1992) and the availability of pure standards isolated from organisms or synthesized chemically, opened the floodgates for easier, less expensive, and more extensive separation of avian integumentary pigments starting in the early 1990s (Hudon 1991, Hamilton 1992) and continuing today (e.g., Prum et al. 2014). Again, the vast majority of research applying HPLC centered on carotenoids. Stradi and colleagues pioneered the use of HPLC in their extensive descriptions of carotenoids in the cardueline finches (Stradi et al. 1995a,b, 1996, 1997) and woodpeckers (Stradi et al. 1998). This foundation of work, coupled with the growing understanding of the diverse biological roles that carotenoids can play in animals (Lozano 1994, Olson and Owens 1998), set the precedent for other groups to apply these methods in other taxa, and to ask specific questions about the ecological, evolutionary, immunological, and behavioral relevance of carotenoid-specific color variation. Nearly 40 different carotenoids have now been described from a few hundred species spanning diverse avian families (reviewed in McGraw 2006b, LaFountain et al. 2015) and including several novel forms (LaFountain et al. 2010). With this large body of information, excellent phylogenetic investigations have been undertaken to trace evolutionary patterns of carotenoid deposition, metabolism, and coloration (Prager and Andersson 2010; Friedman et al. 2014a,b). We now know, for example, that carotenoid pigmentation in plumage has evolved multiple times (as many as 13) across the avian orders, and that over 40% of families have species with carotenoid plumage coloration (Thomas et al. 2014a). Additionally, fine-scale chromatographic work on carotenoids in birds has permitted field ornithological investigations into the dietary limitations of particular carotenoids (McGraw 2006b) as well as physiological experiments on the role of specific carotenoids for boosting immunity (Fitze et al. 2007) or acquiring attractive plumage coloration (Saks et al. 2003).

The HPLC era has also stimulated the development of analytical methods for the two forms of melanin: eumelanin typically creates black and gray tones and pheomelanin typically creates buff and brown colors. This technique was created initially for analysis of mammal skin and hair (Ito and Wakamatsu 2003) and was co-opted for use with bird feathers (Haase et al. 1992, McGraw and Wakamatsu 2004). This preparation method specifically involves the degradation of the two melanin forms into products (pyrrole-2,3,5-tricarboxylic acid and 4-aminohydroxyphenylalanine, respectively) that are analyzed by HPLC (Ito and Wakamatsu 2003). By comparison to carotenoid analyses, these melanin characterizations have been performed on only a few dozen bird species (McGraw 2006a). But from what little has been done, we know that both eumelanin and pheomelanin are present in most melanic plumage colors (McGraw 2004, McGraw et al. 2004). Still, many researchers resort to inferring dominant melanin type from plumage color appearance (Galván and Møller 2013), and this suggests that HPLC techniques need to permeate more studies if we are to attain a deeper understanding of melanin plumage production and evolution at the biochemical level. Some new methods for quantifying melanin have also appeared in recent years (Zhou et al. 2012) that hold promise for more pervasive testing of melanin concentration in feathers, but these methods have only been tested in softer-tissued organisms to date, such as plants and insects (Debecker et al. 2015), and feathers may present a challenge to the extraction procedure.

In the last decade, we have also seen the HPLC-based identification of unique colorants in bird feathers, including both the unique aldehydes (psittacofulvins) that create the red and orange colors of parrot feathers (Stradi et al. 2001, McGraw and Nogare 2005) and the fluorescent, nitrogenous compounds that are responsible for the yellow and orange feathers of penguins (McGraw et al. 2007). HPLC has also recently improved our analyses of iris pterins and purines and facilitated investigation of, for example, sex and age differences in eye colorants of blackbirds (Hudon and Muir 1996). The same is true for eggshell porphyrin pigments as well, including the identification of biliverdin that creates blue and green shell coloration (reviewed in Gorchein et al. 2009).

Benefits and Challenges of Using Specimens

As with museum-based studies of avian morphology, demography, and evolution, for example, museum specimens provide a rich supply of biological material for analysis of pigments across the nearly full range of bird species. This is especially true for the colorants of plumage, as pigments appear to generally be preserved well in feathers of specimens that have been kept in the dark for extended periods of time. Unfortunately, as mentioned earlier for studies of coloration, bare-part colors fade over long periods of time, and this has notably limited our understanding of the distribution and evolution of avian bare-part pigmentation. In a few cases, in fact, pigments and coloration can fade in the feathers of museum skins as well (McNett and Marchetti 2005, Doucet and Hill 2009), so careful attention should be paid, if possible, to learning the preservation history of the skins being analyzed. An important question is, for example, have any specimens ever been used in exhibits and thus exposed to light for some period of time? Careful validation of the specimen colors/pigments under study with those seen in wild birds of the same species (Armenta et al. 2008a) is also critical.

Museum specimens can be precious or delicate, such that destructive sampling should be avoided if possible. Compared to spectrophotometric or photographic studies of coloration in museum specimens, pigment analyses typically incur a greater risk of specimen damage. At present, feathers from a bird skin must be trimmed or plucked for full extraction and characterization of pigments with HPLC. Thus, scientists should be urged to carefully consider the goals of their pigment/coloration study while deciding the best course of action for pigment investigations using museum specimens.

Future Directions

Despite the many recent advances in biochemical analyses of avian integumentary pigments and associated studies of pigment diversity, mechanisms, and evolution in birds, much work still remains to be done. For example, we have not yet identified the integumentary pigments for the vast majority (>90%) of bird species; instead, pigment type has been inferred for a taxon based on reference specimen(s) within that lineage or from either visual estimation or spectral reflectance data (Thomas et al. 2013, Galván and Jorge 2015). Museum specimens could play a key role in a high-throughput pigment screening effort, until one can organize a large call for feather collection from all wild birds currently under study (Smith et al. 2003). For studies where feather removal from a skin is imperative, it may be useful for those preparing specimens to harvest feathers at the time of skinning and separately preserve these alongside the skin with as little modification to the integrity and appearance of the specimen, so that later plucking of feathers from aged/weathered specimens can be avoided. To improve studies of bare-part pigments in bird skins, it would be instructive to consider possible methods that one could employ at capture to preserve/characterize bare-part pigments.

Even among the select group of bird species in which integumentary pigments have been explored, there are many instances of incomplete characterization. The novel pigments in penguin feathers, the yellow psittacofulvins of parrots, and the fluorescent yellow in the down of game bird chicks still require comprehensive elucidation (McGraw et al. 2004). The pigments generating red and yellow plumage colors of adult game birds, such as the Golden and Lady Amherst Pheasant, have also proven particularly challenging analytically. Melanin is the most widespread colorant in the animal kingdom, and is present in all nonwhite structural colors of birds, yet we do not know how types and amounts of eumelanin and pheomelanin

contribute to structural colors or have evolved across a wide range of birds. Last, chromatographically speaking, ultra-performance liquid chromatography (UPLC) has yet to be employed in bird pigment analyses and could enhance the detection of particular integumentary pigment types, for example, isomers and trace levels across Aves.

NONDESTRUCTIVE ANALYSIS WITH RAMAN SPECTROSCOPY

Plumage coloration is a rich source of ecological and evolutionary information and has been studied for many decades using relatively few analytical techniques. As described earlier, the two techniques most commonly used for feather color analysis are spectrophotometry and liquid chromatography, where spectrophotometry provides information about light absorption and reflectance properties (hue, brightness, and saturation), and liquid chromatography can provide deeper insight into pigment chemistry (Kritzler 1943, Dyck 1966). Absorbance and reflectance properties of feathers have proven valuable for analyzing bird health, social behavior, and other ecological and evolutionary parameters (Johnsen et al. 1998, Saks et al. 2003, Andersson and Prager 2006, Montgomerie 2006). Although spectrophotometry has the advantage of being a nondestructive technique (Montgomerie 2006), pigment identification has typically required destructive liquid chromatography analyses (Kritzler 1943, Stradi et al. 1995a). Indeed, liquid chromatography analyses on feather extracts have revealed an array of novel pigments (McGraw 2006b,c). Sample destruction is not always possible for museum specimens however; instead, pigment identification studies in ornithology collections increasingly use nondestructive Raman spectroscopy.

What Is Raman Spectroscopy?

Raman spectroscopy provides information about molecules and minerals in a sample by probing covalent bonds with a laser (Smith and Dent 2005). The instrumentation used for Raman spectroscopy can vary greatly, from a network of open-air mirrors and other equipment, to tiny components nestled inside a scanning electron microscope. More commonly, though, Raman spectroscopy is performed using a modified binocular microscope, which allows for a very simple end-user experience. A sample is placed on the microscope stage and brought into the focal range of the microscope optics. Laser light is channeled through the microscope optics to the sample, and the light that scatters from the sample is channeled back through the optics toward a detector. Information from the detector is interpreted by a computer and used to calculate a Raman spectrum.

Scattered light is the essence of Raman spectroscopy. In brief, laser photons are first focused onto a sample; these photons interact with the sample by stimulating motion, that is, vibrations, between atoms that share covalent bonds. Energy is exchanged between the photons and the vibrating atoms. The photons scatter away from the sample and are channeled to a detector, allowing the energy that the photons have lost to the sample, or gained from the sample, to be calculated. The energy exchanged during the interaction between the laser photons and the sample is presented as a Raman spectrum, where each peak in the spectrum corresponds to a specific motion of covalently bound atoms. Peaks in Raman spectra are identifiable as components of a molecule

or mineral, and chemically distinct structures have characteristic Raman spectra. Hence, Raman spectroscopy is useful for identifying distinct pigments. See Woodward (1967) and Smith and Dent (2005) for nonspecialist introductions to Raman spectroscopy, and Smith and Dent (2005) for a complete technical description with modern equipment.

Comparing Raman Spectroscopy and Spectrophotometry

Raman spectroscopy is still an unfamiliar technique to most avian biologists, whereas ultraviolet-visible spectroscopy (i.e., spectrophotometry) is commonly used. The following description explains the key differences between the two spectroscopy techniques (Figure 3.3).

Regarding spectrophotometry measurements, a light absorption spectrum is often presented as a series of intensity values against wavelength values. Each intensity value reveals the absorption of light at a particular wavelength, and light absorption spectra are intuitive to interpret as they correlate with the visual appearance of an object. A feather may appear orange because it absorbs blue

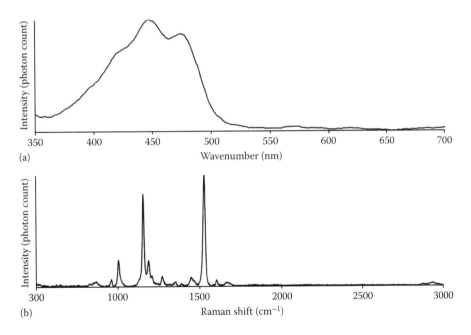

Figure 3.3. (a) UV-visible spectrum and (b) Raman spectrum from a feather pigmented with carotenoids. Spectrophotometry, also known as UV-visible spectroscopy, is routinely used to study the color of feathers and other tissues, whereas Raman spectroscopy is still comparatively rare in ornithology literature. Although spectra are the principal outputs of both spectroscopies, the data from each technique conveys fundamentally different information (see text).

wavelengths of light: a light absorption spectrum of an orange feather would have high intensity values in the blue wavelength range (450–495 nm). Spectrophotometry instruments can also measure reflectance spectra, which are essentially the inverse of an absorption spectrum. A reflectance spectrum of an orange feather would have high intensity values in the orange wavelength range (590–620 nm).

Unlike absorption or reflectance spectra, a Raman spectrum may not correlate with the visual appearance of an object. Instead, interpreting a Raman spectrum requires a moderate understanding of vibrational spectroscopy. A Raman spectrum is often presented as a series of intensity values against wavenumber values (note: not wavelength values), which are measurements of energy. A laser used in Raman spectroscopy will produce photons of a single color (e.g., green, 532 nm), and these photons will therefore all have the same energy value (18,797 wavenumbers, cm^{-1}). When photons interact with a sample they may lose or gain energy; the energy of the photon would become 17,277 cm^{-1} if the photon lost 1,520 cm^{-1} to the sample. The photon would lose 1,520 cm^{-1} if this is the energy "cost" of a vibrational mode within the sample (e.g., stretching of double bonds between carbon atoms). A Raman spectrum reports these energy gains or losses. The wavenumber scale in the Raman spectrum is presented as a Raman shift, where 0 cm^{-1} identifies the energy of the laser (i.e., no shift), and perhaps counterintuitively, positive cm^{-1} values describe energy losses to the sample. Most Raman spectra of molecules have multiple peaks, corresponding to multiple vibrational modes within the molecule. The positions of these peaks are often used as a "chemical fingerprint" for identifying a molecule.

AN OVERVIEW OF RAMAN SPECTROSCOPY AND PLUMAGE PIGMENTS

In the first study to use Raman spectroscopic measurements of pigmented feathers, Stradi et al. (1995b) examined carotenoid pigments in the plumages of eight cardueline finch species, mostly to demonstrate a new method of carotenoid extraction. However, the authors also were interested in the link between the type of carotenoids present in feathers and perceived coloration, and showed that the red and yellow plumage patches of

European Goldfinches (*Carduelis carduelis*) contained the same types of carotenoids. Important color information that is lost during pigment extraction was recovered by studying the pigments *in situ* with Raman spectroscopy (Stradi et al. 1995b). Subtly different Raman spectra were recorded from the red and yellow plumage patches, which both contained canary xanthophyll A and canary xanthophyll B. The Raman spectra from each patch had peaks expected for carotenoids, but the peak positions differed between patches. Stradi et al. (1995b) proposed that the difference in Raman spectra could be explained by "considering the mode of attachment of the carotenoids to different keratin structures." When bound into feathers, the canary xanthophyll molecules likely adopted different conformations, which altered their perceived coloration. Raman spectroscopy is well suited for studying avian pigments *in situ* to understand the relationship between pigment chemistry and plumage color (Stradi et al. 1995b).

"Color-tuning" in plumages refers to chemical interactions that alter the electronic structure of a pigment molecule and therefore change the perceived color. Raman spectroscopy helped reveal that color-tuning could explain the variation in color between red feathers from Scarlet Ibis (*Eudocimus ruber*), orange-red feathers from Summer Tanager (*Piranga rubra*), and violet-purple feathers from White-browed Purpletuft (*Iodopleura isabellae*) (Mendes-Pinto et al. 2012). The same carotenoid (canthaxanthin) was present in the plumages of each bird, and shifts in Raman spectral peaks showed that the pigments were held in subtly different molecular configurations. Likewise, Berg et al. (2013) used Raman spectral data to propose that the interaction between multiple carotenoid molecules (i.e., exciton coupling between chromophores) was a potential explanation for the color variation from "brilliant red to magenta or purple" across rhodoxanthin-pigmented plumages. Color differences between the plumages of two broadbill species studied with Raman spectroscopy (Black-and-red Broadbill, *Cymbirynchus macrorhynchos*; Banded Broadbill, *Eurylaimus javanicus*) has also been attributed to color-tuning: "the polarizing influence of charges nearby the carotenoid, hydrogen bonding, or possibly exciton coupling among neighboring chromophores" (Prum et al. 2014). Although the exact color-tuning mechanism for plumage pigments is still elusive, nondestructive Raman spectroscopy has provided fundamental data that will guide its eventual discovery.

Beyond carotenoids and color-tuning, a diverse range of avian pigments has been analyzed with Raman spectroscopy. New insights into the structures of parrot-specific and penguin-specific pigments have been revealed through in situ analyses, and experiments with biologically ubiquitous melanins have also been reported (Veronelli et al. 1995; Galván et al. 2013a,b; Thomas et al. 2013; Galván and Jorge 2015). Raman spectroscopy has proven to be a useful technique for pigment surveys in ornithology collections.

Applications of Raman Spectroscopy in Museum Collections

Raman spectroscopy was applied to specimens from the National Museum of Natural History, Smithsonian Institution to show that a nondestructive technique could predict the most abundant type of carotenoid in a feather (Thomas et al. 2014b). Fossil feathers preserved in amber were studied with a confocal Raman microscopy instrument in a search for ancient pigments (Thomas et al. 2014c). Raman spectroscopy was also used to expand the known taxonomic distribution of carotenoid plumage pigments (Thomas et al. 2014a). These and other recent reports show the potential of Raman spectroscopy for studies of the specimens found in ornithology collections. Two studies in particular are good platforms for future collections-based research, the color-tuning investigation of Berg et al. (2013), and the carotenoid-type analyses of Thomas et al. (2014b).

Plumage color-tuning mechanisms that involve carotenoid pigments are largely unexplored (Shawkey and Hill 2005). Berg et al. (2013) sought evidence for a color-tuning mechanism in feathers pigmented with the carotenoid rhodoxanthin and presented a set of viable candidates. Spectra in Berg et al. (2013) may help subsequent researchers to find Raman spectral evidence for a particular tuning mechanism, allowing fine-scale selection pressures on plumage color to be studied. Consider that, in addition to the costs of accumulating and displaying carotenoids, some birds invest additional resources to achieve a narrow hue, brightness, and saturation range. The importance of color tuning is well established in many other plant and animal systems (Björn and Ghiradella 2015), and if evidence of a color-tuning mechanism can be discovered in plumage spectra, then it would be possible to study the prevalence of this trait among birds.

Thomas et al. (2014b) also studied Raman spectra collected from carotenoid-pigmented plumage. Carotenoids have a distinctive Raman spectrum that contains three principal peaks that vary slightly in position and intensity for different carotenoids (Veronelli et al. 1995, Thomas et al. 2014b). Raman spectroscopy could therefore be used to taxonomically map plumage carotenoids (Thomas et al. 2014a), revealing associations between lineages and particular types of carotenoids (e.g., Mendes-Pinto et al. 2012).

Raman spectroscopy occupies a valuable analytical niche for pigment research. Like spectrophotometry, Raman spectroscopy is nondestructive and requires no specialized sample preparation. Like liquid chromatography, Raman spectroscopy can be used to chemically identify pigments. The use of Raman spectroscopy for plumage studies has surged recently as researchers have begun to explore the potential of this technique. However, Raman spectroscopy is a potentially valuable technique for all pigmented tissues, not just plumage, and will likely see wider application in coming years.

ADVANCES IN STUDYING STRUCTURAL COLORATION

Avian coloration is produced by a combination of two main mechanisms. The first is the absorption of particular wavelengths of light by pigmentary molecules and analyses of such pigments were described earlier. The second is the differential reflection and refraction of light by biological materials, such as pigmentary and keratin molecules, which is generally referred to as "structural color" (Prum 2006). Structure and pigments cannot operate independently—structural color needs biological molecules like pigments to scatter light (Prum 2006, Shawkey and Hill 2006), and pigmentary color needs structure to reflect the wavelengths of light that are not absorbed (Prum 2006, Shawkey and Hill 2006). However, we can define pigmentary color as color whose reflective properties (e.g., hue) depend primarily on the wavelengths of light not absorbed by a pigment molecule and structural color as color whose reflective properties depend primarily on the light being reflected or refracted by the nanostructures present in the biological material (Prum 2006). Note that it is possible for a biological structure to be made up of the interaction of pigmentary and structural color, as in the case of the plumage of the Budgerigar (*Melopsittacus undulatus*), whose green feathers are a combination of yellow caused by pigments and blue caused by structure (D'Alba et al. 2012). Structural color can be present in avian facial skin, bills, legs, irises, and plumage coloration (Prum and Torres 2003a, Prum 2006).

Mechanisms of Structural Coloration

The mechanisms for the production of structural color can be broadly divided into two types: incoherent and coherent scattering (Prum 2006). Incoherent scattering is characterized by randomly scattered wavelengths of light and is the mechanism behind white, unpigmented feathers (Prum 1999, 2006). For coherent scattering, the light-scattering biological molecules are not randomly distributed, and the wavelengths of light constructively interfere to produce particular colors (Prum et al. 1998; Prum 1999; Prum and Torres 2003a,b), including many of the greens, blues, violet, and ultraviolet colors observed in avian plumage and integumentary structures (Prum 2006; Prum et al. 1998, 2003; Prum and Torres 2003a; Stoddard and Prum 2011). Coherent scattering can be present either in the barbule, producing iridescence, or in the feather barb, generally producing noniridescent colors (Prum et al. 1998, Prum and Torres 2003a, Prum 2006). Within feather barbules, arrays of melanosomes can be arranged in single layers or multilayer crystal-like structures (Prum 2006) and shift the angle of light incidence to produce shifts in color properties, such as hue (Osorio and Ham 2002), which is termed iridescence. One common form of iridescence, that seen on the oil slick-like dark plumage of iridescent members of Icteridae, can be described by thin-film modeling (Shawkey et al. 2006a, Maia et al. 2009), and is produced by a thick layer of keratin on top of a single layer of melanin molecules (Prum 2006). Alternatively, in multilayer arrays, hollow or solid melanosomes can also be arranged in stacks to produce the bright colors such as those observed in hummingbirds (reviewed in Prum 2006).

Noniridescent colors produced in feather barbs are created by quasi-ordered arrays of keratin and air located below the cortex of the feather barb (Prum and Torres 2003a, Prum 2006). The shape of these air-filled channels can be either sphere-like or channel-like (Prum and Torres 2003a, Saranathan et al. 2012), but are uniform in shape and size within a feather barb (Prum and Torres 2003a). The uniform shape of these air-filled channels dictates which wavelength of light will be scattered by any given structure, but, unlike iridescence, the hue will not change with viewing angle (Prum and Torres 2003a). While these arrays produce colors from their physical properties alone, the scattered light can also be partially absorbed by pigments (such as carotenoids or psittacofulvins) that are present in the cortex. This combination of feather structure and pigments can produce hues not created by either alone (D'Alba et al. 2012).

Techniques to Describe Structural Coloration

The colors produced by structural mechanisms can be studied using the photographic or spectrophotometric methods described in previous and subsequent sections. It is also possible to investigate the contributions of different mechanisms to the observed color. To do this, one could remove underlying pigments, like carotenoids, and measure the structure alone (Shawkey and Hill 2005, Jacot et al. 2010). Alternatively, one can saturate feathers in a substance like Cresol (Sigma, St. Louis, MO) that has the same refractive index as keratin to disrupt structural color and measure the pigmentary color alone (Shawkey and Hill 2005).

Additional techniques can be applied to museum specimens to describe the underlying physical structures that produce structural coloration (reviewed by Vukusic and Stavenga 2009). One of the most common techniques is transmission electron microscopy, which can be used to describe the internal nanostructure of feather barbs or barbules (Prum 2006). In the case of iridescent color, the measurements from these images can be applied to single or multilayer thin film models (Prum 2006, Vukusic and Stavenga 2009). The resulting two-dimensional images from quasi-ordered arrays, like those that produce noniridescent structural color in birds, can be described by a Fourier transformation and can predict the shape of the resulting reflectance spectrum (Prum et al. 1998, Prum and Torres 2003a). For example, to obtain a three-dimensional reconstruction of the feather from an Eastern Bluebird (*Sialia sialis*), Shawkey et al. (2009) applied intermediate voltage electron microscopy and more accurately modeled the quasi-ordered structure with a three-dimensional Fourier analysis. Finally, small-angle x-ray scattering can also be used to describe the quasi-ordered structure of noniridescent structural colors (Saranathan et al. 2012).

Applications to Ecology and Evolutionary Biology

Methods used to quantify the nanostructure responsible for the production of structural coloration can be applied to studies of ecology and evolutionary biology. When coupled with museum specimens, these methods have been used to study the anatomical basis for sexual dichromatism (e.g., Shawkey et al. 2005), the mechanism for geographic plumage color differences within a species (e.g., Doucet et al. 2004), and the evolution of iridescent plumage (e.g., Shawkey et al. 2006b) and complex nanostructures (Eliason et al. 2015). Nonetheless, this area remains rife with opportunity.

One study, highlighted here, was conducted by Maia et al. (2013b), and combined transmission electron microscopy, spectrophotometry, and powerful phylogenetic comparative methods to describe how melanosome morphology influenced diversification within the African starlings (Sturnidae). The authors used transmission electron microscopy on feathers from museum specimens to confirm previously described melanosome morphology in at least one species per genus, or more where species were reported to have different morphologies than their closest relatives. They measured reflectance spectra from males and females and analyzed these spectra using the avian color space model (Stoddard and Prum 2008). They then reconstructed the ancestral state of melanosome morphology using reversible-jump Markov chain Monte Carlo, identified the best model of color evolution by comparing models of random evolution (Brownian motion), stabilizing selection (Ornstein-Uhlenbeck processes), or combinations of these models with melanosome type, and estimated diversification rates within the clade. They found that the simple, rod-shaped melanosomes were the ancestral melanosome morphology in the clade, that these melanosomes repeatedly evolved into the more complex morphologies, and that the evolution of these more complex melanosome morphologies not only allowed for a broader area of color space, but that they accelerated the evolution of color differences in coloration between species. Finally, the authors showed that lineages with the more complex melanosome morphologies had faster diversification rates, which has implications for the influence of social signals on lineage diversification. This is just one example of what can be learned by studying a large number of species, an area that is also rife with opportunity, and illustrates the type of project that is greatly facilitated by using the rich resource available from museum skin specimens.

Structural Coloration in Fossil Feathers

While most of what we know about plumage coloration in birds comes from extant species, melanosomes or other pigment molecules can be preserved in fossils. These fossilized molecules can be used to infer likely coloration patterns, including structural coloration (Vinther et al. 2010, Li et al. 2012, Vinther 2015). Structural coloration in fossils was first described in the context of melanosome distribution and morphology to differentiate between eumelanin and pheomelanin (Vinther et al. 2008). In many fossil feathers, the beta-keratin is degraded (Vinther et al. 2010), and so the ability to reconstruct noniridescent structural colors in feather barbs that rely on the organization of keratin and air molecules is limited (Vinther 2015), at least at present. However, the organization of the melanosomes within the barbules can be well-preserved in some cases, and similarities to extant species suggests that some feathers displayed iridescent structural color (Vinther et al. 2010). Using this approach, Li et al. (2012) hypothesized that the feathered dinosaur *Microraptor* likely had predominantly iridescent plumage. The extent of structural coloration is still unknown in early birds and dinosaurs, but is likely to expand in breadth as researchers examine and discover additional well-preserved fossils.

Future Directions

Knowledge of structural coloration has increased dramatically in the last 15 years, but remains a topic open for exploration and study. It will only be through broad surveys of species throughout the avian tree of life, coupled with the ever-increasing knowledge of their genetic relatedness, that we will be able to understand how these mechanisms evolve and to correlate them with life history traits. Museum collections provide a rich resource for completing these surveys by providing the material to broadly sample species throughout the avian tree. Together with spectrophotometer measurements and pigment information, these studies will provide insights into how the gamut of avian coloration evolves in concert with the underlying coloration mechanisms.

Intraspecific differences in feather structure have rarely been studied, but could provide essential information as to how structural coloration might vary within or between populations or individuals. Museum collections also provide a rich resource of material for these types of studies, and can even be used to examine how feather structure might change in a population in historical time.

Finally, many recent advances in the study of structural coloration have come about by collaborations between biologists and physicists. It is only by combining expertise in these areas can we fully understand the basis of these mechanisms and discover previously unknown types of structural coloration.

ADVANCED METHODS
FOR STUDYING AVIAN EGG COLOR

With their striking variation in color and pattern, avian eggs provide a compelling system for investigating the mechanisms and functions of animal coloration. Although they historically have received less research attention than skin collections, egg collections provide a valuable record of life history and behavior, and a rich reservoir of material for researchers (Scharlemann 2001, Kiff and Zink 2005). Consider the collection of eggs at the Natural History Museum in Tring, United Kingdom, which houses over 300,000 clutches and is one of the most comprehensive and actively used egg collections in the world. In recent years, this egg collection has provided the raw material for discoveries related to pigment chemistry (Cassey et al. 2012a), ultraviolet light exposure and solar radiation (Maurer et al. 2015), signal diversity (Cassey et al. 2012b), camouflage (Hanley et al. 2013), egg mimicry (Stoddard and Stevens 2010, 2011), and egg pattern signatures

(Stoddard et al. 2014). New tools for quantifying coloration have helped to usher in a new era of research on avian eggs. Here we briefly introduce the basics of egg coloration and then describe the four main techniques used in museum-based studies—chemical analysis, structural analysis, spectrophotometry, and digital photography—all of which have parallels to the study of plumage and skin coloration.

Egg Coloration: An Overview

The full range of egg coloration appears to stem from just two tetrapyrrole pigments: a red-brown pigment called protoporphyrin and a blue-green pigment called biliverdin (Kennedy and Vevers 1976, McGraw 2006c, Gorchein et al. 2009, Sparks 2011). Both pigments are involved in the biosynthesis of heme, an iron-containing compound important for oxygen transport in the blood stream of vertebrates (Baird et al. 1975). Pigments are deposited on eggshell in the shell gland during the final stages of egg formation, with the bulk of the pigment distributed in the cuticle, an organic layer that typically coats the shell (Hincke et al. 2012). While evidence suggests that biliverdin is produced *de novo* in the shell gland, it is not clear whether the same is true for protoporphyrin, which may be synthesized elsewhere in the body and subsequently mobilized to the shell gland (Sparks 2011). It is also important to note that eggshells that appear white do not necessarily lack pigment, as sometimes protoporphyrin and biliverdin are detected even in white shells (Kennedy and Vevers 1976, Sparks 2011). The glossiness of eggshells, most evident in the highly reflective sheen of many tinamou eggs, results from an extremely smooth cuticle that modifies the appearance of the underlying background color (Igic et al. 2015).

From an evolutionary standpoint, the ancestral egg type was probably white (reviewed in Kilner 2006). However, the detection of both tetrapyrrole pigments in many ratite eggs, including in extinct moa species (Igic et al. 2009), suggests that egg pigments evolved early in birds and are likely to be highly conserved throughout avian evolution. Why are bird eggs colorful? A suite of selective forces likely influences egg appearance, including camouflage, brood parasitism, sexual signaling, thermoregulation, antimicrobial defense, embryonic development, and eggshell strength.

The evolutionary patterns and ecological functions of egg coloration have been reviewed extensively (Underwood and Sealy 2002, Kilner 2006, Reynolds et al. 2009, Cherry and Gosler 2010, Cassey et al. 2011, Maurer et al. 2011, Stoddard et al. 2011, Hauber 2014). We direct readers to the reviews listed here for detailed information, as here we focus on the analysis of egg coloration using museum collections.

Chemical Analysis

As with feather pigments, high-performance liquid chromatography (HPLC) combined with mass spectrometry (MS) is a common technique used for the analysis of eggshell pigments (reviewed in Gorchein et al. 2009). HPLC (usually reverse phase-HPLC) provides a mechanism for separating chemicals along a column; these chemicals are then detected by a mass spectrometer, which provides information about the chemical component's mass and identity. Detailed information about HPLC-MS techniques specific to porphyrins can be found in Lim (2004). With respect to eggshell pigment extraction, researchers typically follow procedures outlined by Mikšík et al. (1996), which involve cutting a small fragment of eggshell and dissolving it in a solution containing methanol and sulfuric acid (Cassey et al. 2012a,b). Some caution should be exercised here, as there are drawbacks to using a methanolic sulfuric acid solution for pigment extraction (Gorchein et al. 2009). However, drawbacks such as the unsuitability of the method for detection of metal compounds may be irrelevant if metal-containing compounds are truly absent from eggshell pigments, which—contrary to the findings of early studies (Kennedy and Vevers 1976)—appears to be the case (Gorchein et al. 2009). Using HPLC-MS, extracted pigments are then identified by comparison to commercially sourced standards, and their concentrations are quantified. Pigment concentration must then be standardized relative to the eggshell fragment's mass or surface area, depending on how the pigment is distributed throughout the shell (Cassey et al. 2012a).

Recent studies on egg pigment chemistry have made exciting contributions to research on breeding behavior, extinct species, and brood parasite-host dynamics. For example, in a broad survey of eggshell pigment concentrations across different groups of British birds, Cassey et al. (2012b) found that, while controlling for phylogenetic relatedness, pigment concentrations are correlated with different ecological and life-history strategies. Specifically, high levels of protoporphyrin are associated with cavity nesting and ground nesting, while biliverdin concentration is associated with noncavity nesting and a greater likelihood of biparental care. Researchers have also demonstrated that eggshell pigments can be extracted from eggshells that are over 500 years old; using HPLC-MS, Igic et al. (2009) successfully recovered biliverdin and protoporphyrin from various extinct species of moa. Finally, comparing the pigment composition of eggshells laid by the brood parasitic Common Cuckoo and its hosts has revealed that cuckoos and their hosts color their eggs using the same pigments and in similar concentrations (Igic et al. 2012).

Chemical analysis of eggshell pigments also holds promise for the study of pesticides and other environmental contaminants. Classic studies exploring the effect of pesticide contamination on eggshells focused on measures of shell thickness (Hickey and Anderson 1968). However, two recent studies suggest that egg appearance itself may be indicative of contaminant load, such that some aspects of egg coloration might provide a rapid and nondestructive means of assessing local pesticide levels (Jagannath et al. 2007, Hanley and Doucet 2012). Though these studies report correlations between measures of egg appearance such as blue-green coloration and pesticide load, neither study quantified pigment concentration. Relating pigment concentration to contaminant load is an important next step for elucidating the mechanisms by which contaminants influence egg coloration. Note that estimates of pigment concentration cannot be made reliably from visual assessment or photographs, at least in some cases (Brulez et al. 2014). For this reason, chemical analyses of eggshell tend to be destructive. Consequently, "shoebox" collections (Russell et al. 2010), which typically lack the high-quality data required for the main collection, are of particular value to researchers, and museum curators should be encouraged to accept such material when the opportunity is presented. Moving forward, it will be important to develop chemical analysis methods that minimize damage to specimens. Promisingly, Raman spectroscopy, a nondestructive spectroscopic method that requires no specialized sample preparation (see earlier section "Nondestructive Analysis with Raman Spectroscopy"), was recently used to detect biliverdin and protoporphyrin from eggshell specimens (Thomas et al. 2015). However, pigment composition may be difficult to determine if both pigments are present in the eggshell, and it is not yet clear whether Raman spectroscopy can provide reliable information about pigment concentration (Thomas et al. 2015).

Structural Analysis

Several nonsignaling hypotheses for the diversity of egg coloration require a detailed understanding of eggshell strength and ultrastructure. Following the initial suggestion from Solomon (1987) that protoporphyrin pigment might contribute to eggshell strength, Gosler et al. (2005) proposed the "structural function" hypothesis for protoporphyrin pigmentation. The hypothesis posits that birds might add pigment to the shells to compensate for shell thinning, which arises from calcium deficiency, and that the added pigment, in turn, affects the egg's mechanical and water vapor conductance properties. Correlative evidence for the hypothesis has been mixed (Gosler et al. 2011, Bulla et al. 2012, Mägi et al. 2012). Another suite of hypotheses suggests that egg pigments interact with the light environment to affect embryonic development (Lahti 2008, Cassey et al. 2011, Maurer et al. 2011). The main ideas here are that shell pigments may protect the developing embryo from overheating, block harmful radiation, provide antimicrobial defense, and serve as wavelength-specific filters, creating a particular light environment that may influence the speed of embryonic growth (Maurer et al. 2011, Maurer et al. 2015).

To fully address these nonsignaling hypotheses, researchers must employ a range of advanced techniques to describe eggshell structure, thickness, strength, water vapor conductance, and light transmission. For analyzing eggshell structure, scanning electron microscopy (SEM) is often used to capture details of the shell's surface texture or of its interior structure (Hincke et al. 2012). Energy-dispersive x-ray spectroscopy (EDS), which provides information about the elemental composition of different eggshell layers, can serve as a useful complement to SEM imaging (Igic et al. 2011). Thickness measurements are typically obtained by a micrometer (Maurer et al.

2012) or from SEM images; the techniques yield similar results (Igic et al. 2010). The mechanical properties of eggshell are commonly assessed using a range of methods, including Vickers hardness testers (Igic et al. 2011) and Instron universal testing frames (Gosler et al. 2011). Water vapor conductance is typically measured by mounting eggshell fragments on test tubes, placing them in a desiccator, and then measuring mass loss, where mass loss is assumed to be water vapor escaping out of the shell (Portugal et al. 2010b). Finally, light transmission is measured using a spectrophotometer, which measures the light that passes through eggshell fragments mounted on plastic cuvettes (Maurer et al. 2015). A goal for future work will be to incorporate these techniques into rigorous, multifaceted studies that address multiple hypotheses about the function of structural variation in eggshells (e.g., Maurer et al. 2015). As with chemical analysis, structural analysis of eggshell material is typically destructive. One exception is the measurement of eggshell thickness, which can sometimes be assessed with measurements through the shell's blowhole or predicted from shell mass and dimensions (see Maurer et al. 2012).

Spectrophotometry

As with feathers, the adoption and then widespread use of spectrophotometry (Andersson and Prager 2006) helped to revolutionize the study of egg coloration. Researchers interested in egg mimicry were quick to use spectrophotometry, demonstrating that cuckoos mimic host eggs across the bird-visible range of wavelengths (Cherry and Bennett 2001, Avilés et al. 2006, Starling et al. 2006). Similarly, many researchers testing the sexually selected eggshell coloration hypothesis (Moreno and Osorno 2003) employed spectrophotometry (reviewed in Reynolds et al. 2009). In these earlier studies, researchers typically used the raw spectra to extract colorimetric variables or to perform PCA (Montgomerie 2006). However, the recent trend has been toward incorporating visual models, which are combined with the raw spectral data to provide estimates of retinal quantal cone catch (see "Analytical Approaches" under "Spectrophotometry" section). In studies that make explicit predictions about signaling (e.g., mimicry, camouflage, sexual signaling, communal breeding), it is important to apply a visual model that is relevant to the signal receiver, which is usually another bird. This approach has been embraced in diverse studies of egg coloration in the field (Spottiswoode and Stevens 2010, Yang et al. 2013) and in museum collections (Stoddard and Stevens 2011, Hanley et al. 2013, Abernathy and Peer 2014), including a broad comparative assessment of egg color variability in museum eggshells representing 251 species (Cassey et al. 2010a). A comprehensive review of avian vision and its application to the study of egg coloration can be found in Stevens (2011), which provides detailed recommendations about spectrophotometry, digital photography, and visual modeling (also see upcoming section "Case Study").

When obtaining spectral data from eggs, there are a few special points to consider. First, it can be challenging to obtain reliable spectra across the egg, especially for maculated (speckled) parts of eggs because speckles can be very small. To help account for speckling, it is advisable to use a custom narrow-ended (1/8-inch diameter) probe (available from Ocean Optics, Dunedin, FL); hold the probe at a constant distance and a fixed angle to the egg surface; and measure reflectance at the top, middle, and base of the egg. Where possible, separate measures should be obtained from the egg background and speckled portions of the egg (Stoddard and Stevens 2011). Second, to avoid contaminated spectra, which overlap with neighboring colors, consult Akkaynak (2014). Third, egg colors can change over time in museum collections (Starling et al. 2006; Cassey et al. 2010b, 2012c), so it is important to consider storage duration, age of the specimen, and measurement device when analyzing egg colors. Recently, Navarro and Lahti (2014) determined that blue-green eggshells decreased in overall reflectance and shifted slightly in terms of spectral shape when exposed to broad-spectrum light under lab conditions for several days; however, these differences happen gradually and extensive handling and exposure of museum eggs would be required to produce large errors. More work on egg fading and photodegradation is needed, particularly for a broader range of egg colors as blue-green egg colors have been the focus so far. Finally, it is worth remembering that egg collections do not always present a random cross-section of wild eggs because some collectors may favor unusual eggs, such as excellent mimics (Starling et al. 2006). Using large sample sizes of clutches from diverse localities, and those acquired by different collectors, can help to counteract these sources of bias and to prevent pseudoreplication.

Digital Photography

Digital photography is especially useful for the study of avian eggs because point-by-point spectrophotometry does not capture the spatial arrangement of egg patterning, which is an important aspect of egg coloration in many avian species (Kilner 2006, Hauber 2014). Already, digital photography, combined with novel ways of quantifying spatial patterns, is helping to shape our understanding of egg camouflage (Lovell et al. 2013), egg mimicry (Spottiswoode and Stevens 2010, Stoddard and Stevens 2010), and egg signatures (Stoddard et al. 2014, Caves et al. 2015). We refer the reader to the earlier section on digital photography and to the following case study for further details.

Case Study

Brood parasites sneak their eggs into the nests of other species, off-loading all parental care to host birds (Davies 2000). The cost of parasitism often triggers an evolutionary arms race between brood parasites and their hosts, with hosts evolving shrewder defenses and parasites evolving better tricks, such as egg mimicry (Rothstein 1990). To study egg mimicry by the Common Cuckoo, Stoddard and Stevens (2010) combined digital image analysis and a model of avian luminance vision with a recently developed method of quantifying spatial patterns called "granularity analysis" (see Figure 3.4b). They found that cuckoo host-races have evolved better egg pattern mimicry for those host species showing the strongest egg rejection. Stoddard and Stevens (2011) next used reflectance spectra to analyze egg color mimicry by the cuckoo. They applied two models of avian color vision to test for egg mimicry: the Vorobyev-Osorio receptor noise-limited discrimination model (Vorobyev and Osorio 1998), which can be used to calculate threshold differences between two colors, and the avian tetrahedral color space model (Goldsmith 1990, Endler and Mielke 2005, Stoddard and Prum 2008), which can be used to represent colors in a three-dimensional chromaticity diagram based on avian spectral sensitivities (Figure 3.4a). Both visual models

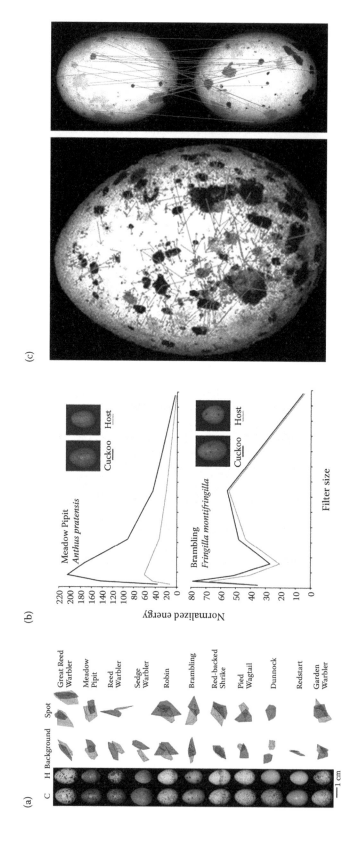

Figure 3.4. New methods for analyzing avian egg color and patterns, applied to cuckoo–host coevolution. (a) Photographs of Common Cuckoo (left) and host (right) eggs are shown next to the color distributions of egg background and spot colors in avian tetrahedral color space, which accounts for the fact that birds have four color cones in their retinas. This illustrates the extent of overlap between cuckoos (red) and their target hosts (blue). The overlap can be used as an estimate of egg color mimicry from a bird's-eye view. (Reproduced with permission from Stoddard and Stevens, 2011. Photographs within the figure are copyright of the Natural History Museum, London [NHM], and the University Museum of Zoology, Cambridge [UMZC].) (b) Average granularity spectra for two hosts (Meadow Pipit and Brambling) and their respective cuckoo gens, or host-race, showing the contribution of different marking sizes to the overall egg pattern. Where the granularity spectra are a good match (as in the Brambling and its cuckoo), the cuckoo egg pattern matches many aspects of the host's egg pattern. (Reproduced with permission from Stoddard and Stevens, 2010. Photographs within the figure are copyright of the NHM.) (c) To study the recognizability of egg patterns, individual features are extracted from a host egg using the SIFT algorithm (left). (Reproduced with permission from Stoddard et al., 2014.) A novel pattern matching algorithm, NaturePatternMatch, then searches for matching features on pairs of eggs, like the pair shown here, laid by the same female (right). (Photographs within the figure are copyright of the NHM.)

confirmed that cuckoo host-races have evolved better egg color mimicry for highly discriminating hosts. Finally, Stoddard et al. (2014) used digital images and a new pattern recognition algorithm (NaturePatternMatch) to demonstrate that host birds have fought back against cuckoo mimicry by evolving individually recognizable patterns on their own eggs (Figure 3.4c). All three of these studies were performed on eggs held in museum collections. Spectrophotometry, digital photography, and advanced models of avian vision and recognition provide outstanding avenues for studying egg coloration in new contexts.

Future Directions

Researchers are now poised to pursue questions about egg coloration with unprecedented rigor and creativity. Museum egg collections will continue to play a vital role in facilitating this work. In the future, it will be fascinating to see how the study of egg coloration will be influenced by new technical advances occurring within museum egg collections, particularly with respect to DNA extraction and whole genome amplification (Lee and Prys-Jones 2008), proteomic analysis (Portugal et al. 2010a), fossil ancient DNA (Oskam et al. 2010), and pollution (Ruuskanen et al. 2013). The critical importance of maintaining and bolstering egg collections cannot be overstated, particularly for use in future studies investigating long-term changes in bird populations that are reflected in eggs (Scharlemann 2001, Kiff and Zink 2005). Finally, museums should be strongly encouraged to digitize their egg collections using photographs. Photographs can be efficiently obtained and easily stored online, and they provide key information about color, pattern, and egg morphology. Care should be taken to use calibrated, UV-sensitive digital cameras, controlled light sources, and appropriate color standards in all images.

CONCLUSIONS

These are exciting times to be studying avian coloration. Just a few decades ago, studies of avian color were mostly limited to human-based assessments of plumage color or labor-intensive biochemical analyses. As outlined in this chapter, recent advances have dramatically expanded the types and nature of approaches that can be used. These advances include quantifying avian coloration using spectrophotometry, high-performance liquid chromatography, new methods for assessing structural color, Raman spectrophotometry, and digital photography and associated analytical techniques. As a result, researchers can now assess coloration from an avian visual perspective, and detailed information on the pigments and structure underlying coloration is known for many species. In addition, these advances have facilitated a better understanding of egg coloration. The growth of data on avian color has greatly expanded our overall understanding of the evolution and ecology of birds. Future directions include more advanced digital photography, the potential to uncover color information from fossils, and more precise assessment of pigments. And how the same pigments can result in different colors. All of these approaches have expanded the role of the traditional study skin, revealing data not typically visible to the naked eye. Thus, they illustrate how the concept of the "extended specimen" is expanding the use of museum collections for the study of avian coloration.

Museum collections represent archives of nature's variation, including variation in color across time and space. These collections allow researchers access to extinct species and extinct populations as well as rare species that cannot be easily sampled in the wild. In addition, they allow researchers to survey densely across taxonomic groups, something that would be prohibitively expensive if all species needed to be sampled in nature. Furthermore, specimens in museum collections preserve the coloration of past populations, allowing for the study of adaptation in coloration across time. For all these reasons, museum collections will continue to provide the essential foundation on which future studies of avian color can be based. These future studies will likely extend the use of specimens even beyond the approaches outlined here.

LITERATURE CITED

Abernathy, V. E., and B. D. Peer. 2014. Intraclutch variation in egg appearance of Brown-headed Cowbird hosts. Auk 131:467–475.

Akkaynak, D. 2014. Use of spectroscopy for assessment of color discrimination in animal vision. Journal of the Optical Society of America A 31:A27–A33.

Akkaynak, D., T. Treibitz, B. Xiao, U. A. Gürkan, J. J. Allen, U. Demirci, and R. T. Hanlon. 2014. Use of commercial off-the-shelf digital cameras for scientific data acquisition and scene-specific color calibration. Journal of the Optical Society of America A 31:312–321.

Andersson, S., J. Örnborg, and M. Andersson. 1998. Ultraviolet sexual dimorphism and assortative mating in blue tits. Proceedings of the Royal Society B 265:445–450.

Andersson, S., and M. Prager. 2006. Quantifying colors. Pp. 41–89 in G. E. Hill and K. J. McGraw (editors), Bird coloration volume 1: mechanisms and measurements, Harvard University Press, Cambridge, MA.

Armenta, J. K., P. O. Dunn, and L. A. Whittingham. 2008a. Effects of specimen age on plumage color. Auk 125:803–808.

Armenta, J. K., P. O. Dunn, and L. A. Whittingham. 2008b. Quantifying avian sexual dichromatism: A comparison of methods. Journal of Experimental Biology 211:2423–2430.

Avilés, J. M., B. G. Stokke, A. Moksnes, E. Røskaft, M. Asmul, and A. P. Møller. 2006. Rapid increase in cuckoo egg matching in a recently parasitized reed warbler population. Journal of Evolutionary Biology 19:1901–1910.

Baird, T., S. E. Solomon, and D. R. Tedstone. 1975. Localisation and characterisation of egg shell porphyrins in several avian species. British Poultry Science 16:201–208.

Barreira, A. S., D. A. Lijtmaer, S. C. Lougheed, and P. L. Tubaro. 2007. Subspecific and temporal variation in the structurally based coloration of the Ultramarine Grosbeak. Condor 109:187–192.

Bennett, A. T. D., I. C. Cuthill, and K. J. Norris. 1994. Sexual selection and the mismeasure of color. American Naturalist 144:848–860.

Bennett, A. T. D., and M. Théry. 2007. Avian color vision and coloration: multidisciplinary evolutionary biology. American Naturalist 169:S1–S6.

Berg, C. J., A. M. LaFountain, R. O. Prum, H. A. Frank, and M. J. Tauber. 2013. Vibrational and electronic spectroscopy of the retro-carotenoid rhodoxanthin in avian plumage, solid-state films, and solution. Archives of Biochemistry and Biophysics 539:142–155.

Björn, L. O., and H. Ghiradella. 2015. Spectral tuning in biology I: Pigments. Pp. 97–117 in L. O. Björn (editor), Photobiology, Springer, New York, NY.

Bridge, E. S., J. Hylton, M. D. Eaton, L. Gamble, and S. J. Schoech. 2007. Cryptic plumage signaling in Aphelocoma Scrub-Jays. Journal of Ornithology 149:123–130.

Brulez, K., P. Cassey, A. Meeson, I. Mikšík, S. L. Webber, A. G. Gosler, and S. J. Reynolds. 2014. Eggshell spot scoring methods cannot be used as a reliable proxy to determine pigment quantity. Journal of Avian Biology 45:94–102.

Brush, A. H. 1967. Pigmentation in the Scarlet Tanager, Piranga olivacea. Condor 69:549–559.

Brush, A. H. 1969. On the nature of "cotingin." Condor 71:431–433.

Brush, A. H. 1970. Pigments in hybrid, variant and melanic tanagers (birds). Comparative Biochemistry and Physiology 36:785–793.

Brush, A. H., and D. M. Power. 1976. House finch pigmentation: carotenoid metabolism and the effect of diet. Auk 93:725–739.

Brush, A. H., and H. Seifried. 1968. Pigmentation and feather structure in genetic variants of the Gouldian Finch, Poephila gouldiae. Auk 85:416–430.

Bulla, M., M. Šálek, and A. G. Gosler. 2012. Eggshell spotting does not predict male incubation but marks thinner areas of a shorebird's shells. Auk 129:26–35.

Burns, K. J., and A. J. Shultz. 2012. Widespread cryptic dichromatism and ultraviolet reflectance in the largest radiation of neotropical songbirds: implications of accounting for avian vision in the study of plumage evolution. Auk 129:211–221.

Cassey, P., M. E. Hauber, G. Maurer, and J. G. Ewen. 2012c. Sources of variation in reflectance spectrophotometric data: a quantitative analysis using avian eggshell colours. Methods in Ecology and Evolution 3:450–456.

Cassey, P., G. Maurer, C. Duval, J. G. Ewen, and M. E. Hauber. 2010b. Impact of time since collection on avian eggshell color: a comparison of museum and fresh egg specimens. Behavioral Ecology and Sociobiology 64:1711–1720.

Cassey, P., G. Maurer, P. G. Lovell, and D. Hanley. 2011. Conspicuous eggs and colourful hypotheses: testing the role of multiple influences on avian eggshell appearance. Avian Biology Research 4:185–195.

Cassey, P., I. Mikšík, S. J. Portugal, G. Maurer, J. G. Ewen, E. Zarate, M. A. Sewell, F. Karadas, T. Grim, and M. E. Hauber. 2012b. Avian eggshell pigments are not consistently correlated with colour measurements or egg constituents in two Turdus thrushes. Journal of Avian Biology 43:503–512.

Cassey, P., S. J. Portugal, G. Maurer, J. G. Ewen, R. L. Boulton, M. E. Hauber, and T. M. Blackburn. 2010a. Variability in avian eggshell colour: a comparative study of museum eggshells. PLoS One 5:e12054.

Cassey, P., G. H. Thomas, S. J. Portugal, G. Maurer, M. E. Hauber, T. Grim, P. G. Lovell, and I. Mikšík. 2012a. Why are birds' eggs colourful? Eggshell pigments co-vary with life-history and nesting ecology among British breeding non-passerine birds. Biological Journal of the Linnean Society 106:657–672.

Caves, E. M., M. Stevens, E. S. Iversen, and C. N. Spottiswoode. 2015. Hosts of avian brood parasites have evolved egg signatures with elevated information content. Proceedings of the Royal Society B 282:20150598.

Charter, M., Y. Leshem, I. Izhaki, and A. Roulin. 2014. Pheomelanin-based colouration is correlated with indices of flying strategies in the Barn Owl. Journal of Ornithology 156:309–312.

Cherry, M. I., and A. T. Bennett. 2001. Egg colour matching in an African cuckoo, as revealed by ultraviolet-visible reflectance spectrophotometry. Proceedings of the Royal Society B 268:565–571.

Cherry, M. I., and A. G. Gosler. 2010. Avian eggshell coloration: new perspectives on adaptive explanations. Biological Journal of the Linnean Society 100:753–762.

Chiao, C.-C., J. K. Wickiser, J. J. Allen, B. Genter, and R. T. Hanlon. 2011. Hyperspectral imaging of cuttlefish camouflage indicates good color match in the eyes of fish predators. Proceedings of the National Academy of Sciences USA 108:9148–9153.

Church, A. H. 1869. Researches on turacin, an animal pigment containing copper. Philosophical Transactions of the Royal Society of London 159:627–636.

Cuthill, I. 2006. Color perception. Pp. 3–40 in G. E. Hill and K. J. McGraw (editors), Bird coloration volume 1: mechanisms and measurements, Harvard University Press, Cambridge, MA.

Cuthill, I. C., A. T. D. Bennett, J. C. Partridge, and E. J. Maier. 1999. Plumage reflectance and the objective assessment of avian sexual dichromatism. American Naturalist 153:183–200.

Czeczuga, B. 1979. Carotenoids in the skin of certain species of birds. Comparative Biochemistry and Physiology B 62:107–109.

D'Alba, L., L. Kieffer, and M. D. Shawkey. 2012. Relative contributions of pigments and biophotonic nanostructures to natural color production: A case study in Budgerigar (Melopsittacus undulatus) feathers. Journal of Experimental Biology 215:1272–1277.

Davies, N. 2000. Cuckoos, Cowbirds and other cheats. T. & A. D. Poyser, London, UK.

Debecker, S., R. Sommaruga, T. Maes, R. Stoks, and G. Davidowitz. 2015. Larval UV exposure impairs adult immune function through a trade-off with larval investment in cuticular melanin. Functional Ecology 29:1292–1299.

Delhey, K., C. Burger, W. Fiedler, and A. Peters. 2010. Seasonal changes in colour: A comparison of structural, melanin- and carotenoid-based plumage colours. PLoS One 5:e11582.

Delhey, K., V. Delhey, B. Kempenaers, and A. Peters. 2014. A practical framework to analyze variation in animal colors using visual models. Behavioral Ecology 26:367–375.

Doucet, S. M., and G. E. Hill. 2009. Do museum specimens accurately represent wild birds? A case study of carotenoid, melanin, and structural colours in Long-tailed Manakins Chiroxiphia linearis. Journal of Avian Biology 40:146–156.

Doucet, S. M., D. J. Mennill, and G. E. Hill. 2007. The evolution of signal design in manakin plumage ornaments. American Naturalist 169:S62–S80.

Doucet, S. M., M. D. Shawkey, M. K. Rathburn, H. L. Mays, and R. Montgomerie. 2004. Concordant evolution of plumage colour, feather microstructure and a melanocortin receptor gene between mainland and island populations of a fairy-wren. Proceedings of the Royal Society B 271:1663–1670.

Dunn, P. O., J. K. Armenta, and L. A. Whittingham. 2015. Natural and sexual selection act on different axes of variation in avian plumage color. Science Advances 1:e1400155.

Dyck, J. 1966. Determination of plumage colours, feather pigments and structures by means of reflection spectrophotometry. Dansk Ornithologisk Forening 60:49–76.

Eaton, M. D. 2005. Human vision fails to distinguish widespread sexual dichromatism among sexually "monochromatic" birds. Proceedings of the National Academy of Sciences USA 102:10942–10946.

Eaton, M. D., and K. P. Johnson. 2007. Avian visual perspective on plumage coloration confirms rarity of sexually monochromatic North American passerines. Auk 124:155–161.

Eliason, C. M., R. Maia, and M. D. Shawkey. 2015. Modular color evolution facilitated by a complex nanostructure in birds. Evolution 69:357–367.

Endler, J. A. 1990. On the measurement and classification of colour in studies of animal colour patterns. Biological Journal of the Linnean Society 41:315–352.

Endler, J. A., and P. W. Mielke. 2005. Comparing entire colour patterns as birds see them. Biological Journal of the Linnean Society 86:405–431.

Endler, J. A., and M. Théry. 1996. Interacting effects of lek placement, display behavior, ambient light, and color patterns in three neotropical forest-dwelling birds. American Naturalist 148:421–452.

Endler, J. A., D. A. Westcott, J. R. Madden, and T. Robson. 2005. Animal visual systems and the evolution of color patterns: sensory processing illuminates signal evolution. Evolution 59:1795–1818.

Fitze, P. S., B. Tschirren, J. Gasparini, and H. Richner. 2007. Carotenoid-based plumage colors and immune function: is there a trade-off for rare carotenoids? American Naturalist 169:S137–S144.

Fox, D. L. 1962. Carotenoids of the Scarlet Ibis. Comparative Biochemistry and Physiology 5:31–43.

Fox, D. L., A. A. Wolfson, and J. W. McBeth. 1969. Metabolism of β-carotene in the American flamingo, *Phoenicopterus ruber*. Comparative Biochemistry and Physiology 29:1223–1229.

Friedman, N. R., C. M. Hofmann, B. Kondo, and K. E. Omland. 2009. Correlated evolution of migration and sexual dichromatism in the New World orioles (*Icterus*). Evolution 63:3269–3274.

Friedman, N. R., K. J. McGraw, and K. E. Omland. 2014a. History and mechanisms of carotenoid plumage evolution in the New World orioles (*Icterus*). Comparative Biochemistry and Physiology B 172-173:1–8.

Friedman, N. R., K. J. McGraw, and K. E. Omland. 2014b. Evolution of carotenoid pigmentation in caciques and meadowlarks (Icteridae): repeated gains of red plumage coloration by carotenoid C4-oxygenation. Evolution 68:791–801.

Galván, I., and A. Jorge. 2015. Dispersive Raman spectroscopy allows the identification and quantification of melanin types. Ecology and Evolution 5:1425–1431.

Galván, I., A. Jorge, K. Ito, K. Tabuchi, F. Solano, and K. Wakamatsu. 2013a. Raman spectroscopy as a noninvasive technique for the quantification of melanins in feathers and hairs. Pigment Cell Melanoma Research 26:917–923.

Galván, I., A. Jorge, F. Solano, and K. Wakamatsu. 2013b. Vibrational characterization of pheomelanin and trichrome F by Raman spectroscopy. Spectrochimica Acta Part A, Molecular and Biomolecular Spectroscopy 110:55–59.

Galván, I., and A. P. Møller. 2013. Pheomelanin-based plumage coloration predicts survival rates in birds. Physiological and Biochemical Zoology 86:184–192.

Gamgee, A. 1895. On the relations of turacin and turacoporphyrin to the colouring matter of the blood. Proceedings of the Royal Society of London 59:339–342.

Gluckman, T.-L., and G. C. Cardoso. 2009. A method to quantify the regularity of barred plumage patterns. Behavioral Ecology and Sociobiology 63:1837–1844.

Goldsmith, T. H. 1980. Hummingbirds see near ultraviolet light. Science 207:786–788.

Goldsmith, T. H. 1990. Optimization, constraint, and history in the evolution of eyes. Quarterly Review of Biology 65:281–322.

Gomez, D., and M. Théry. 2007. Simultaneous crypsis and conspicuousness in color patterns: comparative analysis of a neotropical rainforest bird community. American Naturalist 169:S42–S61.

Gorchein, A., C. K. Lim, and P. Cassey. 2009. Extraction and analysis of colourful eggshell pigments using HPLC and HPLC/electrospray ionization tandem mass spectrometry. Biomedical Chromatography: BMC 23:602–606.

Gosler, A. G., O. R. Connor, and R. H. C. Bonser. 2011. Protoporphyrin and eggshell strength: preliminary findings from a passerine bird. Avian Biology Research 4:214–223.

Gosler, A. G., J. P. Higham, and S. James Reynolds. 2005. Why are birds' eggs speckled? Ecology Letters 8:1105–1113.

Götmark, F. 1987. White underparts in gulls function as hunting camouflage. Animal Behaviour 35:1786–1792.

Haase, E., S. Ito, A. Sell, and K. Wakamatsu. 1992. Melanin concentrations in feathers from wild and domestic pigeons. Journal of Heredity 83:64–67.

Hamilton, P. B. 1992. The use of high-performance liquid chromatography for studying pigmentation. Poultry Science 71:718–724.

Hanley, D., and S. M. Doucet. 2012. Does environmental contamination influence egg coloration? A long-term study in herring gulls. Journal of Applied Ecology 49:1055–1063.

Hanley, D., M. C. Stoddard, P. Cassey, and P. L. R. Brennan. 2013. Eggshell conspicuousness in ground nesting birds: do conspicuous eggshells signal nest location to conspecifics? Avian Biology Research 6:147–156.

Hart, N. S. 2001. Variations in cone photoreceptor abundance and the visual ecology of birds. Journal of Comparative Physiology A 187:685–697.

Hart, N. S., and D. M. Hunt. 2007. Avian visual pigments: characteristics, spectral tuning, and evolution. American Naturalist 169(Suppl. 1):S7–S26.

Hart, N. S., J. C. Partridge, I. C. Cuthill, and A. T. Bennett. 2000. Visual pigments, oil droplets, ocular media and cone photoreceptor distribution in two species of passerine bird: the Blue Tit (*Parus caeruleus* L.) and the Blackbird (*Turdus merula* L.). Journal of Comparative Physiology A 186:375–387.

Hart, N. S., and M. Vorobyev. 2005. Modelling oil droplet absorption spectra and spectral sensitivities of bird cone photoreceptors. Journal of Comparative Physiology A 191:381–392.

Håstad, O., and A. Odeen. 2008. Different ranking of avian colors predicted by modeling of retinal function in humans and birds. American Naturalist 171:831–838.

Hauber, M. E. 2014. The book of eggs: a lifesize guide to the eggs of six hundred of the world's bird species. University of Chicago Press, Chicago, IL.

Hausmann, F., K. E. Arnold, N. J. Marshall, and I. P. Owens. 2003. Ultraviolet signals in birds are special. Proceedings of the Royal Society B 270:61–67.

Hickey, J. J., and D. W. Anderson. 1968. Chlorinated hydrocarbons and eggshell changes in raptorial and fish-eating birds. Science 162:271–273.

Hill, G. E. 1991. Plumage coloration is a sexually selected indicator of male quality. Nature 350:337–339.

Hill, G. E., and K. J. McGraw. 2006a. Bird coloration volume 2: function and evolution. Harvard University Press, Cambridge, MA.

Hill, G. E., and K. J. McGraw. 2006b. Bird coloration volume 1: mechanisms and measurements. Harvard University Press, Cambridge, MA.

Hincke, M. T., Y. Nys, J. Gautron, K. Mann, A. B. Rodriguez-Navarro, and M. D. McKee. 2012. The eggshell: structure, composition and mineralization. Frontiers in Bioscience 17:1266–1280.

Hoi, H., and M. Griggio. 2008. Dual utility of a melanin-based ornament in Bearded Tits. Ethology 114:1094–1100.

Hudon, J. 1991. Unusual carotenoid use by the Western Tanager (*Piranga ludoviciana*) and its evolutionary implications. Canadian Journal of Zoology 69:2311–2320.

Hudon, J., and A. H. Brush. 1992. Identification of carotenoid pigments in birds. Methods in Enzymology 213:312–321.

Hudon, J., A. P. Capparella, and A. H. Brush. 1989. Plumage pigment differences in manakins of the *Pipra erythrocephala* superspecies. Auk 106:34–41.

Hudon, J., and A. D. Muir. 1996. Characterization of the reflective materials and organelles in the bright irides of North American blackbirds (Icterinae). Pigment Cell Research 9:96–104.

Igic, B., K. Braganza, M. M. Hyland, H. Silyn-Roberts, P. Cassey, T. Grim, J. Rutila, C. Moskát, and M. E. Hauber. 2011. Alternative mechanisms of increased eggshell hardness of avian brood parasites relative to host species. Journal of The Royal Society Interface 8:1654–1664.

Igic, B., P. Cassey, T. Grim, D. R. Greenwood, C. Moskat, J. Rutila, and M. E. Hauber. 2012. A shared chemical basis of avian host-parasite egg colour mimicry. Proceedings of the Royal Society B 279:1068–76.

Igic, B., D. Fecheyr-Lippens, M. Xiao, A. Chan, D. Hanley, P. R. L. Brennan, T. Grim, G. I. N. Waterhouse, M. E. Hauber, and M. D. Shawkey. 2015. A nanostructural basis for gloss of avian eggshells. Journal of the Royal Society Interface 12:20141210.

Igic, B., D. R. Greenwood, D. J. Palmer, P. Cassey, B. J. Gill, T. Grim, P. L. R. Brennan, S. M. Bassett, P. F. Battley, and M. E. Hauber. 2009. Detecting pigments from colourful eggshells of extinct birds. Chemoecology 20:43–48.

Igic, B., M. E. Hauber, J. A. Galbraith, T. Grim, D. C. Dearborn, P. L. R. Brennan, C. Moskát, P. K. Choudhary, and P. Cassey. 2010. Comparison of micrometer- and scanning electron microscope-based measurements of avian eggshell thickness. Journal of Field Ornithology 81:402–410.

Ito, S., and K. Wakamatsu. 2003. Quantitative analysis of eumelanin and pheomelanin in humans, mice, and other animals: a comparative review. Pigment Cell Research 16:523–531.

Jacot, A., C. Romero-Diaz, B. Tschirren, H. Richner, and P. S. Fitze. 2010. Dissecting carotenoid from structural components of carotenoid-based coloration: a field experiment with Great Tits (*Parus major*). American Naturalist 176:55–62.

Jagannath, A., R. F. Shore, L. A. Walker, P. N. Ferns, and A. G. Gosler. 2007. Eggshell pigmentation indicates pesticide contamination. Journal of Applied Ecology 45:133–140.

Johnsen, A., S. Andersson, J. Ornborg, and J. T. Lifjeld. 1998. Ultraviolet plumage ornamentation affects social mate choice and sperm competition in Bluethroats (Aves: *Luscinia s. svecica*): a field experiment. Proceedings of the Royal Society B 265:1313–1318.

Johnson, N. K., J. V. Remsen, Jr., and C. Cicero. 1998. Refined colorimetry validates endangered subspecies of the Least Tern. Condor 100:18–26.

Karubian, J., W. R. Lindsay, H. Schwabl, and M. S. Webster. 2011. Bill coloration, a flexible signal in a tropical passerine bird, is regulated by social environment and androgens. Animal Behaviour 81:795–800.

Kelber, A., and D. Osorio. 2010. From spectral information to animal colour vision: experiments and concepts. Proceedings of the Royal Society B 277:1617–1625.

Kelber, A., M. Vorobyev, and D. Osorio. 2003. Animal colour vision—behavioural tests and physiological concepts. Biological Reviews 78:81–118.

Kemp, D. J., M. E. Herberstein, L. J. Fleishman, J. A. Endler, A. T. Bennett, A. G. Dyer, N. S. Hart, J. Marshall, and M. J. Whiting. 2015. An integrative framework for the appraisal of coloration in nature. American Naturalist 185:705–724.

Kennedy, G. Y., and H. G. Vevers. 1976. A survey of avian eggshell pigments. Comparative Biochemistry and Physiology B 55:117–123.

Kiff, L. F., and R. M. Zink. 2005. History, present status, and future prospects of avian eggshell collections in North America. Auk 122:994–999.

Kilner, R. M. 2006. The evolution of egg colour and patterning in birds. Biological Reviews of the Cambridge Philosophical Society 81:383–406.

Kim, M. H., T. A. Harvey, D. S. Kittle, H. Rushmeier, J. Dorsey, R. O. Prum, and D. J. Brady. 2012. 3D imaging spectroscopy for measuring hyperspectral patterns on solid objects. ACM Transactions on Graphics 31:1–11.

Kritzler, H. 1943. Carotenoids in the display and eclipse plumages of bishop birds. Physiological Zoology 16:241–255.

Krukenberg, C. F. W. 1882. Die federfarbstoffe der psittaciden. Vergleichend-physiologische Studien Reihe 2, Abtlg 2:29–36.

LaFountain, A. M., S. Kaligotla, S. Cawley, K. M. Riedl, S. J. Schwartz, H. A. Frank, and R. O. Prum. 2010. Novel methoxy-carotenoids from the burgundy-colored plumage of the Pompadour Cotinga *Xipholena punicea*. Archives of Biochemistry and Biophysics 504:142–153.

LaFountain, A. M., R. O. Prum, and H. A. Frank. 2015. Diversity, physiology, and evolution of avian plumage carotenoids and the role of carotenoid–protein interactions in plumage color appearance. Archives of Biochemistry and Biophysics 572:201–212.

Lahti, D. C. 2008. Population differentiation and rapid evolution of egg color in accordance with solar radiation. Auk 125:796–802.

Lee, P. L., and R. P. Prys-Jones. 2008. Extracting DNA from museum bird eggs, and whole genome amplification of archive DNA. Molecular Ecology Resources 8:551–560.

Li, Q., K.-Q. Gao, Q. Meng, J. A. Clarke, M. D. Shawkey, L. D'Alba, R. Pei, M. Ellison, M. A. Norell, and J. Vinther. 2012. Reconstruction of *Microraptor* and the evolution of iridescent plumage. Science 335:1215–1219.

Lim, C. K. 2004. High-performance liquid chromatography and mass spectrometry of porphyrins, chlorophylls and bilins. World Scientific Publishing Company, Hackensack, NJ.

Lovell, P. G., G. D. Ruxton, K. V. Langridge, and K. A. Spencer. 2013. Egg-laying substrate selection for optimal camouflage by quail. Current Biology 23:260–264.

Lowe, D. G. 1999. Object recognition from local scale-invariant features. Proceedings of the Seventh IEEE International Conference on Computer Vision, 1999.

Lowe, D. G. 2000. Towards a computational model for object recognition in IT cortex. Pp. 20–31 in S.-W. Lee, H. H. Bülthoff, and T. Poggio (editors), Biologically motivated computer vision: First IEEE International Workshop, BMCV 2000 Seoul, Korea, May 15–17, 2000 Proceedings, Springer, Berlin, Germany.

Lozano, G. A. 1994. Carotenoids, parasites, and sexual selection. Oikos 70:309–311.

Maan, M. E., and M. E. Cummings. 2012. Poison frog colors are honest signals of toxicity, particularly for bird predators. American Naturalist 179:E1–E14.

Mägi, M., R. Mänd, A. Konovalov, V. Tilgar, and S. J. Reynolds. 2012. Testing the structural–function hypothesis of eggshell maculation in the Great Tit: An experimental approach. Journal of Ornithology 153:645–652.

Maia, R., J. V. Caetano, S. N. Bao, and R. H. Macedo. 2009. Iridescent structural colour production in male blue-black grassquit feather barbules: the role of keratin and melanin. Journal of the Royal Society Interface 6:S203–S211.

Maia, R., C. M. Eliason, P.-P. Bitton, S. M. Doucet, M. D. Shawkey, and A. Tatem. 2013a. pavo: an R package for the analysis, visualization and organization of spectral data. Methods in Ecology and Evolution 4:906–913.

Maia, R., D. R. Rubenstein, and M. D. Shawkey. 2013b. Key ornamental innovations facilitate diversification in an avian radiation. Proceedings of the National Academy of Sciences USA 110:10687–10692.

Maley, J. M., and K. Winker. 2007. Use of juvenal plumage in diagnosing species limits: an example using buntings in the genus *Plectrophenax*. Auk 124:907–915.

Marchetti, K. 1993. Dark habitats and bright birds illustrate the role of the environment in species divergence. Nature 362:149–152.

Mason, N. A., A. J. Shultz, and K. J. Burns. 2014. Elaborate visual and acoustic signals evolve independently in a large, phenotypically diverse radiation of songbirds. Proceedings of the Royal Society B 281:20140967.

Maurer, G., S. J. Portugal, and P. Cassey. 2011. Review: an embryo's eye view of avian eggshell pigmentation. Journal of Avian Biology 42:494–504.

Maurer, G., S. J. Portugal, and P. Cassey. 2012. A comparison of indices and measured values of eggshell thickness of different shell regions using museum eggs of 230 European bird species. Ibis 154:714–724.

Maurer, G., S. J. Portugal, M. E. Hauber, I. Mikšík, D. G. D. Russell, P. Cassey, and B. Tschirren. 2015. First light for avian embryos: eggshell thickness and pigmentation mediate variation in development and UV exposure in wild bird eggs. Functional Ecology 29:209–218.

McGraw, K. J. 2004. European barn swallows use melanin pigments to color their feathers brown. Behavioral Ecology 15:889–891.

McGraw, K. J. 2006a. Mechanics of melanin coloration in birds. Pp. 243–294 in G. E. Hill and K. J. McGraw (editors), Bird coloration volume 1: mechanisms and measurements, Harvard University Press, Cambridge, MA.

McGraw, K. J. 2006b. Mechanics of carotenoid-based coloration. Pp. 177–242 in G. E. Hill and K. J. McGraw (editors), Bird coloration volume 1: mechanisms and measurements, Harvard University Press, Cambridge, MA.

McGraw, K. J. 2006c. Mechanics of uncommon colours: pterins, porphyrins, and psittacofulvins. Pp. 354–398 in G. E. Hill and K. J. McGraw (editors), Bird coloration volume 1: mechanisms and measurements, Harvard University Press, Cambridge, MA.

McGraw, K. J., and M. C. Nogare. 2005. Distribution of unique red feather pigments in parrots. Biology Letters 1:38–43.

McGraw, K. J., M. B. Toomey, P. M. Nolan, N. I. Morehouse, M. Massaro, and P. Jouventin. 2007. A description of unique fluorescent yellow pigments in penguin feathers. Pigment Cell Research 20:301–304.

McGraw, K. J., and K. Wakamatsu. 2004. Melanin basis of ornamental feather colors in male Zebra Finches. Condor 106:686–690.

McGraw, K. J., K. Wakamatsu, S. Ito, P. M. Nolan, P. Jouventin, F. S. Dobson, R. E. Austic, R. J. Safran, L. M. Siefferman, G. E. Hill, and R. S. Parker. 2004. You can't judge a pigment by its color: carotenoid and melanin content of yellow and brown feathers in swallows, bluebirds, penguins, and domestic chickens. Condor 106:390–395.

McKay, B. D. 2013. The use of digital photography in systematics. Biological Journal of the Linnean Society 110:1–13.

McKay, B. D., H. L. Mays, C.-T. Yao, D. Wan, H. Higuchi, and I. Nishiumi. 2014. Incorporating color into integrative taxonomy: analysis of the Varied Tit (*Sittiparus varius*) complex in East Asia. Systematic Biology 63:505–517.

McNett, G. D., and K. Marchetti. 2005. Ultraviolet degradation in carotenoid patches: live versus museum specimens of Wood Warblers (Parulidae). Auk 122:793–802.

Meadows, M. G., N. I. Morehouse, R. L. Rutowski, J. M. Douglas, and K. J. McGraw. 2011. Quantifying iridescent coloration in animals: a method for improving repeatability. Behavioral Ecology and Sociobiology 65:1317–1327.

Mendes-Pinto, M. M., A. M. LaFountain, M. C. Stoddard, R. O. Prum, H. A. Frank, and B. Robert. 2012. Variation in carotenoid-protein interaction in bird feathers produces novel plumage coloration. Journal of the Royal Society Interface 9:3338–3350.

Mikšík, I., V. Holáň, and Z. Deyl. 1996. Avian eggshell pigments and their variability. Comparative Biochemistry and Physiology B 113:607–612.

Møller, A. P. 1987. Variation in badge size in male House Sparrows *Passer domesticus*: evidence for status signalling. Animal Behaviour 35:1637–1644.

Montgomerie, R. 2006. Analyzing colors. Pp. 90–147 in G. E. Hill and K. J. McGraw (editors), Bird coloration volume 1: mechanisms and measurements, Harvard University Press, Cambridge, MA.

Montgomerie, R., B. Lyon, and K. Holder. 2001. Dirty ptarmigan: behavioral modification of conspicuous male plumage. Behavioral Ecology 12:429–438.

Moreno, J., and J. L. Osorno. 2003. Avian egg colour and sexual selection: does eggshell pigmentation reflect female condition and genetic quality? Ecology Letters 6:803–806.

Murphy, T. G., M. F. Rosenthal, R. Montgomerie, and K. A. Tarvin. 2009. Female American goldfinches use carotenoid-based bill coloration to signal status. Behavioral Ecology 20:1348–1355.

Navarro, J. Y., and D. C. Lahti. 2014. Light dulls and darkens bird eggs. PLoS One 9:e116112.

Nicolaus, M., C. Le Bohec, P. Nolan, M. Gauthier-Clerc, Y. Le Maho, J. Komdeur, and P. Jouventin. 2007. Ornamental colors reveal age in the king penguin. Polar Biology 31:53–61.

Ödeen, A., and O. Håstad. 2013. The phylogenetic distribution of ultraviolet sensitivity in birds. BMC Evolutionary Biology 13:36.

Ödeen, A., S. Pruett-Jones, A. C. Driskell, J. K. Armenta, and O. Håstad. 2012. Multiple shifts between violet and ultraviolet vision in a family of passerine birds with associated changes in plumage coloration. Proceedings of the Royal Society B 279: 1269–1276.

Oliphant, L. W. 1987. Pteridines and purines as major pigments of the avian iris. Pigment Cell Research 1:129–131.

Oliphant, L. W. 1988. Cytology and pigments of non-melanophore chromatophores in the avian iris. Progress in Clinical and Biological Research 256:65–82.

Olson, V. A., and I. P. F. Owens. 1998. Costly sexual signals: are carotenoids rare, risky or required? Trends in Ecology and Evolution 13:510–514.

Oskam, C. L., J. Haile, E. McLay, P. Rigby, M. E. Allentoft, M. E. Olsen, C. Bengtsson, G. H. Miller, J. L. Schwenninger, C. Jacomb, R. Walter, A. Baynes, J. Dortch, M. Parker-Pearson, M. T. Gilbert, R. N. Holdaway, E. Willerslev, and M. Bunce. 2010. Fossil avian eggshell preserves ancient DNA. Proceedings of the Royal Society B 277:1991–2000.

Osorio, D., and A. D. Ham. 2002. Spectral reflectance and directional properties of structural coloration in bird plumage. Journal of Experimental Biology 205:2017–2027.

Osorio, D., and M. Vorobyev. 2005. Photoreceptor spectral sensitivities in terrestrial animals: adaptations for luminance and colour vision. Proceedings of the Royal Society B 272:1745–1752.

Pike, T. 2011. Using digital cameras to investigate animal colouration: estimating sensor sensitivity functions. Behavioral Ecology and Sociobiology 65:849–858.

Portugal, S. J., H. J. Cooper, C. G. Zampronio, L. L. Wallace, and P. Cassey. 2010a. Can museum egg specimens be used for proteomic analyses? Proteome Science 8:40.

Portugal, S. J., G. Maurer, and P. Cassey. 2010b. Eggshell permeability: a standard technique for determining interspecific rates of water vapor conductance. Physiological and Biochemical Zoology 83:1023–1031.

Prager, M., and S. Andersson. 2010. Convergent evolution of red carotenoid coloration in widowbirds and bishops (Euplectes spp.). Evolution 64:3609–3619.

Prum, R. O. 1999. The anatomy and physics of structural colours. Proceedings of the 22nd International Ornithological Congress, Durban, South Africa.

Prum, R. O. 2006. Anatomy, physics, and evolution of structural colors. Pp. 177–242 in G. E. Hill and K. J. McGraw (editors), Bird coloration volume 1: mechanisms and measurements, Harvard University Press, Cambridge, MA.

Prum, R. O., S. Andersson, and R. H. Torres. 2003. Coherent scattering of ultraviolet light by avian feather barbs. Auk 120:163–170.

Prum, R. O., A. M. LaFountain, C. J. Berg, M. J. Tauber, and H. A. Frank. 2014. Mechanism of carotenoid coloration in the brightly colored plumages of broadbills (Eurylaimidae). Journal of Comparative Physiology B 184:651–672.

Prum, R. O., and R. Torres. 2003a. Structural colouration of avian skin: convergent evolution of coherently scattering dermal collagen arrays. Journal of Experimental Biology 206:2409–2429.

Prum, R. O., and R. H. Torres. 2003b. A Fourier tool for the analysis of coherent light scattering by bio-optical nanostructures. Integrative and Comparative Biology 43:591–602.

Prum, R. O., R. H. Torres, S. Williamson, and J. Dyck. 1998. Coherent light scattering by blue feather barbs. Nature 396:28–29.

Pryke, S. R., and S. C. Griffith. 2006. Red dominates black: agonistic signalling among head morphs in the colour polymorphic Gouldian finch. Proceedings of the Royal Society B 273:949–957.

Renoult, J. P., A. Kelber, and H. M. Schaefer. 2015. Colour spaces in ecology and evolutionary biology. Biological Reviews 92:292–315.

Reynolds, S. J., G. R. Martin, and P. Cassey. 2009. Is sexual selection blurring the functional significance of eggshell coloration hypotheses? Animal Behaviour 78:209–215.

Rothstein, S. I. 1990. A model system for coevolution: avian brood parasitism. Annual Review of Ecology and Systematics 21:481–508.

Russell, B. J., and H. M. Dierssen. 2015. Use of hyperspectral imagery to assess cryptic color matching in Sargassum associated crabs. PLoS One 10:e0136260.

Russell, D. G. D., J. White, G. Maurer, and P. Cassey. 2010. Data-poor egg collections: tapping an important research resource. Journal of Afrotropical Zoology 6:77–82.

Ruuskanen, S., T. Laaksonen, J. Morales, J. Moreno, R. Mateo, E. Belskii, A. Bushuev, A. Järvinen, A. Kerimov, I. Krams, C. Morosinotto, R. Mänd, M. Orell, A. Qvarnström, F. Slater, V. Tilgar, M. E. Visser, W. Winkel, H. Zang, and T. Eeva. 2013. Large-scale geographical variation in eggshell metal and calcium content in a passerine bird (Ficedula hypoleuca). Environmental Science and Pollution Research 21:3304–3317.

Saks, L., I. Ots, and P. Horak. 2003. Carotenoid-based plumage coloration of male greenfinches reflects health and immunocompetence. Oecologia 134:301–307.

Santos, S. I. C. O., and J. T. Lumeij. 2007. Comparison of multiple-angle spectrometry of plumage versus individual feathers for the assessment of sexual dichromatism in the Long-tailed Finch (Poephila acuticauda). Journal of Ornithology 148:281–291.

Saranathan, V., J. D. Forster, H. Noh, S. F. Liew, S. G. Mochrie, H. Cao, E. R. Dufresne, and R. O. Prum. 2012. Structure and optical function of amorphous photonic nanostructures from avian feather barbs: a comparative small angle X-ray scattering (SAXS) analysis of 230 bird species. Journal of the Royal Society Interface 9:2563–2580.

Scharlemann, J. P. W. 2001. Museum egg collections as stores of long-term phenological data. International Journal of Biometeorology 45:208–211.

Schmitz-Ornés, A. 2006. Using colour spectral data in studies of geographic variation and taxonomy of birds: examples with two hummingbird genera, Anthracothorax and Eulampis. Journal of Ornithology 147:495–503.

Seddon, N., J. A. Tobias, M. Eaton, A. Ödeen, and B. E. Byers. 2010. Human vision can provide a valid proxy for avian perception of sexual dichromatism. Auk 127:283–292.

Shawkey, M. D., S. L. Balenger, G. E. Hill, L. S. Johnson, A. J. Keyser, and L. Siefferman. 2006a. Mechanisms of evolutionary change in structural plumage coloration among bluebirds (Sialia spp.). Journal of the Royal Society Interface 3:527–532.

Shawkey, M. D., A. M. Estes, L. Siefferman, and G. E. Hill. 2005. The anatomical basis of sexual dichromatism in non-iridescent ultraviolet-blue structural coloration of feathers. Biological Journal of the Linnean Society 84:259–271.

Shawkey, M. D., M. E. Hauber, L. K. Estep, and G. E. Hill. 2006b. Evolutionary transitions and mechanisms of matte and iridescent plumage coloration in grackles and allies (Icteridae). Journal of the Royal Society Interface 3:777–786.

Shawkey, M. D., and G. E. Hill. 2005. Carotenoids need structural colours to shine. Biology Letters 1:121–124.

Shawkey, M. D., and G. E. Hill. 2006. Significance of a basal melanin layer to production of non-iridescent structural plumage color: evidence from an amelanotic Steller's Jay (Cyanocitta stelleri). Journal of Experimental Biology 209:1245–1250.

Shawkey, M. D., V. Saranathan, H. Palsdottir, J. Crum, M. H. Ellisman, M. Auer, and R. O. Prum. 2009. Electron tomography, three-dimensional Fourier analysis and colour prediction of a three-dimensional amorphous biophotonic nanostructure. Journal of the Royal Society Interface 6:S213–S220.

Shultz, A. J., and K. J. Burns. 2013. Plumage evolution in relation to light environment in a novel clade of neotropical tanagers. Molecular Phylogenetics and Evolution 66:112–125.

Smith, E., and G. Dent. 2005. Modern Raman spectroscopy: a practical approach. Wiley, New York, NY.

Smith, I. D., and H. S. Perdue. 1966. Isolation and tentative identification of the carotenoids present in chicken skin and egg yolks. Poultry Science 45:577–581.

Smith, T. B., P. P. Marra, M. S. Webster, I. Lovette, H. L. Gibbs, R. T. Holmes, K. A. Hobson, S. Rohwer, and R. Prum. 2003. A call for feather sampling. Auk 120:218–221.

Solomon, S. 1987. Egg shell pigmentation. Pp. 147–158 in R. G. Wells and C. G. Belyavin (editors), Egg quality: current problems and recent advances, Butterworths, London, UK.

Sparks, N. H. C. 2011. Eggshell pigments—from formation to deposition. Avian Biology Research 4:162–167.

Spottiswoode, C. N., and M. Stevens. 2010. Visual modeling shows that avian host parents use multiple visual cues in rejecting parasitic eggs. Proceedings of the National Academy of Sciences USA 107:8672–8676.

Starling, M., R. Heinsohn, A. Cockburn, and N. E. Langmore. 2006. Cryptic gentes revealed in pallid cuckoos Cuculus pallidus using reflectance spectrophotometry. Proceedings of the Royal Society B 273:1929–1934.

Stevens, M. 2011. Avian vision and egg colouration: concepts and measurements. Avian Biology Research 4:168–184.

Stevens, M., C. A. Párraga, I. C. Cuthill, J. C. Partridge, and T. S. Troscianko. 2007. Using digital photography to study animal coloration. Biological Journal of the Linnean Society 90:211–237.

Stevens, M., M. C. Stoddard, and J. P. Higham. 2009. Studying primate color: towards visual system-dependent methods. International Journal of Primatology 30:893–917.

Stoddard, M. C., R. M. Kilner, and C. Town. 2014. Pattern recognition algorithm reveals how birds evolve individual egg pattern signatures. Nature Communications 5:4117.

Stoddard, M. C., K. L. A. Marshall, and R. M. Kilner. 2011. Imperfectly camouflaged avian eggs: artefact or adaptation? Avian Biology Research 4:196–213.

Stoddard, M. C., and R. O. Prum. 2008. Evolution of avian plumage color in a tetrahedral color space: a phylogenetic analysis of New World buntings. American Naturalist 171:755–776.

Stoddard, M. C., and R. O. Prum. 2011. How colorful are birds? Evolution of the avian plumage color gamut. Behavioral Ecology 22:1042–1052.

Stoddard, M. C., and M. Stevens. 2010. Pattern mimicry of host eggs by the common cuckoo, as seen through a bird's eye. Proceedings of the Royal Society B 277:1387–1393.

Stoddard, M. C., and M. Stevens. 2011. Avian vision and the evolution of egg color mimicry in the common cuckoo. Evolution 65:2004–2013.

Stradi, R., G. Celentano, M. Boles, and F. Mercato. 1997. Carotenoids in bird plumage: the pattern in a series of red-pigmented Carduelinae. Comparative Biochemistry and Physiology B 117:85–91.

Stradi, R., G. Celentano, and D. Nava. 1995a. Separation and identification of carotenoids in bird's plumage by high-performance liquid chromatography-diode-array detection. Journal of Chromatography B: Biomedical Sciences and Applications 670:337–348.

Stradi, R., G. Celentano, E. Rossi, G. Rovati, and M. Pastore. 1995b. Carotenoids in bird plumage—I. the carotenoid pattern in a series of Palearctic Carduelinae. Comparative Biochemistry and Physiology Part B: Biochemistry and Molecular Biology 110:131–143.

Stradi, R., J. Hudon, G. Celentano, and E. Pini. 1998. Carotenoids in bird plumage: the complement of yellow and red pigments in true woodpeckers (Picinae). Comparative Biochemistry and Physiology B 120:223–230.

Stradi, R., E. Pini, and G. Celentano. 2001. The chemical structure of the pigments in *Ara macao* plumage. Comparative Biochemistry and Physiology B 130:57–63.

Stradi, R., E. Rossi, G. Celentano, and B. Bellardi. 1996. Carotenoids in bird plumage: the pattern in three *Loxia* species and in *Pinicola enucleator*. Comparative Biochemistry and Physiology B 113:427–432.

Strain, H. H. 1934. Carotene VIII. separation of carotenes by adsorption. Journal of Biological Chemistry 105:523–535.

Test, F. H. 1969. Relation of wing and tail color of the woodpeckers *Colaptes auratus* and *C. cafer* to their food. Condor 71:206–211.

Thomas, D. B., M. E. Hauber, D. Hanley, G. I. N. Waterhouse, S. Fraser, and K. C. Gordon. 2015. Analysing avian eggshell pigments with Raman spectroscopy. Journal of Experimental Biology 218:2670–2674.

Thomas, D. B., C. M. McGoverin, K. J. McGraw, H. F. James, and O. Madden. 2013. Vibrational spectroscopic analyses of unique yellow feather pigments (spheniscins) in penguins. Journal of the Royal Society Interface 10:20121065.

Thomas, D. B., K. J. McGraw, M. W. Butler, M. T. Carrano, O. Madden, and H. F. James. 2014a. Ancient origins and multiple appearances of carotenoid-pigmented feathers in birds. Proceedings of the Royal Society B 281:20140806.

Thomas, D. B., K. J. McGraw, H. F. James, and O. Madden. 2014b. Non-destructive descriptions of carotenoids in feathers using Raman spectroscopy. Analytical Methods 6:1301–1308.

Thomas, D. B., P. C. Nascimbene, C. J. Dove, D. A. Grimaldi, and H. F. James. 2014c. Seeking carotenoid pigments in amber-preserved fossil feathers. Scientific Reports 4:5226.

Troscianko, J., and M. Stevens. 2015. Image calibration and analysis toolbox—a free software suite for objectively measuring reflectance, colour and pattern. Methods in Ecology and Evolution 6:1320–1331.

Troy, D. M., and A. H. Brush. 1983. Pigments and feather structure of the Redpolls, *Carduelis flammea* and *C. hornemanni*. Condor 85:443–446.

Tswett, M. 1906. Physikalisch-chemische studien über das chlorophyll. Die adsorption. Berichte der Deutschen botanischen Gesellschaft 24:316–326.

Tubaro, P. L., D. A. Lijtmaer, and S. C. Lougheed. 2005. Cryptic dichromatism and seasonal color variation in the Diademed Tanager. Condor 107:648–656.

Underwood, T. J., and S. G. Sealy. 2002. Adaptive significance of egg coloration. Pp. 280–298 in D. C. Deeming (editor), Avian incubation, behaviour, environment and evolution, Oxford University Press, Oxford, UK.

Veronelli, M., G. Zerbi, and R. Stradi. 1995. In situ resonance Raman spectra of carotenoids in bird's feathers. Journal of Raman Spectroscopy 26:683–692.

Vinther, J. 2015. A guide to the field of palaeo colour—melanin and other pigments can fossilise: reconstructing colour patterns from ancient organisms can give new insights to ecology and behaviour. BioEssays 37:643–656.

Vinther, J., D. E. Briggs, J. Clarke, G. Mayr, and R. O. Prum. 2010. Structural coloration in a fossil feather. Biology Letters 6:128–131.

Vinther, J., D. E. Briggs, R. O. Prum, and V. Saranathan. 2008. The colour of fossil feathers. Biology Letters 4:522–525.

Völker, O. 1934. Die abhangigkeit der lipochrombildung bie vogeln von pflanzlichen carotinoiden. Journal für Ornithologie 82:439–450.

Völker, O. 1938. Porphyrin in vogelfedern. Journal für Ornithologie 86:436–456.

Vorobyev, M. 2003. Coloured oil droplets enhance colour discrimination. Proceedings of the Royal Society B 270:1255–1261.

Vorobyev, M., R. Brandt, D. Peitsch, S. B. Laughlin, and R. Menzel. 2001. Colour thresholds and receptor noise: behaviour and physiology compared. Vision Research 41:639–653.

Vorobyev, M., and D. Osorio. 1998. Receptor noise as a determinant of colour thresholds. Proceedings of the Royal Society B 265:351–358.

Vorobyev, M., D. Osorio, A. T. Bennett, N. J. Marshall, and I. C. Cuthill. 1998. Tetrachromacy, oil droplets and bird plumage colours. Journal of Comparative Physiology A 183:621–633.

Vukusic, P., and D. G. Stavenga. 2009. Physical methods for investigating structural colours in biological systems. Journal of the Royal Society Interface 6:S133–S148.

Wackenroder, H. 1831. Ueber das oleum radicis dauci aetherum, das carotin, den carotenzucker und den officinellen succus cauci; so wie auch über das mannit, welches in dem möhrensafte durch eine besondere art der gährung gebildet wird. Geigers Magazin der Pharmazie 33:144–172.

White, T. E., R. L. Dalrymple, D. W. A. Noble, J. C. O'Hanlon, D. B. Zurek, and K. D. L. Umbers. 2015. Reproducible research in the study of biological coloration. Animal Behaviour 106:51–57.

With, T. K. 1978. On porphyrins in feathers of owls and bustards. International Journal of Biochemistry 9:893–895.

Woodward, L. A. 1967. General Introduction. Pp. 1–6 in H. A. Szymanski (editor), Raman spectroscopy: theory and practice, Plenum Press, New York, NY.

Yang, C., Y. Cai, and W. Liang. 2013. Eggs mimicry of Common Cuckoo (Cuculus canorus) utilizing Ashy-throated Parrotbill (Paradoxornis alphonsianus) host. Chinese Birds 4:51–56.

Zhou, J., J. Shang, F. Ping, and G. Zhao. 2012. Alcohol extract from Vernonia anthelmintica (L.) wild seed enhances melanin synthesis through activation of the p38 MAPK signaling pathway in B16F10 cells and primary melanocytes. Journal of Ethnopharmacology 143:639–647.

CHAPTER FOUR

Integrating Museum and Media Collections to Study Vocal Ecology and Evolution*

Nicholas A. Mason, Bret Pasch, Kevin J. Burns, and Elizabeth P. Derryberry

Abstract. Studies of animal vocalizations have generated key insights into the evolutionary and ecological forces that shape behavioral diversity in birds and other animals. Natural history collections and media archives provide a wealth of data that are being incorporated into studies of vocal evolution with increasing frequency and sophistication. Here, we review recent advances regarding the integration of museum and media collections to study vocal evolution and ecology of animals with a special emphasis on birds. We consider how digital archives of bioacoustic data combined with vouchered specimens and other biological collections have improved our understanding of geographic variation in vocalizations, longitudinal studies of cultural evolution, and comparative studies of vocal evolution and diversification, among other topics. We highlight case studies that exemplify the novel approaches and insights gained from studies of animal vocalizations that leverage biological collections. In providing this overview, we encourage the scientific community to further consider how natural history collections can address longstanding questions in ecology and evolutionary biology.

Key Words: archives, bird song, specimen.

Animals vocalize to exchange information with other individuals for tasks essential to reproduction and survival, such as attracting mates and defending territories (Bradbury and Vehrencamp 1998). Bioacoustics and the study of avian vocalizations have deep roots in ornithology, and studies therein have provided important advances that contribute to our broader understanding of animal behavior (e.g., Marler and Slabbekoorn 2004, Dugatkin 2009). Traditionally, studies of animal vocalizations have relied on behavioral observations of live organisms rather than museum specimens from natural history collections; ongoing and recent studies, however, increasingly leverage natural history and technological innovations to enable large-scale investigations of animal vocalizations—both within and among populations and species. Here, by reviewing recent studies that combine bioacoustics and museum specimens, we identify emergent trends in how natural history collections have advanced our understanding of the evolutionary and ecological

* Mason, N. A., B. Pasch, K. J. Burns, and E. P. Derryberry. 2017. Integrating museum and media collections to study vocal ecology and evolution. Pp. 57–74 in M. S. Webster (editor), The Extended Specimen: Emerging Frontiers in Collections-based Ornithological Research. Studies in Avian Biology (no. 50), CRC Press, Boca Raton, FL.

processes that shape the remarkable diversity of animal vocal signals that occur in nature.

Although the field of bioacoustics and the practice of recording birds for scientific purposes have existed since the early 1900s (Gaunt et al. 2005), recent technological innovations have induced a paradigm shift in how natural sounds are collected, archived, and distributed to scientific researchers, educators, and the general public (Ranft 2004, Betancourt and McLinn 2012). The advent of digital audio has transformed large rooms filled with reels upon reels of analog recordings into terabytes—or even petabytes—of digital audio files, which can be accessed by anyone with an Internet connection (Budney et al. 2014, Marques et al. 2014). Large-scale digital archiving initiatives, such as efforts spearheaded by the Macaulay Library (http://macaulaylibrary.org/), the Borror Laboratory of Bioacoustics (https://blb.osu.edu/), and Xeno-Canto (http://www.xeno-canto.org/), among others, have made millions of recordings readily available. This massive amount of bioacoustic data has enabled diverse research initiatives at an unprecedented scale across geographic space, taxonomy, and time.

In this chapter, we review recent developments in how data from ornithological collections—such as audio recordings and physical specimens—have advanced our understanding of the evolution and ecology of bird song, including geographic variation, selective constraints, and rates of vocal evolution and diversification (Figure 4.1). Birds produce a wide variety of sounds, including nonmating calls and mechanical sounds; we focus on songs, which serve as courtship and territorial displays and are typically conspicuous behaviors (Catchpole and Slater 2008). One prominent way in which songs differ among avian lineages is whether vocal displays are learned through cultural transmission or are inherited genetically (Slater 1986, Kroodsma 2005). Learning can have profound effects on the evolution of characters involved in sexual selection

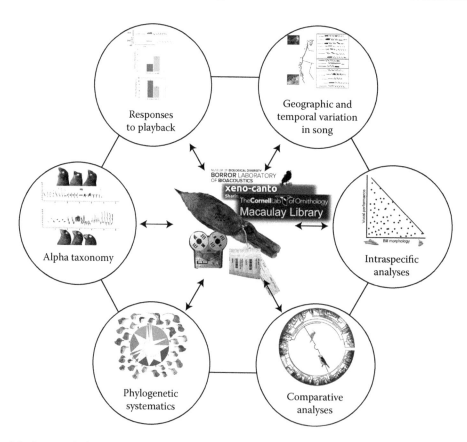

Figure 4.1. Conceptual advances enabled by combining media archives and natural history collections. Each circle on the perimeter represents a research topic that has been advanced by combining data from physical museum specimens and digital bioacoustic archives.

and speciation (Verzijden et al. 2012). In allopatry, learning can accelerate vocal evolution and speciation (Lachlan and Servedio 2004). When distributions overlap, however, learning can diminish the role of songs as premating barriers to gene flow (Seddon and Tobias 2007, Olofsson et al. 2011). Thus, the impacts of learning on vocal evolution and diversification are complex and context dependent (Wilkins et al. 2013). Although we emphasize birds throughout the chapter, we also include a brief section on parallel advances in nonavian taxa. Finally, we consider future directions and goals of the field in leveraging natural history collections to further the study of animal vocalizations.

SPECIES LIMITS AND GEOGRAPHIC VARIATION IN BIRD SONG

Many species exhibit geographic variation in vocalizations (Podos and Warren 2007), and these patterns of vocal variation play an important role in understanding species limits in avian taxonomy (Alström and Ranfft 2003, Tobias et al. 2010). Closely related species are often most differentiated in secondary sexual characteristics (Coyne and Orr 2004), and birds often exhibit stronger premating than postmating reproductive barriers (Grant and Grant 1997, Price 2007). As such, it is not surprising that geographic variation in song can act as a cue and precursor to genetic divergence. Recent studies highlighting the importance of vocalizations in delimiting avian species include the newly described Tropeiro Seedeater (*Sporophila beltoni*; Repenning and Fontana 2013) and Perijá Tapaculo (*Scytalopus perijanus*; Avendaño et al. 2015). Since 1990, more than 25 published species descriptions include analyses of audio recordings that are curated by digital sound archives, such as the Macaulay Library (e.g., Fitzpatrick and Willard 1990, Zimmer et al. 2001, Cuervo et al. 2005, Herzog et al. 2008), and the majority of those publications reference a media specimen as a "vocal type voucher" that can augment the value of traditional museum specimens.

Songs often vary substantially among populations and individuals within a species as well as among species; quantifying geographic variation among populations is another fundamental use of audio recordings of bird song. Until recently, however, most studies exploring geographic variation in bird song have relied on individual efforts to sample vocalizations from different populations

and to characterize variation in songs and vocal repertoires among the sampled individuals and populations. Such early efforts were motivated by attempts to document song learning as well as interspecific and intraspecific patterns in geographic song variation. Early examples include the studies of Luis Baptista from the California Academy of Natural Science. In the 1970s and 1980s, Baptista and colleagues captured individual and geographic measures of song variation across multiple subspecies of White-crowned Sparrows (*Zonotrichia leucophrys*), covering thousands of miles (Baptista 1975, 1977; Orejuela and Morton 1975; Baptista and Morton 1982). Those efforts made possible a series of later studies on geographic and temporal variation in White-crowned Sparrow song (Chilton and Lein 1996a; Harbison et al. 1999; Derryberry et al. 2007; Derryberry 2009, 2011). Many examples exist of similarly impressive early efforts to document geographic variation in the song of birds (see Mundinger 1982 for overview). Unfortunately, many of these early efforts survive only as spectrograms printed in journal articles: the original recordings no longer exist. Baptista's archival efforts reflected his position as a museum curator. Most other early recordists—many of whom were not associated with museums—did not archive their analog materials, and these recordings are often lost or are now too degraded for scientific use (Van Bogart 1995). Historically, gathering recordings to study avian vocalizations was largely an individual endeavor requiring extensive time and effort in the field by the primary investigators rather than a community-based initiative.

In sharp contrast, recent studies of geographic variation in bird song leverage preexisting recordings that are made available by digitized bioacoustics libraries, which has enabled large-scale studies of geographic variation in vocalizations that were previously unfeasible. Direct collection of digital audio recordings, in contrast to analog recordings, which required heavier equipment and digitization time, has facilitated more individuals making recordings (including "citizen science" approaches) and wider dissemination of those recordings (August et al. 2015). This recent boom in material has necessitated new conversations regarding best practices for annotating and vouchering recorded materials (Ranft 2004).

Among songbirds, recent studies on Common Yellowthroat (*Geothlypis trichas*; Bolus 2014), Alder

Flycatcher (*Empidonax alnorum*; Lovell and Lein 2013), Yellowhammer (*Emberiza citronella*; Petrusková et al. 2014), and Ruddy-capped Nightingale-thrush (*Catharus frantzii*; Ortiz-Ramírez et al. 2016) demonstrate how researchers can draw from multiple sources of audio recordings to characterize geographic variation in song. Nonetheless, many studies continue to rely on individual efforts, particularly in geographic regions that have few birders or ornithologists actively collecting animal sounds. Petrusková et al. (2010), for instance, categorized individual and geographic variation from over 2,000 songs of Tree Pipit (*Anthus trivialis*) at multiple spatial scales in eastern Europe. Other examples include Cicero and Benowitz-Fredericks (2000), who recorded over 4,500 songs from more than 50 male Lincoln's Sparrow (*Melospiza lincolnii*) to characterize variation in repertoire size and song types. While the availability of bioacoustics data through media archives and the digitization of analog recordings have enabled large-scale geographic studies, continued individual and collective recording efforts are still required to expand both the breadth and depth of available data for many species and regions.

The studies summarized here do not cover exhaustively the extensive work done by contemporary ornithologists to characterize geographic variation in bird song. Nevertheless, these studies do illustrate general trends in how audio data are collected, digitized, and widely disseminated for modern studies of animal communication. In addition to increased access to bioacoustics data, recent advances in analytical methods, including improved software and algorithms for detailed and automated analyses, have enabled high-throughput processing of numerous recordings (Aide et al. 2013). Together, these approaches have facilitated large-scale studies of avian vocalizations and cultural evolution among many taxa and across impressive geographic and temporal axes.

CULTURAL EVOLUTION AND TEMPORAL VARIATION IN AVIAN VOCAL DISPLAYS

Our understanding of the tempo and mode of song evolution is largely limited to studies of current variation in vocal displays. Whereas contemporary geographic variation in song provides one snapshot of song variation representing the results of different selective pressures acting on song over time, longitudinal studies can provide a more

direct assessment of how sexual and natural selection shape bird song. Nonhistorical approaches to temporal processes are problematic in the study of cultural traits (Payne 1996) as well as genetic traits (e.g., Gillespie 1991). Some common assumptions made by only considering current geographic variation in phenotypes—many of which are probably not met by song—are that traits evolve at a constant rate, that traits diverge gradually over time, and that local dispersal explains the geographic distribution of traits (Lynch et al. 1989, Lynch and Baker 1993). For example, in one of the few studies to carefully follow changes in song over time, changes in song did not reflect a branching tree among Indigo Bunting (*Passerina cyanea*) populations; instead, patterns of song evolution resembled a reticulated network in which song components were shared among individuals from different lineages rather than following a strict pattern of inheritance (Payne 1996). Measures of temporal variation in song are thus essential to a complete understanding of the tempo and mode of song evolution. Although song can evolve within contemporary timescales and can be documented within an investigator's lifetime, many questions remain about temporal variation at longer, multigenerational timescales. Such questions can only be answered with access to archived, historical song samples.

The earliest studies on temporal variation in song documented patterns of cultural transmission. Changes in songs over time were due to innovations, mistakes, or preferential learning. These studies often represented the efforts of individual scientists recording banded individuals within populations over one or a few generations of birds (e.g., Jenkins 1978; Mundinger 1980; Payne et al. 1981, 1988; Payne and Payne 1993; Payne 1996; Lang and Barlow 1997). Building on this foundation, subsequent studies compared songs recorded within the same locality but across decades, often relying on earlier datasets and recordings collected by a different investigator. These types of studies documented both remarkable stability (e.g., Payne et al. 1981, Baker and Jenkins 1987, Harbison et al. 1999, Derryberry 2009) as well as rapid turnover (e.g., Ince et al. 1980, Chilton and Lein 1996b, Payne 1996, Baker et al. 2003) of song features. Such variation in transmission patterns inspired follow-up work to assess the selective pressures shaping bird song (e.g., Nelson et al. 2004, Derryberry 2009). Only a handful of the recordings from earlier studies on temporal variation were archived, thereby limiting the number of species that can be currently studied over multiple generations of birds. Yet continued archiving of natural sounds ensures that future generations of scientists can study how vocal signals evolve using longitudinal datasets.

BEHAVIORAL RESPONSES TO GEOGRAPHIC AND TEMPORAL VARIATION IN SONG

The efficacy of mating signals depends on successful communication, or the transmission and reception of a signal (Searcy and Nowicki 2005). Any breakdown in this process can lead to reduced fitness by suppressing either mate attraction or territoriality (West-Eberhard 1983, Sætre 2000). Less effective signals may reduce an individual's ability to compete successfully against rival males, thereby affecting their ability to obtain a mate (West-Eberhard 1983; Andersson 1994). Interspecific behavioral barriers to gene flow may form if populations diverge in mating signals and/or signal recognition. Grant and Grant (1997) suggest that avian speciation, in particular, is often driven by the evolution of premating barriers, and one source of these barriers may be the diversification of song. Studies of geographic and temporal variation in bird song not only allow determination of the sources of selection driving song diversification, but also permit measurement of the functional consequences of diversification in mating signals. The 60-plus years of work on song indicates that both potential mates and competitors respond differently to geographic and temporal variants of song. Substantial geographic variation in song has been documented in many species (e.g., Baptista and King 1980, Krebs and Kroodsma 1980, Tubaro and Segura 1995, Martens 1996, Chilton et al. 2002). In turn, individuals (both male and female) generally respond more strongly to conspecific rather than heterospecific song (Ratcliffe and Grant 1985, Grant and Grant 1996, Bentley et al. 2000, Soha and Marler 2001, Nelson and Soha 2004; but also see Sætre et al. 1997, Irwin et al. 2001a). Many individuals also prefer songs of their local population over songs of other conspecific populations (reviewed in Andersson 1994, Catchpole and Slater 1995, Searcy and Nowicki 2005). Species often have diverse vocal repertoires that include distinct signals involved in territoriality or mate attraction,

and these different signals frequently evolve under distinct selective pressures among signal producers and receivers (Searcy and Brenowitz 1988, Seddon and Tobias 2010). In certain instances, taxa may converge on similar songs through cultural selection acting on interspecific territorial displays, whereas signals involved in mate attraction may remain diagnostic among species (Podos and Warren 2007, Tobias and Seddon 2009). These behavioral assays of response to heterospecific and geographic song variants demonstrate that the reproductive and territorial efficacy of vocal displays depends on the source population of both the signal producer and the receiver and the ecological context in which signals are produced.

Testing receiver responses to geographic song variants is often used as an approximation of signal evolution over time (e.g., Irwin et al. 2001b). Differences in responses to playbacks of songs from separate populations within a species' range indicate that signal divergence among populations may lead to reproductive isolation (Lachlan and Servedio 2004). Yet patterns of responses to playbacks provide little information on how signal evolution within populations contributes to reproductive isolation. Another approach measures receiver responses to statistical reconstructions of ancestral signals to infer the formation of reproductive barriers between species (Ryan and Rand 1995, Losos 1999).

The most direct test of the effect of signal evolution on signal efficacy within populations is to measure the response of receivers to actual mating signals at different time points within a single population. Measurement of responses to actual mating signals are rarely used because of the obvious difficulty of preserving historical songs in a manner that allows their use in behavioral tests. Stimuli used in behavioral assays must have high signal-to-noise ratios and minimal degradation. In addition, recordings of multiple individuals are needed to avoid pseudoreplication during experiments (Kroodsma 1990). In the 1960s, the use of audio recordings to document bird song became widespread, and recordings that were properly archived are now a primary source of historical songs of sufficient quality to be used in behavioral assays, including territorial playbacks, copulation solicitation, and operant conditioning assays.

Derryberry (2007) used historical recordings of song from a population of White-crowned Sparrows at Tioga Pass, California, to test whether changes in mating signals over time induced a loss in signal efficacy. Derryberry leveraged historical stimuli recorded by Baptista, archived at the California Academy of Sciences, and later curated by the Borror Laboratory of Bioacoustics. This study demonstrated that song evolution affected the efficacy of songs in this population: females performed more solicitations and males approached a simulated intruder more closely during playback of current songs compared to songs recorded 24 years earlier. Historical songs were less effective signals for receivers in the current population, both in the context of female mate choice and male–male competition. These data represented the first direct demonstration in a single natural population that signal evolution can impact the capacity for historical stimuli to elicit conspecific responses. These findings also suggested that signal evolution contributes to the formation of behavioral barriers to gene flow. However, the mechanisms underlying birds' responses to historical songs may not be the same mechanisms underlying responses to geographic variation in song. In order to understand how song evolution within populations contributes to the formation of behavioral barriers between populations, it is important to understand how response to song evolution within populations compares to response to geographic variation in song.

Derryberry (2011) contrasted male responses to temporal and geographic variation in song. Specifically, males' responses to historical local, current nonlocal, and heterosubspecific songs were compared using a series of playback experiments. Historical local songs were as effective as current nonlocal songs and significantly more effective than the songs of another subspecies in eliciting a response from territorial males. In addition, the songs of the local and nonlocal populations had changed in a roughly parallel direction in acoustic space while distinct song types were maintained during the past 35 years (Harbison et al. 1999). Variation in this acoustic space appeared to explain variation in male response to playback songs, such that the more dissimilar a song stimulus was to current variation in local song, the less strongly a male responded (Derryberry 2011). Altogether, these findings suggest that similar mechanisms may explain male response to song along the axes of both time and space. Additional studies are sorely needed to assess whether similar results hold true in other species. Bioacoustics archives offer a starting point for future studies by curating historical and current recordings for the future.

MACROEVOLUTIONARY PATTERNS OF AVIAN VOCALIZATIONS

In parallel with recent improvements in the availability of bioacoustics data and analytical techniques, the systematics and museum communities have generated massive amounts of molecular data, enhancing our understanding of evolutionary relationships among birds. Phylogenies generated from these molecular data provide the necessary framework to consider evolutionary patterns of diversification and trait evolution at deep evolutionary timescales. By examining vocal variation among species in conjunction with additional phenotypic and genetic data gathered from museum specimens, many investigators are answering long-standing questions about how bird song coevolves with other aspects of avian biology. Here, we provide an overview of two research programs as case studies of how phenotypic and genetic data from museum specimens are being combined with vocal data to further our understanding of avian communication and evolution. We focus on recent work on tanagers (Thraupidae) and ovenbirds and woodcreepers (Furnariidae), but note that important advancements regarding interspecific patterns of avian vocal evolution are also being made in other lineages, such as wood warblers (Cardoso 2010, Cardoso and Hu 2011), fringilid finches (Cardoso et al. 2012), fairy-wrens (Greig et al. 2013), blackbirds (Odom et al. 2015), and leaf warblers (Tietze et al. 2015), among many others.

Tanagers (Thraupidae) were traditionally limited to include only a group of bright, frugivorous songbirds in Central and South America that typically produced inconspicuous and simple vocalizations (Isler and Isler 1999). However, recent molecular phylogenies have transformed the taxonomic boundaries of tanagers: many species that were previously considered tanagers are actually members of other avian families, and species that were assumed to be distantly related to tanagers form a monophyletic group with the remaining tanagers (Burns et al. 2014, Barker et al. 2015). The recent revisions of the taxonomic grouping of tanagers now include species that span an impressive array of vocal variation that has been the subject of a recent series of studies on vocal evolution (Mason 2012; Figure 4.2). Mason et al. (2014) used

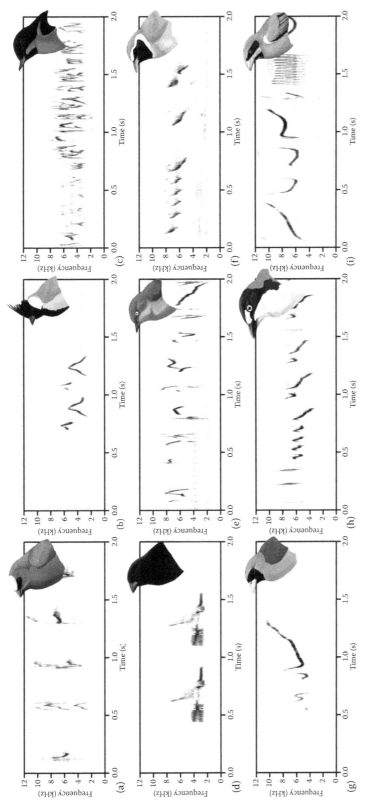

Figure 4.2. Spectrograms illustrating vocal diversity among tanagers. Representative species from 15 different subfamilies of Thraupidae are included. Darker shades of gray in the spectrograms indicate higher decibel levels. Spectrograms are on equal scales for frequency and time, demonstrating the diversity in frequency and temporal aspects of thraupid songs. These vocal data were used in combination with museum specimen data to study vocal evolution in a series of publications. (a) Catamblyrhynchus diadema, (b) Charitospiza eucosma, (c) Diglossa brunneiventris, (d) Geospiza fuliginosa, (e) Hemispingus xanthophthalmus, (f) Hemithraupis guira, (g) Incaspiza personata, (h) Nemosia pileata, and (i) Parkerthraustes humeralis. (Portraits are courtesy of Mary Margaret Ferraro. Adapted from Figure 1 of Mason and Burns 2015 with permission from Biological Journal of the Linnean Society.)

(Continued)

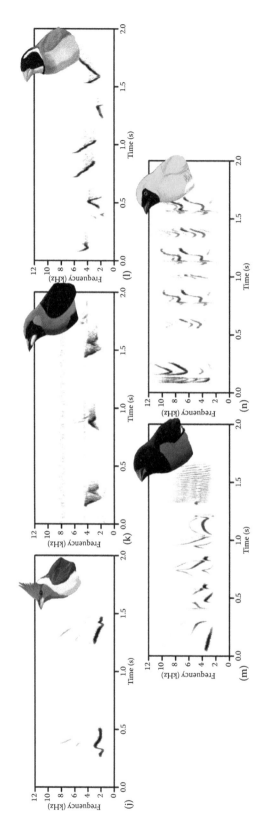

Figure 4.2. (Continued) Spectrograms illustrating vocal diversity among tangers. Representative species from 15 different subfamilies of Thraupidae are included. Darker shades of gray in the spectrograms indicate higher decibel levels. Spectrograms are on equal scales for frequency and time, demonstrating the diversity in frequency and temporal aspects of thraupid songs. These vocal data were used in combination with museum specimen data to study vocal evolution in a series of publications. (j) *Parouria coronata*, (k) *Ramphocelus nigrogularis*, (l) *Saltator aurantiirostris*, (m) *Sporophila angolensis*, and (n) *Tersina viridis*. Adapted from Figure 1 of Mason and Burns 2015 with permission from *Biological Journal of the Linnean Society*. (Portraits are courtesy of Mary Margaret Ferraro. Adapted from Figure 1 of Mason and Burns 2015 with permission from *Biological Journal of the Linnean Society*.)

a dataset of over 300 species of tanagers with archived recordings in natural sounds collections and spectrophotometry data from museum specimens. This study tested whether there is an evolutionary trade-off between song and plumage elaboration, which has been coined the "transfer hypothesis" (Darwin 1871, Gilliard 1956). The transfer hypothesis posits that elaborate plumage and song are both costly to produce and maintain in males, and that female-driven sexual selection acts on just a single trait such as song or plumage, but not both (Iwasa and Pomiankowski 1994). Despite the extensive variation in both song and plumage complexity among tanagers, there was no generalized relationship between song and plumage complexity, suggesting that elaborate vocal displays and complex plumage evolve independently in tanagers (Mason et al. 2014). Using the same dataset of tanager vocalizations, Mason and Burns (2015) found that body mass data, which were also collected from museum specimens, were strongly correlated with multiple aspects of tanager song, whereas broad categorizations of habitat type were not. These studies illustrate the capacity for contemporary studies to quantify macroevolutionary patterns across many hundreds of species and deep evolutionary timescales. Studies of this nature are not possible without digital sound archives, the availability of decades of bioacoustics recordings, and the ongoing recording efforts of the ornithological community. Moreover, comparisons among vocalization, coloration, and body mass data, as well as molecular phylogenies, all depend heavily on museum specimens, recapitulating the importance of museum specimens in modern, comparative studies of avian vocalizations.

Parallel studies in other avian lineages have similarly leveraged museum collections to produce new insights into avian vocal evolution. Among suboscines, the Neotropical radiation of woodcreepers and ovenbirds (Furnariidae) has provided an important system for understanding signal evolution in taxa with innate, or unlearned, vocalizations (Tobias et al. 2012). Derryberry et al. (2012) examined the tempo and mode of song evolution within woodcreepers and evaluated morphological constraints on song diversification. Dendrocolaptinae exhibit remarkable diversity in bill morphology, in which species with larger bills are limited in their ability to produce rapid trills over large frequency ranges, which emphasizes the role of indirect selection on song via morphological adaptation (Derryberry et al. 2012).

Currently, Derryberry and colleagues are using 276 Furnariid taxa to test the relative roles of sensory drive and "magic traits" in the evolution of innate song (unpubl. data). Sensory drive is the process by which acoustic differences among signaling environments influence the evolutionary divergence of mating signals by optimizing signal transmission (Wilkins et al. 2013). So-called magic traits are phenotypes that are under natural selection and also function in mate choice, such that ecological divergence can influence sexual selection and possibly speciation (Servedio et al. 2011). Sensory drive is often assumed to be more pervasive than magic traits, yet our understanding of how these direct and indirect processes interact during signal evolution is limited. Derryberry et al. (unpubl. data) compared the extent to which vocal evolution was related to the direct influence of habitat characteristics and the indirect effect of beak size, a well-established magic trait in birds. They found that sensory drive is an important factor explaining the structure of acoustic signals across this diverse radiation, but when the effects of beak size were included, it became clear that a balance between these processes drives signal evolution. These results confirm that two independent mechanisms shape bird songs over evolutionary time, and also suggest that the primary origin of signal diversification is ecological selection on magic traits (beaks) driving correlated evolution of mating signals (songs).

Similar to the series of tanager studies, this research on Furnariids relied heavily on museum collections for access to vouchered DNA samples, study skins to take accurate and detailed morphological measurements, and song samples with detailed and accurate metadata from various bioacoustics archives, including the Macaulay Library and Xeno-Canto. In many cases, morphological data could be taken from the same study skin used to voucher a DNA sample, creating a powerful link between morphological and genetic data. Such a link exists between some song and museum specimens as occurs when the audio recording is of the same bird that was subsequently collected, but these links are rare (see next section). Comparative analyses on the tempo and mode of vocal evolution will be most powerful when information linking museum vouchers to song data are readily available.

LINKING MUSEUM VOUCHERS AND AUDIO RECORDINGS

Curated links between physical specimens and behavioral data, such as bird song, strengthen studies on the tempo and mode of signal evolution and recognition. For example, if a recording is associated with a specific physical specimen, then any question as to the species or population identity of that vocalization can be addressed by returning to the physical specimen of interest (Winker et al. 2010). This is similar to the traditional use of a physical specimen as a permanent record with potentially unforeseen utility, but the vouchered data extended to include curated vocalizations linked to the specimen (Palmer et al. 2013). The ability to make links between physical specimens and curated data is especially important when experts disagree over the identity of a song or simply when behavioral data are needed by someone other than the original collector. Population-level vouchers or curated links between song data and specimens collected from the same population at the same time (but not the same individual) can also facilitate the correct association of genetic and behavioral data (see Chapter 13, this volume).

Curating behavioral data in association with physical specimens should provide an important means of (1) validating behavioral data, (2) facilitating studies comparing behavior to other phenotypic traits (e.g., bill morphology), and (3) lowering the expense and time associated with studies on signal evolution. For example, incorrectly associating a song phenotype with a terminal taxon of a molecular phylogeny may yield false estimates of the rate of behavioral change. Further, building cross-referenced molecular, morphological, and behavioral datasets, given the separate collection of those data, is extremely difficult and time consuming. These issues are avoided when analyzing genetic or morphological data by restricting sampling to tissues associated with a physical specimen or by taking morphological data only from physical specimens. Similarly, issues of validation and building cross-referenced datasets can be minimized by the curation and use of song data associated with physical specimens.

Research on song evolution in the Neotropical radiation of woodcreepers and ovenbirds (see earlier) is a case study that illustrates the time involved in building cross-referenced datasets due to a lack of vouchered behavioral data. Several datasets were required for this research project: morphological data on bill size, molecular data for phylogenetic analysis, and behavioral song data. The morphological dataset was assembled by taking measurements from study skins archived in natural history museum collections around the world. The molecular dataset was assembled by sequencing DNA extracted from tissues frozen at the time of specimen collection and stored in cryo facilities at various institutions. These "vouchered tissues" were linked in the field to specific physical specimens later identified and curated into a collection. If, at any time, a question arose regarding results associated with the molecular or morphological dataset, sequences and traits could be traced directly back to a physical specimen to verify its identity. Although representative songs for most of the 290-plus species in Furnariidae could be found in existing scientific or online song databases (Figure 4.3), no means existed to link these recordings directly to specimens used in the molecular dataset. Although a population-level voucher would suffice in this comparative study, there also were no curated links between acoustic and physical collections at the population level. Instead, it took more than 2 years to associate the geographic origin of the song data with the geographic origin of the physical specimens and then to validate that one was representative of the other. Such efforts would be streamlined by curated song data, particularly when associated with physical specimens.

The best method to align behavioral and molecular data for comparative analyses of trait evolution is to associate both sources of data based on collection locality. However, many collections of acoustic and physical specimens do not provide full locality information in their databases. For physical vouchers, it is often necessary to look at the actual specimen tags attached to the study skin to identify exact location. For acoustic data, it is necessary to return to the notes associated with the song or to contact the recordist directly to determine the exact location of a particular recording. Once locality data are obtained for both datasets for each taxon, it is then necessary to determine whether the two locations for each

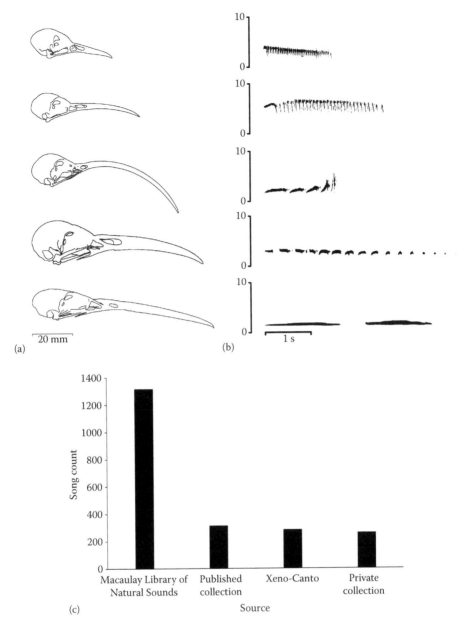

(a)

20 mm

(b)

1 s

(c)

Figure 4.3. Illustration of datasets linked by specimens (morphological and genetic data) and not linked by specimens (song) used in analysis of the evolution of ecological and social traits in the Furnariidae. (a) Outlines of bill profiles and (b) sound spectrograms of typical loud songs produced by five species representing variation in bill and song structure in Dendrocolaptinae (Aves: Furnariidae): (top to bottom) *Certhiasomus stictolaemus, Lepidocolaptes albolineatus, Campylorhamphus trochilirostris, Xiphocolaptes promeropirhynchus,* and *Nasica longirostris.* (Adapted from Figure 1 in Derryberry et al. 2012 with permission from *Evolution*.) (c) Histogram of the sources for songs used in the study, including the Macaulay Library of Natural Sounds, collections published on CDs, Xeno-Canto, and individual private collections. (Details on sources and data can be found in Tobias et al. 2014.)

taxon are representative of one another. To do this requires extensive knowledge about the scale at which species vary in their song phenotype as well as the degree to which populations within species are genetically differentiated. Often, both of these types of information are not known or are held by different experts. For example, are *Asthenes dorbignyi* songs recorded in Peru in the Departamento de Tacna representative of the genetic population of *A. dorbignyi* sampled in the Departamento de Arequipa? As songs are not available from the Departamento de Arequipa, it is necessary to find an expert who can identify the song of *A. dorbignyi* and has been to both locations. Failing that, we have to assume that the variation between songs from these two locations is less than the variation found between songs of *A. dorbignyi* and its sister species, *A. baeri*. Although working assumptions are valid, noise is added to the dataset, which might obscure important patterns of signal evolution.

Building curated links between natural history collections and collections of behavioral data will facilitate many important research studies on the tempo and mode of signal evolution. A number of museums with active natural history collections often make a vocal recording of an individual prior to collecting the voucher, including the Louisiana State University Museum of Natural Science and the Kansas University Biodiversity Institute and Natural History Museum among others. As more major natural history collections actively collect behavioral data in the field, the need for standard acquisition and curation of these ancillary data is pressing. Use of a standard approach to curate behavioral data improves the long-term quality of these data, consistency in these data across collections, and relevance to scientists over time. Similar approaches could be applied to banding stations, in which banders could prioritize audio recordings from the birds that they band, measure, and take blood or feather samples. The studies summarized here, and discussion of the challenges associated with linking behavioral and genetic data, indicate that the labor involved in collecting and curating these recordings is commendable and justifiable.

PARALLEL ADVANCES IN NONAVIAN TAXA

Acoustic signals offer a powerful lens into the evolution of communication in multiple taxa, not just birds. Similarly, physical specimens provide a starting point for studying mechanisms of signal evolution and recognition in nonavian taxa. Here, we briefly highlight a few recent studies linking museum specimens with sound archives to better understand the biology of anurans, insects, and mammals.

Morphology and allometry play important roles in shaping vocal evolution at broad evolutionary scales. Frequently, the physical dimensions and

properties of peripheral structures involved in sound production explain the majority of variation in signal form. As illustrated by Mason and Burns (2015), a long-standing inverse relationship between body size and dominant frequency has been documented in other birds (Wallschager 1980, Ryan and Brenowicz 1985), mammals (Hauser 1993, Jones 1999, Fitch and Hauser 2003, García-Navas and Blumstein 2016), and across many distantly related taxa (Gillooly and Ophir 2010). In anurans, similar correlations have been inferred among datasets with less geographic or taxonomic sampling (Cocroft and Ryan 1995, Hoskin et al. 2009). To explore whether such patterns generalize to larger phylogenies, Gingras et al. (2013) examined vocal parameters extracted from sound archives with data on body size based in part on museum specimens from 136 frog species distributed worldwide. The authors confirmed a strong association between body size and call frequency in anurans, and posited that laryngeal allometry and vocal fold dimensions mediate this relationship in most vertebrate taxa (Gingras et al. 2013).

Similar to investigations of putative evolutionary constraints between elaborate plumage and avian song complexity (e.g., Mason et al. 2014), Santos et al. (2014) explored the relationship between aposematic coloration and acoustic diversification in poison frogs (Dendrobatidae). Although some poison frogs advertise their unpalatable skin alkaloids to predators with visually conspicuous colors, most are cryptic and rely on camouflage (Santos et al. 2003). In addition, males of all species produce acoustic advertisement signals to attract mates (Zimmerman 1990). If predators associate mating signals with unpalatability, then release from predation pressure may allow aposematic species to evolve more elaborate mating signals. Santos et al. (2014) combined data on visual conspicuousness from live-caught animals, molecular data from museum specimens, and acoustic data from their own recordings and sound archives to test the hypothesis that aposematism enhances signal diversification. Indeed, changes in spectral and temporal features of acoustic signals were associated with aposematism, suggesting that relaxed predation facilitated the diversification of vocal signals, presumably via sexual selection.

Evolutionary constraints in acoustic parameters may also arise through morphometric and/or energetic trade-offs inherent in sound production mechanisms. For example, many crickets stridulate by rubbing a scraper on one wing against a file on the opposite wing, and females are sensitive to temporal aspects of male calls (Gerhardt and Huber 2002). Given constant wing closure velocities, longer files produce longer pulse durations, but also increase the time between each stridulation. Until recently, however, empirical evidence for such trade-offs did not exist. Walker and Funk (2014) combined morphological and molecular data from physical specimens with acoustic recordings from sound archives to investigate the evolution of calls in North American brown sword-tailed crickets (*Anaxipha* spp.). As expected, pulse duration was associated with files that were longer and the density of file teeth, but was inversely related to pulse rate. Another recent study verified a similar morphological and energetic trade-off in a closely related subfamily (*Oecanthus*; Symes et al. 2015), reflecting the importance of biomechanical trade-offs in constraining acoustic parameters across insects and disparate taxa, including birds (Podos 2001) and singing mice (Pasch et al. 2011).

Finally, the efficacy of acoustic communication relies on coevolution of senders and receivers, whereby diversification in signal production is complemented by adaptations in perception (Bradbury and Vehrencamp 1998). Thus, understanding receiver anatomy can provide important insight into the evolution of signal structure. In bats, structural modifications of the inner ear are thought to accompany production of high frequency echolocation signals. To explore the origins of ultrasonic hearing and echolocation, Davies et al. (2013) compared microcomputed tomography scans of cochleae from museum specimens of 16 bat families, encompassing echolocating and nonecholocating suborders, with call parameters obtained from the primary literature and sound archives. The researchers found strong correlations between echolocation call frequencies and cochlear morphology, with ancestral reconstructions of basilar membrane length and cochlear coiling suggesting a common ancestor capable of ultrasonic hearing. However, variation in rates of basilar membrane evolution between suborders indicated that laryngeal echolocation evolved independently (Davies et al. 2013). Finer sampling of museum specimens and sound archives, along with functional experiments that determine hearing sensitivities, will help resolve this complex evolutionary puzzle. Integrating media collections with museum specimens holds great prospects for exploring similar evolutionary and ecological links in birds and other taxa (e.g., Walsh et al. 2009).

CONCLUSIONS AND FUTURE DIRECTIONS

It is an exciting time to study the ecology and evolution of vocalizations in birds and other taxa. Recent advances in museum science and the technologies used to acquire and disseminate natural sound recordings have enabled new research initiatives. Increasingly, natural history collections are used to study multiple aspects of the biology of avian vocalizations. Yet the wealth and diversity of new sources of bioacoustic data present certain challenges for ornithologists and curators of bioacoustics archives. These challenges include creating efficient platforms for large-scale collection of acoustic recordings (and other media) and powerful search algorithms that allow researchers to locate the recordings needed for a particular project. Looking ahead, another key challenge will be to standardize the content and availability of data associated with bioacoustic recordings, especially those with links to vouchered specimens. By combining traditionally disparate data from bioacoustics archives and museum specimens, ornithologists continue to gain insight into the ecological and evolutionary forces that shape acoustic signals in nature.

ACKNOWLEDGMENTS

We thank M. S. Webster and the organizers of the symposium "The Extended Specimen: Emerging Frontiers in Collections-based Ornithological Research" for inspiring this chapter. We also thank A. H. Dalziell for helpful feedback and constructive criticism on an earlier version of this chapter. NAM was supported by an EPA STAR fellowship (F13F21201). KJB was supported by the National Science Foundation (DEB-1354006).

LITERATURE CITED

Aide, T. M., C. Corrada-Bravo, M. Campos-Cerqueira, C. Milan, G. Vega, and R. Alvarez. 2013. Real-time bioacoustics monitoring and automated species identification. PeerJ 1:e103.

Alström, P., and R. Ranfft. 2003. The use of sounds in avian systematics and the importance of bird sound archives. Bulletin of the British Ornithological Club 123A (Suppl.):114–135.

Andersson, M. 1994. Sexual selection. Princeton University Press, Princeton, NJ.

August, T., M. Harvey, and P. Lightfoot. 2015. Emerging technologies for biological recording. Biological Journal of the Linnean Society 115: 731–749.

Avendaño, J. E., A. M. Cuervo, J. P. López-O, N. Gutiérrez-Pinto, A. Cortés-Diago, and C. D. Cadena. 2015. A new species of tapaculo (Rhinocryptidae: Scytalopus) from the Serranía de Perijá of Colombia and Venezuela. Auk 132:450–466.

Baker, A. J., and P. F. Jenkins. 1987. Founder effect and cultural evolution of songs in an isolated population of chaffinches, Fringilla coelebs, in Chatham Islands. Animal Behaviour 35:1793–1803.

Baker, M. C., M. S. A. Baker, and E. M. Baker. 2003. Rapid evolution of a novel song and an increase in repertoire size in an island population of an Australian songbird. Ibis 145:465–471.

Baptista, L. F. 1975. Song dialects and demes in sedentary populations of the White-crowned Sparrow (Zonotrichia leucophrys nuttalli). University of California Publications in Zoology 105:1–52.

Baptista, L. F. 1977. Geographic variation in song and dialects of the Puget Sound White-crowned Sparrow. Condor 79:356–370.

Baptista, L. F., and J. R. King. 1980. Geographic variation in song and song dialects of montane White-crowned Sparrows. Condor 82:267–284.

Baptista, L. F., and M. L. Morton. 1982. Song dialects and mate selection in montane White-crowned Sparrows. Auk 99:537–547.

Barker, F. K., K. J. Burns, J. Klicka, S. M. Lanyon, and I. J. Lovette. 2015. New insights into New World biogeography: an integrated view from the phylogeny of blackbirds, cardinals, sparrows, tanagers, warblers, and allies. Auk 132:333–348.

Bentley, G. E., J. C. Wingfield, M. L. Morton, and G. F. Ball. 2000. Stimulatory effects on the reproductive axis in female songbirds by conspecific and heterospecific male song. Hormones and Behavior 37:179–189.

Betancourt, I., and C. M. McLinn. 2012. Teaching with the Macaulay Library: an online archive of animal behavior recordings. Journal of Microbiology and Biology Education 13:86–88.

Bolus, R. T. 2014. Geographic variation in songs of the Common Yellowthroat. Auk 131:175–185.

Bradbury, J. W., and S. L. Vehrencamp. 1998. Principles of animal communication. Sinauer Associates, Sunderland, MA.

Budney, G., W. McQuay, and M. Webster. 2014. Transitioning the largest archive of animal sounds from analogue to digital. Journal of Digital Media Management 2:212–220.

Burns, K. J., A. J. Shultz, P. O. Title, N. A. Mason, F. K. Barker, J. Klicka, S. M. Lanyon, and I. J. Lovette. 2014. Phylogenetics and diversification of tanagers (Passeriformes: Thraupidae), the largest radiation of Neotropical songbirds. Molecular Phylogenetics and Evolution 75:41–77.

Cardoso, G. C. 2010. Loudness of birdsong is related to the body size, syntax and phonology of passerine species. Journal of Evolutionary Biology 23:212–219.

Cardoso, G. C., and Y. Hu. 2011. Birdsong performance and the evolution of simple (rather than elaborate) sexual signals. American Naturalist 178:679–686.

Cardoso, G. C., Y. Hu, and P. G. Mota. 2012. Birdsong, sexual selection, and the flawed taxonomy of canaries, goldfinches and allies. Animal Behaviour 84:111–119.

Catchpole, C. K., and P. J. B. Slater. 2008. Bird song: biological themes and variations. Cambridge University Press, Cambridge, UK.

Chilton, G. 2003. Cultural evolution and song stereotypy in an isolated population of montane White-crowned Sparrows (Zonotrichia leucophrys oriantha). Bird Behavior 15:53–63.

Chilton, G., and M. R. Lein. 1996a. Song repertoires of Puget Sound White-crowned Sparrows Zonotrichia leucophrys pugetensis. Journal of Avian Biology 27:31–40.

Chilton, G., and M. R. Lein. 1996b. Long-term changes in songs and song dialect boundaries of Puget Sound White-crowned Sparrows. Condor 98:567–580.

Chilton, G., M. O. Wiebe, and P. Handford. 2002. Large-scale geographic variation in songs of Gambel's White-crowned Sparrows. Condor 104:378–386.

Cicero, C., and M. Benowitz-Fredericks. 2000. Song types and variation in insular populations of Lincoln's Sparrow (Melospiza lincolnii), and comparisons with other Melospiza. Auk 117:52–64.

Cocroft, R. B., and M. J. Ryan. 1995. Patterns of advertisement call evolution in toads and chorus frogs. Animal Behaviour 49:283–303.

Coyne, J. A., and H. A. Orr. 2004. Speciation. Sinauer Associates, Sunderland, MA.

Cuervo, A. M., C. D. Cadena, N. Krabbe, and L. M. Renjifo. 2005. *Scytalopus stilesi*, a new species of tapaculo (Rhinocryptidae) from the Cordillera Central of Colombia. Auk 122:445–463.

Darwin, C. 1871. The descent of man and selection in relation to sex. John Murray, London, UK.

Davies, K. T., I. Maryanto, and S. J. Rossiter. 2013. Evolutionary origins of ultrasonic hearing and laryngeal echolocation in bats inferred from morphological analyses of the inner ear. Frontiers in Zoology 10:2.

Derryberry, E. P. 2007. Evolution of bird song affects signal efficacy: an experimental test using historical and current signals. Evolution 61:1938–1945.

Derryberry, E. P. 2009. Ecology shapes birdsong evolution: variation in morphology and habitat explains variation in White-crowned Sparrow song. American Naturalist 174:24–33.

Derryberry, E. P. 2011. Male response to historical and geographic variation in bird song. Biology Letters 7:57–59.

Derryberry, E. P., N. Seddon, S. Claramunt, J. A. Tobias, A. Baker, A. Aleixo, and R. T. Brumfield. 2012. Correlated evolution of beak morphology and song in the Neotropical woodcreeper radiation. Evolution 66:2784–2797.

Dugatkin, L. A. 2009. Principles of animal behavior. W. W. Norton and Company, New York, NY.

Fitch, W. T., and M. D. Hauser. 2003. Unpacking "honesty": vertebrate vocal production and the evolution of acoustic signals. Pp. 65–137 in A. M. Simmons, A. N. Popper, and R. R. Fay (editors), Acoustic communication. Springer, New York, NY.

Fitzpatrick, J. W., and D. E. Willard. 1990. *Cercomacra manu*, a new species of antbird from southwestern Amazonia. Auk 107:239–245.

García-Navas, V., and D. T. Blumstein. 2016. The effect of body size and habitat on the evolution of alarm vocalizations in rodents. Biological Journal of the Linnean Society 118:745–751.

Gaunt, S. L., D. A. Nelson, M. S. Dantzker, G. F. Budney, J. W. Bradbury, and R. M. Zink. 2005. New directions for bioacoustics collections. Auk 122:984–987.

Gerhardt, H. C., and F. Huber. 2002. Acoustic communication in insects and anurans. Chicago University Press, Chicago, IL.

Gillespie, J. H. 1991. The causes of molecular evolution. Oxford University Press, Oxford, UK.

Gilliard, E. 1956. Bower ornamentation versus plumage characters in bower-birds. Auk 73: 450–451.

Gillooly, J. F., and A. G. Ophir. 2010. The energetic basis of acoustic communication. Proceedings of the Royal Society B 277:1325–1331.

Gingras, B., M. Boeckle, C. T. Herbst, and W. T. Fitch. 2013. Call acoustics reflect body size across four clades of anurans. Journal of Zoology 289:143–150.

Grant, B. R., and P. R. Grant. 1996. Cultural inheritance of song and its role in the evolution of Darwin's finches. Evolution 50:2471–2487.

Grant, P. R., and B. R. Grant. 1997. Genetics and the origin of bird species. Proceedings of the National Academy of Sciences USA 94:7768–7775.

Greig, E. I., J. J. Price, and S. Pruett-Jones. 2013. Song evolution in Maluridae: influences of natural and sexual selection on acoustic structure. Emu 113:270–281.

Harbison, H., D. A. Nelson, and T. P. Hahn. 1999. Long-term persistence of song dialects in the mountain White-crowned Sparrow. Condor 101:133–148.

Hauser, M. D. 1993. The evolution of nonhuman primate vocalizations—effects of phylogeny, body weight, and social context. American Naturalist 142:528–542.

Herzog, S. K., M. Kessler, and J. A. Balderrama. 2008. A new species of tyrannulet (Tyrannidae: *Phyllomyias*) from Andean foothills in northwest Bolivia and adjacent Peru. Auk 125:265–276.

Hoskin, C. J., S. James, and G. C. Grigg. 2009. Ecology and taxonomy-driven deviations in the frog call-body size relationship across the diverse Australian frog fauna. Journal of Zoology 278:36–41.

Ince, S. A., P. J. B. Slater, and C. Weismann. 1980. Changes with time in the songs of a population of chaffinches. Condor 82:285–290.

Irwin, D. E., P. Alström, and U. Olsson. 2001a. Cryptic species in the genus *Phylloscopus* (Old World leaf warblers). Ibis 143:233–247.

Irwin, D. E., S. Bensch, and T. D. Price. 2001b. Speciation in a ring. Nature 409:333–337.

Isler, M., and P. Isler. 1999. The tanagers. Smithsonian, Washington, DC.

Iwasa, Y., and A. Pomiankowski. 1994. The evolution of mate preferences for multiple sexual ornaments. Evolution 48:853–867.

Jenkins, P. F. 1978. Cultural transmission of song patterns and dialect development in a free-living bird population. Animal Behaviour 26:50–78.

Jones, G. 1999. Scaling of echolocation call parameters in bats. Journal of Experimental Biology 202:3359–3367.

Krebs, J. R., and D. E. Kroodsma. 1980. Repertoires and geographic variation in bird song. Advances in the Study of Animal Behavior 11:143–177.

Kroodsma, D. 2005. The singing life of birds. Houghton Mifflin Harcourt, Boston, MA.

Kroodsma, D. E. 1990. Using appropriate experimental-designs for intended hypotheses in song playbacks, with examples for testing effects of song repertoire sizes. Animal Behaviour 40:1138–1150.

Lachlan, R. F., and M. R. Servedio. 2004. Song learning accelerates allopatric speciation. Evolution 58:2049–2063.

Lang, A. L., and J. C. Barlow. 1997. Cultural evolution in the Eurasian tree sparrow: divergence between introduced and ancestral populations. Condor 99:413–423.

Losos, J. 1999. Uncertainty in the reconstruction of ancestral character states and limitations on the use of phylogenetic comparative methods. Animal Behaviour 58:1319–1324.

Lovell, S. F., and M. R. Lein. 2013. Geographic variation in songs of a suboscine passerine, the Alder Flycatcher (*Empidonax alnorum*). Wilson Journal of Ornithology 125:15–23.

Lynch, A., and A. J. Baker. 1993. A population memetics approach to cultural evolution in chaffinch song: meme diversity within populations. American Naturalist 141:597–620.

Lynch, A., G. M. Plunkett, A. J. Baker, and P. F. Jenkins. 1989. A model of cultural evolution of chaffinch song derived with the meme concept. American Naturalist 133:634–653.

Marler, P., and H. Slabbekkoorn. 2004. Nature's music: the science of birdsong. Elsevier Academic Press, Amsterdam, Netherlands.

Marques, P. A. M., D. M. Magalhães, S. F. Pereira, and P. E. Jorge. 2014. From the past to the future: natural sound recordings and the preservation of the bioacoustics legacy in Portugal. PLoS One 9:e114303.

Martens, J. 1996. Vocalizations and speciation of palearctic birds. Pp. 221–240 in D. E. Kroodsma and E. H. Miller (editors), Ecology and evolution of acoustic communication in birds. Cornell University Press, Ithaca, NY.

Mason, N. A. 2012. Song complexity and its evolutionary correlates across a continent-wide radiation of songbirds. M.Sc. thesis, San Diego State University, San Diego, CA.

Mason, N. A., and K. J. Burns. 2015. The effect of habitat and body size on the evolution of vocal displays in Thraupidae (tanagers), the largest family of songbirds. Biological Journal of the Linnean Society 114:538–551.

Mason, N. A., A. J. Shultz, and K. J. Burns. 2014. Elaborate visual and acoustic signals evolve independently in a large, phenotypically diverse radiation of songbirds. Proceedings of the Royal Society B 281:20140967.

Mundinger, P. C. 1980. Animal cultures and a general theory of cultural evolution. Ethology and Sociobiology 1:183–223.

Mundinger, P. C. 1982. Microgeographic and macrogeographic variation in the acquired vocalizations of birds. Pp. 147–208 in D. E. Kroodsma and E. H. Miller (editors), Acoustic communication in birds. Academic Press, New York, NY.

Nelson, D. A., K. I. Hallberg, and J. A. Soha. 2004. Cultural evolution of Puget Sound White-crowned Sparrow song dialects. Ethology 110:879–908.

Nelson, D. A., and J. A. Soha. 2004. Male and female White-crowned Sparrows respond differently to geographic variation in song. Behaviour 141:53–69.

Odom, K. J., K. E. Omland, and J. J. Price. 2015. Differentiating the evolution of female song and male-female duets in the New World blackbirds: can tropical natural history traits explain duets? Evolution 69:839–847.

Olofsson, H., A. M. Frame, and M. R. Servedio. 2011. Can reinforcement occur with a learned trait? Evolution 65:1992–2003.

Orejuela, J. E., and M. L. Morton. 1975. Song dialects in several populations of mountain White-crowned Sparrows (*Zonotrichia leucophrys oriantha*) in the Sierra Nevada. Condor 77:145–153.

Ortiz-Ramírez, M. F., M. J. Andersen, A. Zaldívar-Riverón, J. F. Ornelas, and A. G. Navarro-Sigüenza. 2016. Geographic isolation drives divergence of uncorrelated genetic and song variation in the Ruddy-capped Nightingale-Thrush (*Catharus frantzii*; Aves: Turdidae). Molecular Phylogenetics and Evolution 94:74–86.

Palmer, C. L., N. M. Weber, T. Munoz, and A. H. Renear. 2013. Foundations of data curation: the pedagogy and practice of "purposeful work" with research data. Archive Journal 3.

Pasch B., A. S. George, P. Campbell, and S. M. Phelps. 2011. Androgen-dependent male vocal performance influences female preference in Neotropical singing mice. Animal Behaviour 82:177–183.

Payne, R. B. 1996. Song traditions in indigo buntings: origin, improvisation, dispersal, and extinction in cultural evolution. Pp. 198–220 in D. E. Kroodsma and E. H. Miller (editors), Ecology and evolution of acoustic communication in birds. Cornell University Press, Ithaca, NY.

Payne, R. B., and L. L. Payne. 1993. Song copying and cultural transmission in indigo buntings. Animal Behaviour 46:1045–1065.

Payne, R. B., L. L. Payne, and S. M. Doehlert. 1988. Biological and cultural success of song memes in indigo buntings. Ecology 69:104–117.

Payne, R. B., W. L. Thompson, W. L., and K. L. Fiala. 1981. Local song traditions in Indigo Buntings: cultural transmission of behavior patterns across generations. Behaviour 77:199–221.

Petrusková, T., L. Diblíková, P. Pipek, E. Frauendorf, P. Procházka, and A. Petrusek. 2014. A review of the distribution of Yellowhammer (Emberiza citrinella) dialects in Europe reveals the lack of a clear macrogeographic pattern. Journal of Ornithology 156:263–273.

Petrusková, T., T. S. Osiejuk, A. Petrusek, and A. M. Dufty. 2010. Geographic variation in songs of the Tree Pipit (Anthus trivialis) at two spatial scales. Auk 127:274–282.

Podos, J. 2001. Correlated evolution of morphology and vocal signal structure in Darwin's finches. Nature 409:185–188.

Podos, J., and P. Warren. 2007. The evolution of geographic variation in birdsong. Advances in the Study of Behavior 37:403–458.

Price, T. 2007. Speciation in birds. Roberts and Company, Greenwood Village, CO.

Ranft, R. 2004. Natural sound archives: past, present and future. Anais da Academia Brasileira de Ciências 76:455–465.

Ratcliffe, L. M., and P. R. Grant. 1985. Species recognition in Darwin's finches (Geospiza, Gould). III. Male responses to playback of different song types, dialects and heterospecific songs. Animal Behaviour 33:290–307.

Repenning, M., and C. S. Fontana. 2013. A new species of gray seedeater (Emberizidae: Sporophila) from upland grasslands of southern Brazil. Auk 130:791–803.

Ryan, M. J., and E. A. Brenowitz. 1985. The role of body size, phylogeny, and ambient noise in the evolution of bird song. American Naturalist 126:87–100.

Ryan, M. J., and A. S. Rand. 1995. Female responses to ancestral advertisement calls in túngara frogs. Science 269:390–392.

Sætre, G. P. 2000. Sexual signals and speciation. Pp. 237–257 in Y. Epsmark, T. Amundsen, and G. Rosenqvist (editors), Animal signals: signalling and signal design in animal communication. Tapir Academic Press, Trondheim, Norway.

Sætre, G. P., M. Král, and S. Bureš. 1997. Differential species recognition abilities of males and females in a flycatcher hybrid zone. Journal of Avian Biology 28:259–263.

Santos, J. C., M. Baquero, C. Barrio-Amoros, L. A. Coloma, L. K. Erdtmann, A. P. Lima, and D. C. Cannatella. 2014. Aposematism increases acoustic diversification and speciation in poison frogs. Proceedings of the Royal Society B 281:20141761.

Santos, J. C., L. A. Coloma, and D. C. Canatella. 2003. Multiple, recurring origins of aposematism and diet specialization in poison frogs. Proceedings of the National Academy of Sciences USA 100: 12792–12797.

Searcy, W. A., and E. A. Brenowitz. 1988. Sexual differences in species recognition of avian song. Nature 332:152–154.

Searcy, W. A., and S. Nowicki. 2005. The evolution of animal communication: reliability and deception in signaling systems. Princeton University Press, Princeton, NJ.

Seddon, N., and J. A. Tobias. 2007. Song divergence at the edge of Amazonia: an empirical test of the peripatric speciation model. Biological Journal of the Linnean Society 90:173–188.

Seddon, N., and J. A. Tobias. 2010. Character displacement from the receiver's perspective: species and mate recognition despite convergent signals in suboscine birds. Proceedings of the Royal Society B 277:2475–2483.

Servedio, M. R., G. S. van Doorn, M. Kopp, A. M. Frame, and P. Nosil. 2011. Magic traits in speciation: "magic" but not rare? Trends in Ecology and Evolution 26:389–397.

Slater, P. J. 1986. The cultural transmission of bird song. Trends in Ecology and Evolution 1:94–97.

Soha, J. A., and P. Marler. 2001. Cues for early discrimination of conspecific song in the White-crowned Sparrow (Zonotrichia leucophrys). Ethology 107:813–826.

Symes, L. B., M. P. Ayres, C. Cowdery, and R. A. Costello. 2015. Energetic, morphometric, and acoustic trade-offs define signal diversification ridges in Oecanthus tree crickets. Evolution 69:1518–1527.

Tietze, D. T., J. Martens, B. S. Fischer, Y. H. Sun, A. Klussmann-Kolb, and M. Päckert. 2015. Evolution of leaf warbler songs (Aves: Phylloscopidae). Ecology and Evolution 5:781–798.

Tobias, J. A., J. D. Brawn, R. Brumfield, E. P. Derryberry, A. N. Kirschel, and N. Seddon. 2012. The importance of suboscine birds as study systems in ecology and evolution. Ornitología Neotropical 23:259–272.

Tobias, J. A., and N. Seddon. 2009. Signal design and perception in *Hypocnemis* antbirds: evidence for convergent evolution via social selection. Evolution 63:3168–3189.

Tobias, J. A., N. Seddon, C. N. Spottiswoode, J. D. Pilgrim, L. D. C. Fishpool, and N. J. Collar. 2010. Quantitative criteria for species delimitation. Ibis 152:724–746.

Tubaro, P. L., and E. T. Segura. 1995. Geographic, ecological and subspecific variation in the song of the Rufous-browed Peppershrike (*Cyclarhis gujanensis*). Condor 97:792–803.

Van Bogart, J. W. C. 1995. Magnetic tape storage and handling: a guide for libraries and archives. National Media Laboratory, Saint Paul, MN.

Verzijden, M. N., C. ten Cate, M. R. Servedio, G. M. Kozak, J. W. Boughman, and E. I. Svensson. 2012. The impact of learning on sexual selection and speciation. Trends in Ecology and Evolution 27:511–519.

Walker, T. J., and D. H. Funk. 2014. Systematics and acoustics of North American *Anaxipha* (Gryllidae: Trigonidiinae). Journal of Orthoptera Research 23:1–38.

Wallschlager, D. 1980. Correlation of song frequency and body weight in passerine birds. Cellular and Molecular Life Sciences 36:412.

Walsh, S. A., P. M. Barrett, A. C. Milner, G. Manley, L. M. Witmer. 2009. Inner ear anatomy is a proxy for deducing auditory capability and behaviour in reptiles and birds. Proceedings of the Royal Society B 276:1355–1360.

West-Eberhard, M. J. 1983. Sexual selection, social competition, and speciation. Quarterly Review of Biology 58:155–183.

Wilkins, M. R., N. Seddon, and R. J. Safran. 2013. Evolutionary divergence in acoustic signals: causes and consequences. Trends in Ecology and Evolution 28:156–166.

Winker, K., J. M. Reed, P. Escalante, R. A. Askins, C. Cicero, G. E. Hough, and J. Bates. 2010. The importance, effects, and ethics of bird collecting. Auk 127:690–695.

Zimmer, K. J., A. Whittaker, and D. C. Oren. 2001. A cryptic new species of flycatcher (Tyrannidae: *Suiriri*) from the Cerrado region of central South America. Auk 118:56–78.

Zimmermann, E. 1990. Behavioral signals and reproduction modes in the Neotropical frog family Dendrobatidae. Pp. 61–73 in W. Hanke (editor), Biology and physiology of amphibians. Gustav Fischer Verlag, Stuttgart, Germany.

CHAPTER FIVE

Leveraging Diverse Specimen Types
to Integrate Behavior and Morphology*

Kimberly S. Bostwick, Todd Alan Harvey, and Edwin Scholes III

Abstract. Biological specimens can hold a surprising wealth of information, and different specimen types hold different, but complementary, sets of data. This is true not only of physical specimens, such as study skins and skeletal preparations, but also of media "specimens," such as an audio recording of an animal's voice, a video of its display, or a photograph of its nest. When diverse specimen types are taken from the same species (species-level vouchering) and especially the same individual (individual-level vouchering), they can be leveraged to extract ever more complete insights into evolution, ecology, behavior, and functional morphology. In this chapter, we present two case studies that combine data obtained from analyses of both physical and media specimens. These case studies illustrate the diverse approaches undertaken using diverse specimen resources, approaches that allow us to address challenging questions and explore new areas of inquiry. Modern collecting techniques,

such as behavioral vouchering using high-speed video and audio recordings, and advanced digital techniques, including several types of anatomical, acoustic, and optical analyses, were applied to extract information from specimens that previously would have been impossible to obtain. Results include surprising behavioral, functional, and evolutionary insights into two fascinating groups of birds: the manakins (Pipridae) and the birds-of-paradise (Paradisaeidae). Similar approaches can be employed to gain insights into other taxa. Importantly, these insights were only possible through an integrated approach that combined information gleaned from multiple specimen types, thereby highlighting the complementary nature of diverse specimen types.

Key Words: BRDF, computer graphics, courtship phenotype, CT scan, functional morphology, high-speed video, model, museum specimen, Paradisaeidae, photogrammetry, Pipridae, reflectance, sonation.

For centuries, scientists have relied on experimentation to tease apart conflicting interpretations of nature. For biologists, bringing organisms into a lab setting has facilitated the control of targeted variables, a practice that is essential for discerning between competing hypotheses. One productive offshoot of this approach has been to take physical models into the real world, as when introducing stuffed or robotic specimens into territories to observe behavioral

* Bostwick, K. S., T. A. Harvey, and E. Scholes III. 2017. Leveraging diverse specimen types to integrate behavior and morphology. Pp. 75–88 in M. S. Webster (editor), The Extended Specimen: Emerging Frontiers in Collections-based Ornithological Research. Studies in Avian Biology (no. 50), CRC Press, Boca Raton, FL.

responses (e.g., Patricelli et al. 2002, Baldassarre et al. 2016). Today, museum specimens are being used to take this process of model and experimentation one step further: using cutting-edge technologies and advanced analytics together with physical and media specimens, scientists can now infer the functional and evolutionary processes that have shaped biodiversity. This chapter seeks to illustrate the groundbreaking potential of studies that integrate morphological and behavioral data derived from traditional physical museum specimen types, and newer, usually digital, media specimen types (such as photos, video, and audio recordings). Combining such "extended specimen" resources greatly expands the scope of research questions open to field biologists.

The approaches illustrated in this chapter were not possible just a few years ago, and yet represent just the tip of the iceberg for what is now possible. The diversity of new technologies and analytic methods have exploded, especially in nonbiological fields. By bringing these developments into our own field, biologists can get new insights into old questions and can ask entirely new questions. Indeed, right now the expanding opportunities created by new technological developments seem to be outpacing our ability to perceive—let alone employ—them; this is a rich time for new approaches, perspectives, and discoveries. In addition, the rich and productive fields of animal behavior and functional morphology have both explored their respective interests in relative isolation for many decades. The relatively unexplored "space" between the two disciplines, given their respective independent advancements, will undoubtedly offer fertile ground for new discovery. It is time to reexamine the fascinating interface between morphology and behavior.

The two case studies outlined in this chapter illustrate how extended specimens create the potential for new discoveries at the interface between morphology and behavior. This intensive multidisciplinary approach can be extremely rewarding in allowing access to species, traits, behaviors, systems, and questions that hitherto have been logistically inaccessible. Simply getting access to a given species, and/or studying that species in the wild where their behaviors are performed, often can be difficult or even impossible. Moreover, many of the classic laboratory techniques that have been used to study

organisms are simply not feasible in the field. Thus, the approaches illustrated in this chapter free the researcher to pursue important questions that have been inaccessible in the past, and further may suggest new lines of inquiry that we have not previously considered.

INSIGHTS INTO THE ORIGIN OF NOVELTY IN THE MANAKINS (PIPRIDAE)

Evolutionary novelties have long fascinated biologists, in large part because these unique features, as deviations from "typical" ones, hold unique potential for insights into evolutionary processes (e.g., Wagner 2010). Evolutionary novelties typically concern (1) a morphology-based trait, often with a behavioral counterpart, that (2) exists in one or very few species, (3) seems to have arisen *de novo*, and (4) functions in some way that does not exist commonly, if it all, elsewhere in nature (Wagner 2010).

The Club-winged Manakin, *Machaeropterus deliciosus* (Passeriformes: Pipridae), exhibits one such behavioral and morphological novelty: males of this species produce an unmodulated, harmonically rich, tone with modified wing feathers while perched, that is, not while flying (Willis 1966, Prum 1998, Bostwick 2000). The anatomy of the feathers, muscles, and bones has been modified to allow production of this mechanically produced sound, hereafter referred to as a "sonation" (Darwin 1874, Prum 1998, Bostwick 2000). No other species of bird is known to produce a sonation of this sort, nor to exhibit the morphological modifications found in *M. deliciosus* (Bostwick 2006). Two obvious questions can be asked about this trait: (1) How does the morphology function in order to produce the sound? (2) How did this complex morphobehavioral feature arise through the ancestors of *M. deliciosus*? or, more simply, How did this novelty evolve?

MANAKIN METHODS

One approach to answering the preceding functional morphological questions would be to bring a male *M. deliciosus* into the lab in order to manipulate and record the bird performing its wing-sound behavior. This approach, though, is not feasible: *M. deliciosus* is endemic to northwestern Ecuador and southwestern Colombia, and has never been kept in captivity. Moreover, obtaining permission to capture and maintain these birds, and setting up a suitable lab, would be expensive, time-consuming, and risky, given that many manakin species do not do well in captivity.

Accordingly, an alternative approach was needed that did not involve bringing live animals into captivity. Fortunately, the two basic questions to be addressed—how does the modified morphology of M. *deliciosus* function to produce the male's sonation, and how did the morphology, behaviors, and acoustics evolve—could both be addressed through complementary analyses of morphological, behavioral, and acoustic datasets obtained from museum specimens. To address the functional questions, physical museum specimens of M. *deliciosus* were examined and compared to those of other, more passerine-typical manakins, to characterize in detail their distinctive morphology. For behavioral and acoustic data, audio and video recordings were made of individuals producing their sonations under natural conditions in the field. For comparative purposes, similar data were obtained for manakin species known to make simpler sonations than M. *deliciosus*. For these acoustic data, audio recordings were obtained from colleagues, from biological sound archives, or spectrograms were obtained from previous publications as available.

Characterizing the variation among species in various traits was useful not only in forming hypotheses about how novel morphology enables novel function, but also, when analyzed in a phylogenetic context, how function, morphology, and behaviors have been modified over time. For this reason, the relationships among the piprids (as hypothesized in Bostwick 2000; S. Hackett, unpubl. data; and later by Ohlson et al. 2013) were explicitly considered when choosing focal study species. In general, five levels of phylogenetic and functional comparative "controls" were examined as available: (Level 1) genus *Machaeropterus* (three species); (Level 2) genus *Ceratopipra* (five species, formerly the "higher" *Pipras*); (Level 3) the genus *Pipra* (three species); (Level 4) the monotypic species Dixiphia pipra (a secondarily nonsound-producing relative embedded within the clade that includes *Pipra*, *Machaeropterus*, and *Ceratopipra*); and (Level 5) more distantly related nonsonating manakins, such as those in the genera *Lepidothrix* and *Cryptopipo* (formerly *Chloropipo*) (see Figure 5.1).

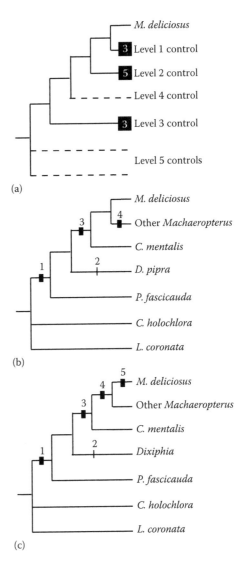

(a)

(b)

(c)

Figure 5.1. Phylogenetic hypothesis and character mappings for a reduced Pipridae phylogeny, as considered in these studies, modified from Ohlson et al. (2013). (a) The five levels of comparative/functional controls and the occurrence of the sonation character. Dashed lines indicate lack of wing-produced sonations; numerals in the boxes indicate the number of species represented by this clade. Control clade memberships are as follows: Level 1, other members of the genus *Machaeropterus*; Level 2, members of the genus *Ceratopipra*; Level 3, members of the genus *Pipra*; Level 4, *Dixiphia pipra* (a secondarily nonsound-producing species); Level 5, more distantly related nonsonating manakins, such as those in the genera *Lepidothrix* and *Crytopipo*. (b) Distribution of sonation-related backward dance behaviors in the study clade. Heavy bars indicate likely origin of character, thin bars indicate loss: (1) backward dance; (2) loss of backward dance; (3) addition of sonation to backward dance; (4) vertical backward dance. (c) Distribution of sonation types within the study clade: (1) simple single percussive *snaps*; (2) sonation lost; (3) double- or triple-pulsed *snaps*; (4) multi-pulsed percussive *peent* sound; (5) stridulatory, resonant *ting*.

Physical Specimens

Traditional museum skins, including round skin and spread wing preparations, were initially used for qualitative examination of the shapes of wings and wing feathers in multiple manakin species. In one study, modified secondary feathers from the round skins of male Club-winged Manakins were removed to allow quantification of their resonant properties using laser vibrometry (for another approach using laser vibrometry, see Elias et al. 2003). The resonant properties of these feathers were measured both in isolation and in the context of the other wing feathers, and compared to homologous secondary feathers of two other manakin species—*Pipra fascicauda* (Level 3 control) and *Lepidothrix coronata* (Level 5 control)—which likely represent the more plesiomorphic feather form. For more details on these methods, see Bostwick et al. (2009).

To identify those aspects of the musculoskeletal system in the Club-winged Manakin that deviated most from the "typical passerine condition," dried skeletal specimens of three manakin species, including *M. deliciosus*, were compared, described, and illustrated in detail. In addition, the soft tissues of the wings of fluid-preserved specimens of seven manakin species were thoroughly dissected and described with special emphasis on variation in wing myology (Bostwick 2002).

For a comparative morphometric analysis, the lengths and widths of four major wing bones were measured in 458 individual specimens across 36 piprid species. An independent contrast analysis was conducted by regressing the phylogenetically corrected components of the relative proportions of wing segments in each species against the relative complexity of a given species' sonation (Bostwick 2002).

Fluid-preserved specimens were also used for high-resolution microcomputed tomography (microCT) scans of the wing's musculoskeletal elements. Eight species (including at least one each from the comparative levels listed above) were scanned at Cornell University's MicroCT imaging facility. Scans were made with a GE CT120 microCT scanner (GE Healthcare, London, ON). The scan data allowed determination and interspecific comparisons of qualitative measures of bone volume, shape, solidness, and tissue mineral density (see Bostwick et al. 2012 for more details on methods).

Tissue samples collected routinely with museum specimens were borrowed from several natural history museums and used to extract nuclear and mitochondrial DNA, and ultimately to generate a phylogenetic hypothesis of the relationships among the various species of manakins (Bostwick et al., unpubl. data). This phylogenetic hypothesis informed all comparative inferences and was ultimately used in the independent contrast analysis of wing morphometrics described earlier.

Digital Media Specimens

Audio and video specimens were also used for both functional and evolutionary analyses. Video recordings of behavior were used in two basic ways. First, standard rate videos (30 frames per second) were recorded and analyzed to determine the behavioral repertoire of *M. deliciosus*, focusing especially on those behaviors during which sonations were produced. Using previously published behavioral repertoire characterizations of 11 other species of manakins, phylogenetic analyses of behavior and morphology (separate and combined) were conducted to generate hypotheses of behavioral homology. These analyses included heavier sampling among those species most closely related to *M. delicious* (9 of the 13 possible species in the top four levels of comparative controls), and allowed for detailed assessment of behavioral and signal homology between species (Bostwick 2000).

Second, high-speed video recordings (250, 500, and 1,000 frames per second, as permitted by ambient lighting conditions) were made and analyzed frame-by-frame to allow more detailed analysis of the wing, feather, and other body motions—or *kinematics*—involved in sound production for three species: *Machaeropterus deliciosus*, *Ceratopipra mentalis* (a Level 2 control), and *Manacus candei* (a more distantly related species of manakin that has likely evolved the ability to sonate independently from the *Pipra/Machaeropterus* clade). Such analyses enabled the development of more detailed hypotheses for how motion and morphology were ultimately translated into sound. For more details on methods, see Bostwick and Prum (2003).

Finally, audio recordings of the sonations of the Club-winged Manakin were analyzed to characterize their basic temporal and frequency characteristics (as were recordings of the simpler sonations produced by other species). For one analysis, a single audio recording was corecorded with a high-speed video during sonation in *M. deliciosus*. This allowed for a one-to-one alignment of that males' sonation with the wing motions used to produce the sound, and these digital assets were essential for testing competing hypotheses for how the sonation was produced (Bostwick and Prum 2005).

Meta-Analyses

Functional Morphology

The aforementioned studies established the ways in which the wing morphology of M. *deliciosus* differed from that of "normal" manakins. Once such morphological deviations had been characterized, hypotheses could be generated, in the context of the acoustic and video analyses, of the chains of causality between the associated kinematics and the physical mechanisms that might generate the observed sounds.

Evolutionary Analyses

As referenced earlier, an independent contrast analysis (Garland et al. 1992, Bostwick 2002) and phylogenetic analyses (Bostwick 2000) of cross-species' morphological, behavioral, and acoustic characters were used to determine character polarity and homology, as well as evolutionary patterns of association between morphological and acoustic characters.

MANAKIN RESULTS

How It Works: The Functional Morphology of the Club-winged Manakin Sonation

Detailed examination and measurement of several physical specimen types (including round skins, skeletons, and fluid specimens), combined with acoustic and kinematic analysis of audio and video recordings, respectively, yielded numerous insights into the functional morphology of the novelty found in M. *deliciosus*. Specifically, M. *deliciosus* males' inner secondary wing feathers exhibit a suite of modifications including, most prominently, a "pick" shaped 5th secondary (with a thin and bent distal tip) and a subtle "washboard-like" ribbed 6th secondary (Figure 5.2a,b, respectively; Bostwick and Prum 2005). The rachi of both the 6th and 7th secondary feathers are hypertrophied and twisted, and were further found to have highly developed resonating properties relative to the two control species (Figure 5.3; Bostwick et al. 2009).

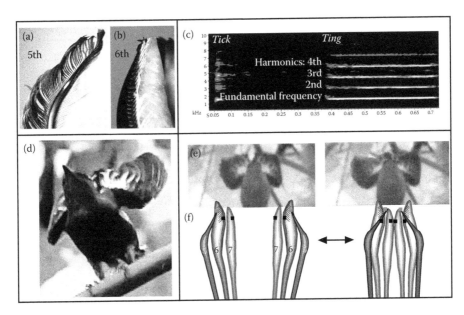

Figure 5.2. Modified feathers, sonation acoustics, and posture of *Machaeropterus deliciosus* during sonation. (a, b) Distal tips of modified 5th and 6th secondary feathers, respectively, showing hypertrophied, twisted rachi, and ridges on shaft. (c) Spectral structure of tick ting sonation showing fundamental frequency and integer harmonics. (d) Body posture during sonation, and (e) two frames taken from high-speed video recordings showing separation of feathers over back (left frame) and subsequent contact (right frame). (f) Diagram indicating hypothesized stridulatory feather movements corresponding to video frames in part (e). Notice ridges crossed on 6th secondary by tip of 5th (red dot). (Modified from Bostwick and Prum 2005.)

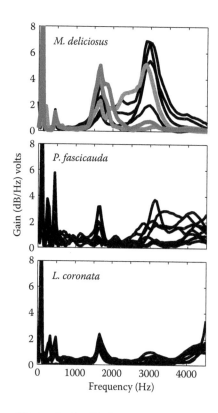

Figure 5.3. Amplitude of feather resonant responses across a broad sweep of frequencies in three species of Pipridae. Each line in each graph corresponds to a different individual secondary feather. All three species have the higher responsiveness near the frequency of the sonation produced by M. *deliciosus* in nature (~1.5 kHz); this indicates the basal (unmodified) resonant response of normal feathers of approximately this size and composition. M. *deliciosus* exhibits a significantly stronger response that includes more feathers (especially the 6th and 7th secondaries), and the higher level harmonics are also visible. (Modified from Bostwick et al. 2009.)

The wing musculature of male M. *deliciosus* also is highly modified from a "typical" passerine wing form, specifically with respect to the shapes and sizes of individual muscles, and also in the placements of their origins and insertions (Bostwick 2002), implying use of wing motions not typical of other species. Further, the shapes of the articulated surfaces of the elbow and shoulder joints are modified in M. *deliciosus* in ways that likely increase joint mobility during the wing pronation that is used prominently during sound production (Bostwick 2002; also see below). In addition to modified musculature, M. *deliciosus* exhibit distinctively shaped ulnae and humeri, both of which are enlarged in diameter, and the ulnae have a

uniquely triangular cross-section and exceptionally pronounced quill knobs (Figure 5.4a). Overall size, solidness, and bone density are greatly increased in male M. *deliciosus* relative to other manakins (Figure 5.4b; Bostwick et al. 2013).

The sonation itself is a sustained (~1/3 second) note composed of integer harmonics with a fundamental frequency of 1.49kHz (Figure 5.2c; see http://macaulaylibrary.org/audio/187177). Video recordings documented that sound production occurs when males pronate their wings, that is flip their dorsal surfaces anteriorly (Figure 5.2d; see http://macaulaylibrary.org/video/458497). High-speed video recordings revealed that, during wing pronation and sound production, the inner margins of the wings meet repeatedly over the back, as the male strikes the modified secondary feathers against one another (Figure 5.2e).

The sonation is produced continuously during the period of wing knocking (Bostwick and Prum 2005). Based on the acoustic and kinematic analysis of our audio and video recordings, we hypothesize that the sonation is produced when the resonance of the 6th and 7th secondaries is excited, primarily by the repeated feather collisions above the back, and is maintained when the distal tip of the 5th secondary stridulates the ribbed surface of the 6th secondary (Figure 5.2f). The number of ribs on the feather shaft (7), rubbed twice for each cycle of wing movement (for 14 stridulatory events), at the rate of collision shown by the video analysis (107 knock/second), exactly matches the fundamental frequency produced (1.49 kHz) during sonation for the male recorded (Bostwick and Prum 2005, Bostwick et al. 2009). The massive solidified ulna reported above is hypothesized to improve the emission of sound from the attached resonating wing feathers (Bostwick et al. 2013).

Overall, the joint analyses of both physical specimens and digital recordings revealed a unique and tightly integrated complex of morphological and behavioral traits that function together to produce M. *deliciosus*'s unique sonation. Such analyses highlight the power of using physical and digital specimens in concert.

How It Evolved: The Integrated Evolutionary Patterns of Character Change

Behavioral, acoustic, and morphological comparisons between species of manakins, in the context of a phylogeny, provided a perspective on the

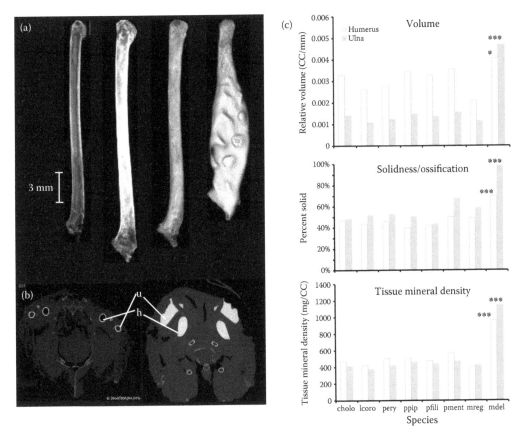

Figure 5.4. Characterization of the distinctive wing bones of M. *deliciosus*. (a) CT scan image of the relative size and shape of three control species (leftmost) compared to M. *deliciosus*. (b) Cross-sectional view of L. *coronata* versus M. *deliciosus* (u indicates the ulna, h indicates the humerus). (c) Data comparing the relative size, solidness, and tissue mineral density of species of piprid humeri and ulnae relative to M. *deliciosus*. (Modified from Bostwick et al. 2012.)

evolutionary history of the behavioral, acoustic, and morphological novelties found in M. *deliciosus*. For instance, several display behaviors found in M. *deliciosus*, such as the backward dance and to-and-fro displays (see http://macaulaylibrary.org /video/458510), appear to be basal to the Pipra/ *Machaeropterus*/*Ceratopipra* study clade, and persist through many speciation events in recognizable form (Figure 5.1b). The backward dance is the main and most complex courtship display reserved for visiting females, and the male's sonation is prominently used in this display. In addition, the ancestral forms of the sonation found in M. *deliciosus* not only arose prior to the species itself, but likely evolved as many as three speciation nodes prior, such that simpler forms of sonations are common among several of the closest relatives of the Club-winged Manakin (Figure 5.1c). Simple single percussive sounds appear to have increased in complexity through

doubled, tripled, and multipulsed percussive sounds. Further, other species incorporate their simpler sonations into their backward dance displays (e.g., *Pipra mentalis*).

Across the family in general, the wing bones of those manakin species that make sounds with their wings are significantly shorter and stockier than those of nonsonating species. M. *deliciosus* represents the most extreme case of this correlation (Bostwick 2002). The extreme morphological modifications observed in the Club-winged Manakin evolved in the context of a trend of increasing sonation complexity. Thus, in this case, the morphological modifications found in M. *deliciosus* seem to have been precipitated by pressure on males to produce increasingly complex sonations. Additionally, the placement of males' sonations in varying parts of evolving repertoires suggests these behaviors have evolved in a modular fashion (Scholes 2008a).

THE COURTSHIP PHENOTYPE OF THE
BIRD-OF-PARADISE (PARADISAEIDAE)

The birds-of-paradise are a signature, sexually selected, radiation of species, renowned for their exotic appearances and bewildering behavioral diversity (Frith and Beehler 1998, Irestedt et al. 2009, Laman and Scholes 2012). The extremity and diversity of sexually selected characters exhibited by the birds-of-paradise are particularly promising for insights into how sexual selection operates (Scholes 2008a).

To look at a female, it is clear that birds-of-paradise are fairly typical, large, passerine birds (Frith and Beehler 1998). However, examining the plumage of the males reveals the existence of a distinctive and extreme *courtship phenotype* (Scholes 2008a). The diverse and unusual shape, texture, and color components of male plumage make them strikingly different from both the females and males of more "typical" birds. The courtship phenotype, though, is not simply the male's plumage; instead it is the portion of the overall phenotype that is perceived and evaluated by females during courtship. A species' courtship phenotype can include, and often combines, visual and/or acoustic components, as well as the morphological and behavioral components that underlie them (Scholes 2008a). Thus, the courtship phenotype is that portion of the male's phenotype that has been shaped primarily by female mating preferences.

Because the courtship phenotype is essentially an emergent property of behavior and morphology, its structure cannot be investigated from physical specimens or media specimens, for example, video, alone. Media specimens are useful for ethographic measurement (duration and frequency, gross movement/posture) and physical specimens are useful for morphological measurement (size, shape, and some qualities like color). But media recorded outside of controlled conditions are typically very poor for morphological measurement, 3D shape, analysis of color, and precise analysis of movement over space and time. Similarly, although physical specimens present the bodily parts that comprise the courtship phenotype, that does not necessarily mean they are observable or measurable *as they are actually used* when courting females. Thus, there is an inherent "gap" between physical and digital specimens for understanding complex components of the courtship phenotype.

In most birds-of-paradise, courtship only occurs under very specific conditions: The location and layout of the court and location of the female on it are carefully managed by the male (Scholes 2006, 2008b,c). Further, several birds-of-paradise prominently use iridescent plumage ornaments in their displays, and an essential property of iridescence is its directional nature (Stavenga et al. 2011, Wilts et al. 2014). The courtship phenotype of most birds-of-paradise have not been observed, let alone recorded, from the female perspective. In order to fully understand the courtship phenotype of males, we need to visualize and analyze male displays from the female's perspective.

Yet studying the courtship phenotype of male birds-of-paradise *in situ* is challenging. Most birds-of-paradise are restricted to the island of New Guinea and nearby smaller islands (Frith and Beehler 1998, Laman and Scholes 2012). New Guinea is one of the most underdeveloped landmasses in the world, largely because of its pervasive steep, mountainous terrain. This same terrain makes both accessing the ranges of many species and following individual birds in the wild exceptionally difficult. Although captive birds-of-paradise exist, we know too little about the context of display of these birds in the wild to adequately reproduce their display environments in the lab.

BIRDS-OF-PARADISE:
A MODELING APPROACH

In order to better understand the courtship phenotypes of birds-of-paradise, we are modeling components of the *Parotia wahnesi* courtship phenotype. *Parotia wahnesi* is a species endemic to the midmontane forests of the Huon Peninsula in northeastern Papua New Guinea (Scholes 2008c). The male's multistage "ballerina dance" is his most complex courtship display, and the one that typically precedes copulation (Scholes 2008c). Several of the male's iridescent ornaments may be visible to the female during this display, and the male manages her viewing behavior by predetermining his display and her viewing positions. However, how his courtship phenotype appears to her during this display is not fully known (see Figure 5.5a).

We are currently using museum specimens of male *Parotia wahnesi*, in conjunction with field-generated media specimens (images and video) of

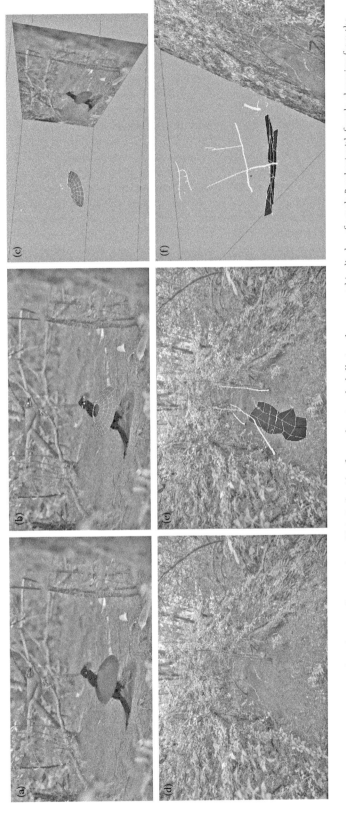

Figure 5.5. Real and virtual male and court of *Parotia wahnesi*. (a) A single video frame showing the ballerina dance courtship display of a male *P. wahnesi* with female observing from the main court perch. (b) View through virtual camera match-moved to real camera with 3D surface models of motion tracked structures including the feather "skirt," breast patch ornament, hind crown patch ornament, bill, and eye of both the male and female. (c) The same moment in time from a different perspective and with frustum and image plane of match-moved video camera shown. (d) A single photograph of the *P. wahnesi* display court used to create 3D model of the courtship environment (photogrammetric analysis). (e) Same photograph and camera perspective with 3D model of key display court features in place. (f) 3D surface model of the court environment viewed from an alternate perspective, which shows the frustum and image plane of the match-moved video camera.

displaying males, to create 3D computer graphics (CG) simulations of the courtship phenotype and courtship environment. Our goal is to bridge the gap between the appearance of static physical specimens and the inherently limited interpretations possible from media specimens alone. Our 3D CG visualization/modeling approach is allowing us to create a virtual world through use of existing specimens and associated media. The resulting simulations will allow us to conduct experiments and address questions that would otherwise not be possible. Specifically, our models, with simulated courtship behavior and environment, will permit us to experimentally manipulate critical aspects of the *Parotia wahnesi* courtship phenotype, including plumage structure, ornaments, male and female position and orientation, and other environmental factors, such as lighting.

Because our goal is to understand how all components of the courtship phenotype of male *Parotia* function with respect to how females observe and evaluate them, it has been important to investigate how light interacts with ornamental structures across a hierarchy of scale from feather microanatomy to whole ornament in the context of the courtship environment. Using a round skin specimen of an adult male *Parotia wahnesi* (Ornithology Collection at the University of Kansas, KU:Birds:93603), we have dissected individual contour feathers from various body regions of the specimen (e.g., forehead, crown, nape, and breast) and imaged individual feathers using x-ray computed tomography (CT). Feathers were imaged using the Advanced Photon Source (APS) at the U.S. Department of Energy's Argonne National Laboratory (Lemont, Illinois). Resulting CT data were analyzed and reconstructed into 3D volumetric models using industrial CT analysis software (Argonne National Laboratory Tomopy and Volume Graphics VG Studio, to reconstruct and analyze CT, respectively). From the 3D feather reconstructions, we were then able to perform virtual extractions and create 3D surface models of key feather structures at various structural scales, for example, individual barbs and barbules, and clusters of barbs and barbules (Figure 5.6a–d).

Another analysis of the physical specimen involved characterizing the color and directional reflectance of the iridescent ornaments. Such characterizations, called bidirectional reflectance

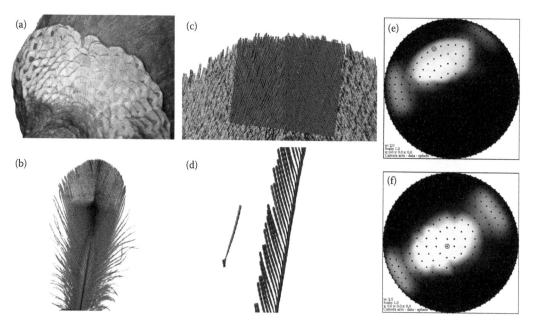

Figure 5.6. Examples of feather imaging from *P. wahnesi* round skin specimen showing morphological features across a hierarchy of scale and their signaling potential. (a) 3D surface model of male's breast shield ornament. (b) Photograph of an individual feather showing unobscured region of iridescence in red. 3D surface model of the breast feather extracted from CT data showing (c) intermediate and (d) high resolution of barbs and barbule. Two BRDF hemispherical plots from the shown region of the single breast feather from two different illumination directions: (e) normal to the plane of the vane and (f) inclined 30 degrees.

distribution functions (BRDFs; Nicodemus et al. 1977), are being made according to the methods outlined in Harvey et al. (2013a,b). This approach images an object (in this case the plumage ornaments of the *Parotia* specimen) from hundreds of different combinations of camera and incident light directions to yield a comprehensive and quantifiable view of how light (and hence color) is reflected over the surface of the object. So far we have found patterns of color reflectance from the feathers of the iridescent breast shield ornament to be similar to those identified by Stavenga et al. (2011) for another species of the genus *Parotia*, *P. lawesii* (see plots in Figure 5.6e,f).

Video documentation of the behavioral component of the *P. wahnesi* courtship phenotype, curated by the Macaulay Library at the Cornell Lab of Ornithology (CLO:ML:469255), was used for modeling movement during display. Because the sequence of behavioral events captured in the media specimens included simultaneous recordings by three video cameras, approximating the front, side, and top perspectives, the videos could be processed with 3D camera tracking/match-moving analysis software (such as Andersson Technologies SynthEyes, Agisoft PhotoScan, and Autodesk Maya). This allows moving objects to be tracked within a 3D environment. Because we were primarily interested in shape transformation, movement, and position in 3D space of specific courtship ornaments during specific phases of the ballerina dance display, motion tracking allowed us to create an animated 3D model of key focal structures, for example, the breast shield ornament, hind-crown ornament, flank-plume "skirt," and other anatomical landmarks like the male's eyes and bill. In addition, because the 3D geometry of the male's ornaments in relation to the female observer were important, we also tracked the position of the eye of an observing female (Figure 5.5a–c).

As male *Parotia* select, modify, and manage the environment from which they display to females (i.e., the court), we also constructed a 3D model for key components of the courtship environment. To do this, we used a series of digital photographs taken from multiple camera perspectives surrounding the actual display court in the wild and processed them with photogrammetric analysis software to generate a 3D model of the courtship environment (again using Andersson Technologies SynthEyes, Agisoft PhotoScan, and

Autodesk Maya). Key features modeled were the basic court geometry, including the main court perch used by females to observe male display, and the primary features impacting the light environment (Figure 5.5d–f).

The 3D models made from the physical and media specimens, as described earlier, are now being imported into 3D CG software and used for virtual simulations representing different aspects of the *Parotia wahnesi* courtship phenotype. Although these analyses are just now underway and results are not ready for publication, we include this case study here to illustrate the power of combining analyses of physical specimens with analyses of digital media that capture the courtship phenotype, as well as the sophisticated analytical tools that are now available for doing both. The sophistication and power of such approaches will undoubtedly grow in the near future.

GENERAL CONCLUSIONS

A new era for specimen-based biology is opening to scientists whose research interests fall at the intersection between organisms' signaling traits and trait use. These two case studies highlight some of the different ways in which combining analyses of physical and digital specimens—extensions beyond the typical museum specimen—can yield results otherwise impossible to obtain from either specimen type by itself, or, in many cases, by any other means. In the first case, using specimens (held in museum collections) together with natural history observations in the field (formalized in the form of media specimens), allowed many surprising insights into the history and function of *M. deliciosus*'s behavioral and morphological novelty. Specifically, the details of how *M. deliciosus*'s unusual wing morphology contributes to sound production are now much more clearly understood. Further, the intermediate behavioral, acoustic, and morphological states that likely led to the evolution of this novelty are also now mapped out in much greater detail than ever before.

Similarly, for *P. wahnesi*, traditional skin specimens, field-based video recordings of courtship displays, and photographs of the court environment together allowed for the generation of accurate, detailed, and complete models. Through such models, a virtual male courtship phenotype can now be placed "within" a realistic display environment that has accurate court and viewing geometries. This 3D framework allows for virtual experiments to be performed, which will allow for a better understanding of the functional morphology of *Parotia wahnesi*'s courtship phenotype. Specifically, we can now "view" any modeled aspect of the courtship phenotype as experienced by the arbiters of sexual selection themselves, the visiting females.

Despite almost nonoverlapping analytical methods, both of the aforementioned studies are essentially studies of functional morphology. But we note that, for both, there are some important departures between these studies and more classic functional morphology. First, rather than examining what are primarily naturally selected traits, we were examining sexually selected traits. Therefore, rather than examining traits that function primarily by mechanical means, both studies were examining *signal* functions, which have not only essential morphological components, but also significant auditory and visual components. This means our focus was less on mechanical laws and more on optical or acoustic laws, and therefore acoustic and visual information gathered in the field—and captured in the media specimens that we collected—were as critical to the success of these projects as the physical specimens. We conclude that, for birds at least, one very fruitful place to generate evolutionary and/or functional insights into behavior, communication, and sexual selection is at the psychosensory/signal–morphology interface, for which both physical and media specimens become essential research materials. However, although many sexually selected traits may provide obvious subjects for similar approaches combining physical and media specimens, no reason exists for why naturally selected functional traits could not be analyzed in similar ways. For instance, images and videos of a cryptically plumaged bird, like a woodcock, in various real-life environments could be combined with optical analyses of plumage patterns based on specimens to reveal in greater detail how the camouflage functions, and how birds do or do not adjust their behaviors to match their backgrounds.

From the curation side of managing specimens, it is worth noting that these studies also highlight the ways in which the lines dividing physical specimens, media specimens, and the materials derived therefrom, have, and will increasingly become, more blurred. That is, for parts of the manakin research (like the analysis of bone structure), and for the entirety of the *Parotia* study, physical specimen use entailed the creation of digital images as an intermediate state of analysis (through photography, photogrammetry, x-ray computed tomography, BRDF, etc.). The resulting images and datasets represent new specimen assets, such as CT scan data, or a 3D model of a specimen or behavior. Although not themselves biological specimens, these assets are nonetheless expensive and time-consuming to produce and have the potential to increase the value of the specimens from which they are derived. For this reason, as curators increasingly secure the connections between their physical specimens and associated media specimens, as well as other related datasets such as gene sequences and research papers, they will likely be motived to maintain connections between these novel digital assets and the curated specimens from which they were derived. It is also worth noting here that no single, one-size-fits-all digitizing solution is available for any given specimen for either study. Instead, knowledge of which subsequent analyses were desired informed the scale, resolution, and viewing geometry of any digital image that was captured. *A priori* digitizing of specimens without knowledge of the specific research questions to be addressed is likely to be of limited utility.

Both of these studies also make it clear that a multiplicity of specimen types, in addition to a greater number of representations of any given individual or species, greatly increases the research value of all associated specimens. For example, in the *Parotia* study, from relatively modest samples (typical museum specimens, videos, and photos), we were able to strategically extract and mobilize large amounts of information to model the system at a wide variety of spatial scales and then to conduct virtual experiments. Such modeling allows virtual experiments for species and/or conditions that are difficult and/or impossible to study in the field. Further, once the model is constructed, the multidimensional depth of data in the model can allow researchers to perform *ad hoc* experiments as desired.

The 2013 AOU symposium, from which this volume and chapter originated, was motivated, in part, by recent and ongoing efforts to reconnect various media specimens with their physical museum specimen counterparts. Such media specimens have largely remained uncurated over the last 30-plus years. Case studies such as these demonstrate timeliness and importance of such efforts. Media specimens, whether curated in traditional museum collections or in dedicated media collections, are critical to future studies of this sort. Such materials enable comparative and modeling studies like these to be done by saving the researcher the considerable investment of going to the field (which may be impossible or at least challenging for some species), and thereby enable studies on difficult-to-study species, larger-scale studies, and so on.

Finally, two important ideas that these case studies also raise are those concerning different levels of vouchers (i.e., species level, population level, and individual level), and how such vouchers relate to co-collected physical and media specimens. These case studies illustrate the importance of species-level vouchering: Both studies were enabled by having specimens of the correct species that could be analyzed alongside the collected media specimens. It would be an improvement to use specimens derived from the same population, or population-level vouchers, from which the media were collected. Such vouchering reduces potential issues in analysis, such as geographic variation in morphology and/or behavior. Better yet would be to have one-to-one vouchers, where there are physical specimens of the very same individuals that were recorded acoustically or behaviorally, as this would also eliminate issues introduced by individual-level variation and allow for stronger tests of mechanistic hypotheses. For example, in the manakin study, the fundamental frequency and the amplitude, or loudness, of each male's sonation is related to the resonant properties on his individual secondary feathers, the ridges on his feathers, and the rate at which he oscillates his wings. The secondary-feather stridulation hypothesis, therefore, predicts that a given male's variation in pitch should be directly correlated with these feather variables. Studies of female choice would require characterizing such tight one-to-one associations between the male's morphology and the signal he produces. Such analyses can only be made by having media and morphological specimens from the same individuals. Prioritization of such individual-level vouchers has the potential to generate additional value to the already invaluable materials held in traditional museum collections, making them newly relevant to a whole new generation of ornithologists.

ACKNOWLEDGMENTS

All authors are, of course, indebted to the many museums from which they have borrowed specimens, and consequently to the individual collectors who work to make these collections possible. Most critical, the University of Kansas Museum of Natural History, the Peabody Museum, the Cornell University Museum of Vertebrates, the University of Michigan Museum of Zoology, the Louisiana State University Museum of Natural History, the American Museum of Natural History, and the Macaulay Library of Natural Sounds supported these various research projects with unique specimens.

For the *Parotia* research, all digital specimens are archived in the Macaulay Library, Cornell Lab of Ornithology. CT data of *Parotia* feathers were collected and processed with the help of X. Xiao and D. Gursoy at beam line 2-BM of the Advanced Photon Source at Argonne National Labs, and supported under proposal 39343. The Richard O. Prum Lab at Yale University provided CT computing resources. D. Milewicz of Yale University segmented *Parotia* CT data. S. Marschner of Cornell University provided access to his lab for imaging the *Parotia* specimen required for computing directional light scattering and photogrammetric surface reconstructions. D. Young produced 3D surface models of the specimen and its environment.

LITERATURE CITED

Baldassarre, D. T., E. I. Greig, and M. S. Webster. 2016. The couple that sings together stays together: duetting, aggression and extra-pair paternity in a promiscuous bird species. Biology Letters 12:20151025.

Bostwick, K. S. 2000. Display behaviors, mechanical sounds, and evolutionary relationships of the Club-winged Manakin (*Machaeropterus deliciosus*). Auk 117:465–478.

Bostwick, K. S. 2002. Phylogenetic analyses of the evolution of behavior, wing morphology, and the kinematics of mechanical sound production in the Neotropical manakins (Aves: Pipridae). Ph.D. dissertation, University of Kansas, Lawrence, KS.

Bostwick, K. S. 2006. Mechanisms of feather sonation in Aves: unanticipated levels of diversity. Acta Zoologica Sinica 52:68–71.

Bostwick, K. S., and R. O. Prum. 2003. High-speed video analysis of wing-snapping in two manakin clades (Pipridae: Aves). Journal of Experimental Biology 206:3693–3706.

Bostwick, K. S., and R. O. Prum. 2005. Courting bird sings with stridulating wing feathers. Science 309:736.

Bostwick, K. S., D. O. Elias, A. Mason, and F. Montealegre-Z. 2009. Resonating feathers produce courtship song. Proceedings of the Royal Society B 277:835–841.

Bostwick, K. S., M. L. Riccio, and J. M. Humphries. 2012. Massive, solidified bone in the wing of a volant courting bird. Biology Letters 8:760–763.

Darwin, C. 1874. The descent of man, and selection in relation to sex (2nd ed.). John Murray, London, UK.

Elias, D. O., A. C. Mason, W. P. Maddison, and R. R. Hoy. 2003. Seismic signals in a courting male jumping spider (Araneae: Salticidae). Journal of Experimental Biology 206:4029–4039.

Frith, C. B., and Beehler, B. M. 1998. The Birds of Paradise: Paradisaeidae. Oxford University Press, Oxford, UK.

Garland, T., P. H. Harvey, and A. R. Ives. 1992. Procedures for the analysis of comparative data using phylogenetically independent contrasts. Systematic Biology 41:18–32.

Harvey, T. A., K. S. Bostwick, and S. Marschner. 2013a. Measuring spatially- and directionally-varying light scattering from biological material. Journal of Visualized Experiments e50254.

Harvey, T. A., K. S. Bostwick, and S. Marschner. 2013b. Directional reflectance and milli-scale feather morphology of the African Emerald Cuckoo, *Chrysococcyx cupreus*. Journal of the Royal Society Interface 10:20130391.

Irestedt, M., K. A. Jønsson, J. Fjeldså, L. Christidis, and P. G. Ericson, P.G. 2009. An unexpectedly long history of sexual selection in birds-of-paradise. BMC Evolutionary Biology 9:235–246.

Laman, T., and E. Scholes. 2012. Birds of paradise: revealing the world's most extraordinary birds. National Geographic Books, Washington, DC.

Nicodemus, F. E., J. C. Richmond, J. J. Hsia, I. W. Ginsberg, and T. Limperis. 1977. Geometrical considerations and nomenclature for reflectance (Vol. 160). U.S. Department of Commerce, National Bureau of Standards, Washington, DC.

Ohlson, J. I., J. Fjeldså, and P. G. P. Ericson. 2013. Molecular phylogeny of the manakins (Aves: Passeriformes: Pipridae), with a new classification and the description of a new genus. Molecular Phylogenetics and Evolution 69:796–804.

Patricelli, G. L., J. A. C. Uy, G. Walsh, and G. Borgia. 2002. Sexual selection: male displays adjusted to female's response. Nature 415:279–280.

Prum, R. O. 1990. Phylogenetic analysis of the evolution of display behavior in the Neotropical manakins (Aves: Pipridae). Ethology 84:202–231.

Prum, R. O. 1998. Sexual selection and the evolution of mechanical sound production in manakins (Aves: Pipridae). Animal Behaviour 55:977–994.

Scholes, E., III. 2006. Courtship ethology of Carola's Parotia (*Parotia carolae*). Auk 123:967–990.

Scholes, E., III. 2008a. Evolution of the courtship phenotype in the bird of paradise genus *Parotia* (Aves: Paradisaeidae): homology, phylogeny, and modularity. Biological Journal of the Linnean Society 94:491–504.

Scholes, E., III. 2008b. Structure and composition of the courtship phenotype in the bird of paradise *Parotia lawesii* (Aves: Paradisaeidae). Zoology 111:260–278.

Scholes, E., III. 2008c. Courtship ethology of Wahnes' Parotia *Parotia wahnesi* (Aves: Paradisaeidae). Journal of Ethology 26:79–91.

Stavenga, D. G., H. L. Leertouwer, N. J. Marshall, and D. Osorio. 2011. Dramatic colour changes in a bird of paradise caused by uniquely structured breast feather barbules. Proceedings of the Royal Society B 278:2098–2104.

Wagner, G. P., and V. J. Lynch. 2010. Evolutionary novelties. Current Biology 20:R48–R52.

Willis, E. O. 1966. Notes on a display and nest of the club-winged manakin. Auk 83:475–476.

CHAPTER SIX

Emerging Techniques for Isotope Studies of Avian Ecology*

Anne E. Wiley, Helen F. James, and Peggy H. Ostrom

Abstract. Stable isotope analysis has expanded the usefulness of avian specimen collections in ways that were generally unanticipated by the original collectors. This technique enables aspects of avian ecology to be studied using very small samples of feather or other tissue taken from specimens. Stable isotope values can be informative about diet, trophic ecology, spatial habitat use, and the migratory connectivity of breeding and nonbreeding populations. They can also provide access to understudied time periods in a bird's annual cycle, particularly periods of molt. Use of museum specimens offers advantages for these studies: museum collections generally span longer time periods than field samples and represent broad geographic ranges. In this chapter, we review established and emerging isotopic techniques and their applications in avian ecology, with an aim to inspire and inform research using museum collections. We focus on analyses of bulk tissues and proteins, and we discuss ways in which sampling strategies

for museum specimens can be adjusted to suit the research question, including considerations such as the timing of feather molt and the length of time during the life of a bird that is reflected by a sample. We also discuss general caveats for interpreting isotope data, the variance of which is usually influenced by multiple factors. In greater detail, we present recommendations for sampling museum specimens, such as avoiding a potential confounding influence of melanin concentration on carbon isotope values in feather, and minimizing damage to museum specimens when sampling. Stable isotope data from museum specimens have great potential to inform ornithologists about the trophic and spatial ecology of birds, and to provide long-term baseline data for studying how these attributes may be changing in the Anthropocene.

Key Words: bird, carbon isotope, diet, feather, foraging, hydrogen isotope, migration, nitrogen isotope, seabird.

S table isotope analysis is now used to address a broad range of research topics in avian ecology and physiology, from the migratory connectivity of songbird populations and the source of nutrients that females provision to eggs, to the spatial and trophic ecology of seabird communities (Rubenstein and Hobson 2004, Inger and Bearhop 2008). Stable isotope-based research has been used in ornithological research since the 1980s (Minagawa and Wada 1984, Schoeninger and DeNiro 1984, Hobson 1987) and has grown steadily in abundance since the

---

* Wiley, A. E., H. F. James, and P. H. Ostrom. 2017. Emerging techniques for isotope studies of avian ecology. Pp. 89–109 in M. S. Webster (editor), The Extended Specimen: Emerging Frontiers in Collections-based Ornithological Research. Studies in Avian Biology (no. 50), CRC Press, Boca Raton, FL.

late 1990s (Figure 6.1). While a small proportion of those studies have incorporated samples from museum specimens, we see great potential to enhance ornithological knowledge by more fully employing museum collections.

In the most general terms, stable isotope analysis is useful in the study of birds because the isotopic composition of avian tissues can reflect aspects of diet, geographic location, habitat, or physiology, depending upon the chemical element studied. This approach is based on the observations that (a) the isotopic composition of bird tissues is similar to, or differs predictably from, diet or drinking water; and (b) that there are consistent geographic and physiologically driven patterns in stable isotope ratios. As intrinsic markers, stable isotopes can be used to study aspects of bird ecology that are difficult or impossible to observe directly, including the migratory patterns of rare or otherwise difficult-to-track populations and the former diets and habitats of extinct species (Hobson and Montevecchi 1991, Chamberlain et al. 1997, Cherel et al. 2006, Clark et al. 2006). Stable isotope analysis also holds practical appeal due to its relatively low per-sample cost, in terms of both time and money, and because samples required for most isotope analyses are very small (e.g., 1 mg) and can thus be taken with minimal effects on live birds or museum specimens. For these reasons, stable isotopes are often useful companions to more expensive and invasive techniques, such as satellite tracking and dietary studies that rely on forced regurgitation or sacrifice of the animal for stomach content analysis (Meckstroth et al. 2007, Votier et al. 2010, González-Solís et al. 2011).

Museums can be a critical sample resource for isotope-based studies of avian ecology. Museums often hold specimens collected from across a species' geographic range, which can be particularly valuable to developing an understanding of spatial isotope patterns, for example, to understand connectivity between widely dispersed breeding and nonbreeding populations of migrants (Kelly et al. 2002, Lott et al. 2003). In addition, museum collections often hold specimens that are difficult or impossible to sample in the field because they are from rare, logistically difficult to access, or extinct species. Indeed, historically collected or archaeological and paleontological specimens provide particularly exciting samples to isotope researchers, who can use them to reconstruct historical trends in ecology (e.g., Thompson et al. 1995, Becker and Beissinger 2006, Wiley et al. 2013). Ornithologists today are faced with a pressing need to characterize and preserve diminishing avian biodiversity. This task involves understanding the nature and scope of human impacts on bird populations over timescales ranging from decades to millennia. Museum specimens are a source of samples that span these broad temporal scales, and the stable isotope ratios of organic matter preserved within these specimens provide a means of documenting shifts in foraging ecology, environmental biogeochemistry, and physiology. Stable isotope data can therefore make valuable contributions to the extended specimen

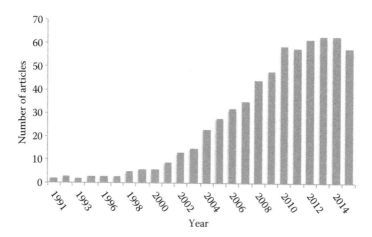

Figure 6.1. Published articles with key words "stable isotope" and "bird" within the Web of Science, by year. (Modified from a Web of Science citation report created December 15, 2015.)

(Chapter 1, this volume); they can both elaborate on the phenotype of individuals represented in museum collections and describe the ecological context from which birds were sampled.

This chapter is written to both inspire and inform the use of museum specimens in stable isotope-based studies of birds. To date, these studies have predominantly focused on isotope data from proteins and whole tissues; this chapter is similarly focused. For readers new to the field of isotope ecology, we include a brief section introducing stable isotope methodology and a section exploring applications of stable isotopic data in ornithology. We do not attempt to cover all studies in this large and growing field, but rather provide selected examples to capture its breadth, complications, and possibilities. With respect to museum specimens, we include a section on the challenges of using stable isotope techniques and sources of interpretation errors, and another section on sampling considerations and strategies, highlighting examples from our own research on procellariiform seabirds. This latter section places a heavy emphasis on the feather, as it is the most readily available tissue from museum specimens and is by far the most commonly used for isotope studies. Finally, we discuss challenges and recommendations for future isotope studies, and we propose ways in which the utility of museum collections could be increased for isotope researchers.

STABLE ISOTOPES: BASIC THEORY AND METHODOLOGY

Stable isotopes are unique forms of an element that differ in neutron number and do not radioactively decay. For example, ^{12}C and ^{13}C are stable isotopes, but not ^{14}C, which decays into ^{14}N through time. Due to the difference in neutron number, stable isotopes of a given element differ from each other in mass and relative abundance. For example, about 98.9% of all carbon atoms on earth are isotope ^{12}C, while only 1.1% are ^{13}C. ^{12}C is therefore referred to as the abundant isotope, while ^{13}C is considered the rare isotope. Although the stable isotopes of an element co-occur in natural substances and participate in the same chemical reactions, the reaction rate constant, or the rate at which a reaction proceeds, is higher for the lighter, typically more abundant, isotope than for the heavier, typically rare, one. Because of its higher reaction rate constant, the light isotope is

often enriched in the product of reaction, as compared with the heavy isotope. This phenomenon is referred to as fractionation and is described by a fractionation factor (α), which is the ratio of two reaction rate constants:

$$\alpha = k2/k1$$

where k1 and k2 are the rate constants for the light and heavy isotope, respectively.

Fractionation causes isotope ratios to vary among different organs, organisms, and nonliving substances. For example, plants using C3 photosynthesis incorporate less ^{13}C during CO_2 fixation than do plants with C4 photosynthesis, and so they have less ^{13}C in their tissues than do plants using C4 photosynthesis (Marshall et al. 2007). Biologists have learned to exploit naturally occurring isotopic variation in order to study the flow of elements through ecosystems and organisms, in the process gaining knowledge of food web structure, habitat use, and physiology for a broad range of habitats and species (Fry 2006). Just as C3 and C4 plants differ from each other in the relative abundance of ^{13}C, entire food webs based on these photosynthetic pathways are also isotopically distinct. Similarly, differences between fresh and seawater in the ratio of oxygen's most abundant isotopes, ^{18}O and ^{16}O, are incorporated into animal tissues and have been used to infer populations' relative use of those water sources (Schaffner and Swart 1991, Clark et al. 2006).

Stable isotopes of the "light" elements—C, N, H, O, and S—are most commonly used to study bird ecology. Isotope data are reported in delta notation as $\delta^{13}C$, $\delta^{15}N$, δD (or δ^2H), $\delta^{18}O$, and $\delta^{34}S$ values, expressed as ‰ (per mil). Delta notation represents the ratio of isotopes within a substance, as compared to that of an internationally accepted standard:

$$\delta X = (R_{sample}/R_{standard} - 1) \times 1000$$

where $X = {}^{13}C$, ^{15}N, D, ^{18}O, or ^{34}S; and R is the corresponding ratio of the element of interest: $^{13}C/^{12}C$, $^{15}N/^{14}N$, $^2H/^1H$, or $^{34}S/^{32}S$ (Coplen et al. 1983, Coplen 1994).

Stable isotope measurements are typically made using stable isotope mass spectrometers, which require that the sample be turned into a gas, and then the gas of interest (e.g., CO_2) be purified from other gas species in the sample. For example, the measurement of $\delta^{13}C$ from bone collagen requires combustion of collagen and purification of CO_2 from other combustion products such as N_2, H_2O, O_2, and SO_x. Purification can be achieved cryogenically, chemically, and/or chromatographically. Within the source of a mass spectrometer, the gas of interest is ionized and then accelerated into a magnetic field. Within the magnetic field, ions are deflected and separated according to their mass to charge ratio (m/z) based on the principle that ions of higher m/z have a greater radius of deflection than those of lower m/z.

In ecological stable isotope studies, researchers are interested in either the analysis of whole or "bulk" organic material such as tissues, proteins, or sediments, or else they are focused on individual compounds such as amino acids and alkanes. The majority of bulk isotope data used in ecological studies is now produced with the aid of an elemental analyzer that combusts and purifies gases (e.g., N_2, CO_2) from a small aliquot of liquid or solid. A stream of helium (He) serves to move the sample gas through the elemental analyzer and into the mass spectrometer for isotope analysis. The analysis of individual compounds or compound specific isotope analysis (CSIA) first requires the compound class of interest (i.e., amino acids, hydrocarbons, fatty acids) to be purified from a mixture. In the case of amino acids derived from collagen, the collagen is purified from the bone mineral and other proteins, and then hydrolyzed to produce free amino acids. If the compound class is not volatile, its constituents must be chemically modified to a volatile form, as is the case with amino acids. The individual compounds in the compound class must be separated on a gas chromatograph (GC) prior to combustion and isotope analysis. Thus, the sample is injected into a GC where it is carried onto a column and, ultimately, to the mass spectrometer by a stream of He. After separation on the GC column, compounds move, individually, into a furnace where they are combusted. The combustion gases are then cryogenically purified into the gas of interest prior to entering the mass spectrometer for isotope analysis.

AVIAN SPECIMENS IN STABLE ISOTOPE ECOLOGY

Stable isotope studies in ornithology can be considered a subdiscipline within a field often referred to as stable isotope ecology, which is itself a large and expanding discipline. Several excellent reviews are available for readers interested in the broader field of stable isotope ecology and further discussion of ornithological applications (Rubenstein and Hobson 2004, Fry 2006, West et al. 2006, Inger and Bearhop 2008, del Rio et al. 2009, Boecklen et al. 2011). Here, we discuss the well-established applications of stable isotopes in ornithology and highlight those that are most likely to involve museum specimens.

Two of the most productive areas of isotope-based ornithological research include studies of diet composition and spatial habitat use. Such studies are founded on the principle that the isotope value of a bird's tissues typically reflects the isotope value of its diet (or diet and drinking water, in the case of H and O), plus a predictable offset. This offset is often referred to as a discrimination factor. For example, the offset between local precipitation and a given animal's tissue is referred to as a precipitation-to-tissue discrimination factor, whereas the difference in isotope value between an organism, or its tissue, and its diet is called a trophic discrimination factor (TDF). Controlled feeding experiments can delineate the TDF for different tissues and demonstrate a predictable relationship between a bird and its diet, facilitating dietary studies for wild bird populations (e.g., Hobson and Clark 1992a,b, Hobson and Bairlein 2003, Cherel et al. 2005, Federer et al. 2010). Although isotope-based studies of bird diet can only differentiate between foods that are isotopically distinct, there are a range of variables that cause avian diets to differ from one another, including trophic level, ecosystem or habitat type, and spatial position; spatial patterns in isotope values are often referred to as isoscapes. In Table 6.1, we detail sources of stable isotope variation that have been used as the basis for avian studies. Although nonexhaustive, this list demonstrates the diversity of factors contributing to stable isotope variation in birds and their diets, and the corresponding diversity of ornithological research that is possible with stable isotope techniques.

The types of questions that isotope-based studies can address expands further when you consider that different bird tissues provide information from distinct timescales (see Table 6.2; Dalerum and Angerbjörn 2005). Among metabolically active tissues, the rate of turnover varies such that isotope values can reflect periods as short as

TABLE 6.1
A nonexhaustive list of sources or interpretations of stable isotope variations in bird tissues.

δ value	Environment	Source of variation	Example studies
$\delta^{13}C$	Marine	Foraging latitude	Cherel and Hobson (2007), Jaeger et al. (2010)
		Benthic versus pelagic food webs	Hobson et al. (1994)
	Terrestrial	C3 versus C4 versus CAM-based food webs; mesic versus xeric habitats	Marra et al. (1998), Bearhop et al. (2004), Clark et al. (2006)
	All	Incorporation of dietary lipids into tissue protein	Thompson et al. (2000)
	Mixed	Marine versus terrestrial C3/freshwater	Tietje and Teer (1988), Hobson and Sealy (1991), Gunnarsson et al. (2005)
$\delta^{15}N$	Marine	Foraging location (spatial variation in N cycling, e.g., denitrification)	Cherel et al. (2000), Wiley et al. (2012)
	Terrestrial	Mesic versus xeric habitats	Sealy et al. (1987), Ambrose (1991)
	All	Anthropogenically fixed N additions (fertilizer use)	Hebert and Wassenaar (2001)
		Trophic level	Kelly (2000), Becker and Beissinger (2006)
		Food restriction/starvation	Cherel et al. (2005) (but see Graves et al. 2012)[a]
	Mixed	Marine versus terrestrial/freshwater diet	Chamberlain et al. (2005)
$\delta D/\delta^{18}O$	Terrestrial	Geographic patterns in meteoric water (associated with altitude, latitude, and distance from ocean)	Hobson et al. (2001), Hobson et al. (2004), Bearhop et al. (2005), Greenberg et al. (2007)
$\delta^{18}O$	Mixed	Marine versus freshwater use	Shaffner and Swart (1991)
δD	All	Evaporative water loss	McKechnie et al. (2004), Powell and Hobson (2006)
$\delta^{34}S$	Marine	Benthic versus pelagic food webs	Thode (1991)
	Mixed	Marine versus estuarine/marsh versus terrestrial/freshwater diet	Hobson et al. (1997), Thode (1991)

[a] The tendency of food restriction and starvation to elevate $\delta^{15}N$ values appears to vary based on tissue type, with tissues that are actively maintained or synthesized during periods of fasting being most likely to show increased $\delta^{15}N$ (see Martinez del Rio et al. 2008).

days, as in blood plasma, or as long as years, as in bone collagen (Hobson and Clark 1992a, Pearson et al. 2003, Hedges et al. 2007). Isotope data from metabolically inert tissues such as feathers and toenails record information over the period of tissue synthesis, after which time they remain isotopically unchanged (Hobson and Clark 1992a). Understandably, feathers have become a favorite of isotope researchers as they can be sampled with minimal damage to live birds and museum specimens. Additionally, feathers are sometimes grown on distant nonbreeding grounds when birds typically are more difficult to study directly.

Researchers have successfully used the isotopic composition of multiple tissues, including feathers, to study different time periods within the annual cycle and to study intraindividual variation in resource use at a variety of timescales (Dalerum and Angerbjörn 2005). An important caveat is that it can be difficult to directly compare isotope data from multiple tissues: they will differ not only as a function of time, but also in accordance with their unique TDFs. For example, different tissues have distinct amino acid compositions and can be synthesized from different portions of the diet; the disparity in isotope values between diet and tissues (tissue-specific TDFs) tends to vary. Notably, TDFs have been measured for a variety of avian tissues and species, but because TDFs vary among species

TABLE 6.2
Bird tissues commonly used for stable isotope analysis and the time periods they reflect, described by half-life (metabolically active tissues) and growth rates (metabolically inert tissues).

Tissue	Species	Half-life (days)[a]		Source
		$\delta^{13}C$	$\delta^{15}N$	
Blood plasma	American Crow (*Corvus brachyrhynchos*)	2.9	0.7–1.7	Hobson and Clark (1992a)
	Yellow-rumped Warbler (*Setophaga coronata*)	0.4–0.7		Pearson et al. (2003)
Red blood cells	American Crow	29.8		Hobson and Clark (1992a)
Whole blood	Japanese Quail (*Coturnix japonica*)	11.4		Hobson and Clark (1992a)
	Great Skua (*Stercorarius skua*)	14	15.7	Bearhop et al. (2002)
	Garden Warbler (*Sylvia borin*)	11	5–5.7	Hobson and Bairlein (2003)
	Yellow-rumped Warbler	4–6	7.5–27.7	Pearson et al. (2003)
	Dunlin (*Calidris alpina*)	11.2	10	Ogden et al. (2004)
Liver	Japanese Quail	2.6		Hobson and Clark (1992a)
Muscle	Japanese Quail	12.4		Hobson and Clark (1992a)
Bone collagen	Japanese Quail	173.3[b]		Hobson and Clark (1992a)
		Growth rate		
Feather	Various[c]	1.7–11.0mm/day		Prevost (1983)[d]
Toenail	Five Palearctic passerine species	0.04 ± 0.01 mm/day growth rate		Bearhop et al. (2003)

[a] The turnover rate of a particular tissue can vary between species, generally slowing in larger species and species or individuals with lower protein intake (Martinez del Rio et al. 2009).

[b] Although Hobson and Clark (1992a) is the only study documenting bone collagen turnover in birds, other studies suggest that carbon in collagen can reflect decades or even the lifetime of an adult (e.g., Hedges et al. 2007).

[c] Lower range of feather growth rates was reported for the Ashy Storm-Petrel (*Oceanodroma homochroa*), and upper range for the Red-crowned Crane (*Grus japonensis*).

[d] Reports a range of growth rates, as reviewed by Rohwer et al. (2009).

(e.g., due to different dietary protein intake), there is some degree of error when TDFs are extrapolated beyond the original study organisms where they were measured (Cherel et al. 2005, Robbins et al. 2005).

Taking advantage of both a suite of available study variables and timescales, isotope researchers have built a substantial body of literature on avian diet and habitat use. For example, marine and terrestrial organisms generally differ in $\delta^{13}C$ and $\delta^{15}N$, differences that have been used to estimate the marine contribution to the diets of insular Northern Saw-whet Owl (*Aegolius acadicus*) and ancient California Condors (*Gymnogyps californianus*), and to infer a temporal increase in the use of terrestrial food by Herring Gulls (*Larus argentatus*) in the Great Lakes (Hobson and Sealy 1991, Chamberlain et al. 2005, Hebert et al. 2008).

$\delta^{15}N$ values are used widely to study the trophic level of birds and food web structure (Hobson et al. 1994, Sydeman et al. 1997, Beaudoin et al. 2001). This tool is the basis of a growing body of literature suggesting that seabirds in both coastal and oceanic regions may have shifted to lower trophic level prey after the onset of industrialized fishing and whaling within their foraging ranges (Becker and Beissinger 2006, Emslie and Patterson 2007, Farmer and Leonard 2011, Wiley et al. 2013). Archived bird specimens were sampled in most of the aforementioned studies, highlighting the value of museum collections in addressing these questions. Isotope research delineating changes in foraging habits through time can play a critical role in understanding avian population declines, and we predict this approach will have a rich future that is closely tied to museum specimen availability (Norris et al. 2007, Blight et al. 2014).

Stable isotopes also provide insight into bird migration and other spatial movement patterns (Hobson 1999, Webster et al. 2002, Rubenstein and Hobson 2004). The use of hydrogen isotopes to study migration is particularly common. This technique is based on the existence of continental-scale patterns in the δD value of precipitation, which is related to latitude, distance from the ocean, and altitude, and on the observation that birds typically incorporate the δD value of local precipitation into their growing tissues with a predictable offset, or precipitation-to-tissue discrimination factor (Chamberlain et al. 1997, Hobson and Wassenaar 1997, Hobson et al. 2004, Inger and Bearhop 2008). Geographic patterns of precipitation δD values are particularly well-described across continental North America, where feathers collected on the breeding grounds, but grown on the wintering grounds, are now commonly used to infer general wintering location and to help link wintering and breeding populations (Hobson et al. 2001, Bearhop et al. 2005). The utility of this approach is nicely demonstrated in the study by Greenberg et al. (2007), which used δD values to estimate the latitude of previously unknown Coastal Plain Swamp Sparrow (*Melospiza georgiana nigrescens*) wintering locations. With additional insight on the brackish habitat type of the wintering location inferred from feather $\delta^{13}C$ and $\delta^{15}N$, the authors were able to physically locate the wintering population (Greenberg et al. 2007).

Although geographic patterns, or isoscapes, of δD are absent in the ocean, $\delta^{13}C$ isoscapes have been commonly used to infer relative foraging latitude among marine birds (Cherel et al. 2006, Cherel and Hobson 2007, Jaeger et al. 2010). The broad-scale, inverse relationship between $\delta^{13}C$ and latitude throughout the world's oceans results from temperature-induced variation in the concentration of CO_2, which changes the extent to which phytoplankton incorporate ^{13}C during CO_2 uptake (Goericke and Fry 1994). A variety of other spatial patterns in $\delta^{13}C$, $\delta^{15}N$, and $\delta^{34}S$ have been recognized in marine, freshwater, and terrestrial environments, and these isoscapes can and have been used for studies of movement patterns in a variety of birds (Kelly 2000, Rubenstein and Hobson 2004, Braune et al. 2005, Hebert and Wassanaar 2005, Coulton et al. 2010, Wiley et al. 2012). In some cases, samples from museum

specimens have helped to describe isoscapes by increasing their spatial coverage and the rapidity of their development (e.g., Kelly et al. 2002).

As direct electronic tracking of birds becomes more feasible, stable isotopes have often entered into a partnership with tracking data, where isotopes provide increased sample size and the ability to include archived samples from birds that cannot be captured and tagged (Knoche et al. 2007). Although isotope analyses cannot provide the level of spatial resolution that is available through electronic tracking (Robinson et al. 2010, Hedd et al. 2012), they are much less expensive and labor-intensive. Importantly, isotope values often reflect diet in addition to location, meaning they can add a perspective on foraging that is impossible to obtain through tracking. We suggest that, after study systems are well-described by joint isotope-tracking research, stable isotopes may provide a means of cheaply monitoring for changes in a population's movements and foraging habits into the future.

Stable isotopes have provided important advances on a variety of other topics, such as the balance of endogenous versus exogenous nutrient allocation to egg and feather formation (Hobson et al. 1997, Klaassen et al. 2001, Fox et al. 2009, DeVink et al. 2011), the transport of marine-derived nutrients to terrestrial ecosystems by breeding seabirds (Wainright et al. 1998, Stapp et al. 1999), and potential carryover effects on fitness between the period of molt and the breeding season (Marra et al. 1998, Norris and Marra 2007, Kouwenberg et al. 2013). Stable isotopes have also proven useful in studies of habitat segregation between the sexes during the, often understudied, non-breeding season (Marra and Holmes 2001) and in investigations of variable contaminant and disease exposure (Bearhop et al. 2000, Braune and Hobson 2002, Hobson et al. 2002, Hebert et al. 2014). Each of these studies relied on one or more of the sources of isotope variation listed in Table 6.1.

EMERGING ANALYTICAL TECHNIQUES

Before moving on, we highlight two emerging methods in stable isotope ecology that are growing in influence. First, Bayesian analyses of mass balance models are becoming common in food web studies. Mass balance models, such as Isosource, have long been used in food web studies to evaluate the relative importance of dietary items to consumers (Phillips and Greg 2003). Bayesian analysis of mass balance models produces estimates of the relative contributions of sources (often prey) to isotope values (often of consumers) as true probability density functions rather than the simple range of possible values that is produced by mass balance models (Parnell et al. 2010). Bayesian methodology also has the advantage that the analyst can incorporate prior knowledge about source contributions to the diet of the animal under investigation, which can increase precision and accuracy of diet estimates (Parnell et al. 2013).

Standard ellipsoid models are an important approach derived from a Bayesian framework (Jackson et al. 2011). The mean isotope values for the population are the center of the ellipse, whereas eigenvalues of a covariance matrix generated from a multivariate normal distribution are used to derive the area of the ellipse (SEA_b). The ellipse defines an area on a bivariate isotope plot and is a reflection of isotopic variation for a population. The ellipse area can also be interpreted as a measure of "isotopic niche width" if isotope values can be clearly related to foraging attributes such as habitat (Newsome et al. 2010, Jackson et al. 2011). Importantly, Bayesian methods can also provide estimates of the probability that two means may be different, rather than confidence intervals provided by frequentist analysis of variance (ANOVA) approaches.

Advances are also being made in the use of compound-specific isotope analysis. This technique generates isotope data from individual compounds such as fatty acids or amino acids, instead of whole tissues or proteins; this latter class of analyses is often referred to as "bulk" isotope analysis. Compound-specific isotope analyses offer the ability to differentiate between otherwise confounded sources of isotope variation. For example, $\delta^{15}N$ values are commonly used to infer relative trophic position, but spatial variability in $\delta^{15}N$ values (e.g., from differences in the relative use of nitrate versus atmospheric N_2 at the base of marine food webs) can confound interpretations of trophic position. Some amino acids, such as phenylalanine (Phe), are incorporated from diet into consumer tissues with negligible TDFs and have $\delta^{15}N$ values that predominantly reflect the nitrogen source, that is, the $\delta^{15}N$ of nitrogen used by primary producers at the base of the food web. Other amino acids, such as glutamic acid (Glu), have large TDFs and therefore show isotopic shifts due to trophic transfers, in addition to reflecting source nitrogen $\delta^{15}N$ (McClelland and Montoyo 2002, Chikaraishi et al. 2009). By comparing the $\delta^{15}N$ value of Phe and Glu (or other so-called source and trophic amino acids), researchers are able to distinguish between variation in trophic level and variation in nitrogen source (Choy et al. 2012, Steffan et al. 2013). This amino acid-specific $\delta^{15}N$ technique offers another exciting prospect: the potential to generate reliable estimates of trophic level for individual birds in the absence of prey or primary producer samples. However, accurate trophic level estimates of birds will require further experimental work that clearly defines the TDF for amino acids in avian tissues (Lorrain et al. 2009, McMahon et al. 2015).

POTENTIAL CONFOUNDING FACTORS AND OTHER SOURCES OF ERROR

Stable isotopes can be powerful tools, but interpretation of isotope data can be challenging, as the ratio of isotopes for any given element can vary as a function of multiple factors. It is the responsibility of every thoughtful isotope researcher to understand the potential sources of isotope variation within their study system as well as the assumptions of isotope methods that, when violated, may be sources of interpretive error. For example, many isotope studies assume that birds are in isotopic equilibrium with their food sources, such that newly formed tissues will reflect the present diet and location. This assumption is not always met. Birds may retain metabolic pools derived from previous locations and may catabolize previously built tissues in order to synthesize new ones; this has been highlighted by isotope studies of exogenous versus endogenous reserves in egg and feather formation (Klaassen et al. 2001, Fox et al. 2009).

We do not attempt to summarize all potential sources of error here, but note that careful consideration of spatial and temporal scale is imperative. For example, δD isoscapes can be powerful indicators of long-distance movements, but at small spatial scales, δD variation may predominately derive from factors other than geographical location because of isotopic influences from local hydrology or water lost from birds' bodies through evaporative cooling (Powell and Hobson

2006, Wunder et al. 2012). Similarly, isoscapes do not always preclude the possibility that bird tissues with similar isotope values were formed in very distant locations. Just as researchers should carefully consider the limitations of isoscapes, they should also acknowledge limits in temporal resolution. For example, the precise timing of feather growth is often unknown, even when molt sequence and season have been described, and intraspecific variation in the timing of feather growth can sometimes explain significant isotope differences (e.g., Wunder et al. 2012).

The occasional discovery of new sources of isotope variation in bird tissues highlights our sometimes-incomplete understanding, but, at the same time, it can inspire the development of new isotopic research tools. For example, Ostrom et al. (2014) discovered a range of >100‰ in δD values among seabirds foraging in the open ocean, even though δD values of water and potential prey are relatively invariant. They hypothesized that isotopic discrimination associated with salt excretion via nasal salt glands is the main cause of this variability, and that birds that eat salty invertebrates as opposed to less salty teleost fish tend to have higher δD values. Further work is needed to determine if δD is a useful tool for studying diet in birds with well-developed salt glands.

Studies uncovering new sources of isotope heterogeneity may also alert us to potentially confounding factors that can be better controlled in future research. For instance, Michalik et al. (2010) and Wiley et al. (2010) found that feather containing the pigment eumelanin tends to be depleted in ^{13}C relative to melanin-free feather (Figure 6.2). This effect may result from the synthesis of eumelanin from the amino acid tyrosine, which is incorporated into avian tissues from diet with little or no biochemical alteration, and is itself depleted in ^{13}C relative to bulk keratin (McCullagh et al. 2005). Similarly, Grecian et al. (2015) showed an offset in $\delta^{13}C$ between feather vane and feather rachis, which they attributed to a difference in amino acid composition between these two feather components. Regardless of cause, these findings demonstrate that consistency in sampling technique can be important, and isotope researchers should continue to explore isotopic heterogeneity within bird populations, individuals, and tissues to test assumptions of their techniques.

When sampling museum specimens, researchers should also consider the possibility that preservation methods may have altered the original isotope values of avian tissues. Most published research on this topic has concentrated on the

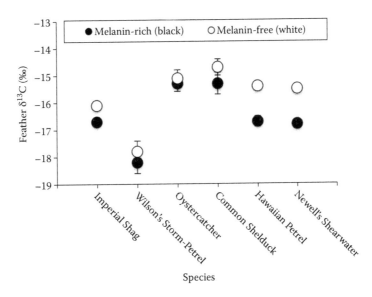

Figure 6.2. Stable carbon isotope values of melanin-rich and melanin-free feathers in six species. Data points represent group means ± standard errors. (Imperial Shag [*Phalacrocorax atriceps*], Wilson's Storm-Petrel [*Oceanites oceanicus*], Oystercatcher [*Haematopus ostralegus*], and Common Shelduck [*Todorna tadorna*] data are from Michalik et al. 2010. Hawaiian Petrel [*Pterodroma sandwichensis*] and Newell's Shearwater [*Puffinus auricularis newelli*] data are from Wiley et al. 2010.)

use of ethanol and formalin-preserved specimens. Several studies found no impact of ethanol preservation on $\delta^{13}C$ and $\delta^{15}N$ values, and a consistent, and therefore correctable, impact of formalin on $\delta^{13}C$ (Sarakinos et al. 2002 and citations therein, Barrow et al. 2008). However, Kaehler and Pakhomov (2001) and Kelly (2006) found that ethanol can sometimes alter $\delta^{13}C$ values of preserved organisms, and the effects of ethanol and formalin can apparently vary between species and tissue types (e.g., Sweeting et al. 2004). Taken together, these studies suggest that investigators should carefully consider methods of sample preservation and may need to carry out experiments to test the equivalency of preservation methods (e.g., drying versus freezing versus ethanol and formalin preservation) for the particular taxa and tissues of interest.

The list of potentially confounding factors in stable isotope studies may be daunting to new isotope users. However, we note that multiple isotopes can often be used to constrain potential sources of variation and afford more definitive conclusions (Lott et al. 2003). In other cases, ground-truthing with conventional, nonisotopic techniques can provide a foundation for isotope-based inferences (González-Solís et al. 2011).

Knowledge of a species' natural history can also be a means of discounting potential sources of isotope variation. Finally, we echo a conclusion made by Inger and Bearhop (2008): although potential sources of error in isotope-based studies of birds exist, many can be overcome simply by thoughtful study design.

STRATEGIES FOR SAMPLING MUSEUM SPECIMENS

For any isotope-based study of birds, one of the first decisions about sampling will be the choice of tissue type. Avian museum specimens are most commonly sampled for feather and bone. Bone has generally been used in studies of zooarchaeological and paleontological samples, which tend to rely on isotope data from collagen, the dominant bone protein. A much larger number of studies have used feathers. To illustrate the different utilities of these two tissues, we present data from the Hawaiian Petrel (*Pterodroma sandwichensis*) as a case study.

Thanks to the abundant ancient bone record of the Hawaiian Petrel, Wiley et al. (2013) were able to construct isotope timelines reaching back circa 4,000 years (Figure 6.3). These timelines reflect

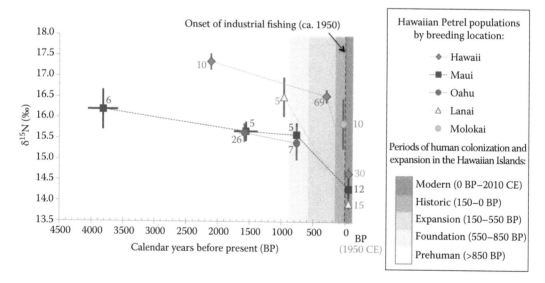

Figure 6.3. Isotope data from bone collagen provide time-averaged signals of diet and the possibility of constructing historic and prehistoric timelines. Here, $\delta^{15}N$ values of modern and radiocarbon-dated bone collagen are shown for five Hawaiian Petrel populations. The average age and isotopic composition of each time bin, $\pm$ standard error, is plotted with sample size noted. Gray shading indicates time bins. Modern samples were unavailable from Oahu and Molokai due to population extirpation. Stippled lines connecting data points are for visualization purposes; isotopic shifts between time bins may have occurred nonlinearly. CE, Common Era. (Reprinted with permission from Wiley et al. 2013.)

an integrated measure of diet over the course of the annual cycle, and likely over months to years in the life of each individual bird. Such time-averaged signals can be ideal for studies covering large temporal scales. Hawaiian Petrel $\delta^{15}N$ values show a striking pattern of decline through time, with all significant change isolated to the interval between the two most modern time bins in three populations. Wiley et al. (2013) interpreted this $\delta^{15}N$ decline as most likely reflecting a decline in trophic level: perhaps a shift in petrel diet caused by the onset of industrial fishing. Such a decline was previously undocumented for a tropical, open ocean predator, and suggests potentially widespread changes in Hawaiian Petrel food webs, which extend from the equator to near the Aleutian Islands. Importantly, the Hawaiian Petrel isotope chronology includes data from well before the advent of humans in the breeding grounds or foraging range of the species. These data therefore establish a prehuman baseline for an endangered seabird, against which modern and future data can be compared.

In contrast, Wiley et al. (2014) use stable carbon and nitrogen isotope data from Hawaiian Petrel feathers to detail modern foraging diversity (Figure 6.4). Whereas collagen data do not show significant differences among modern petrel populations, data from flight feathers grown during the chick-rearing period (by hatch-year birds) and nonbreeding season (adults) are consistent with differences in foraging location between populations and also between periods of the annual cycle (Wiley et al. 2014). These data provide concrete guidance on how to collect electronic tracking data that represent the species as a whole. For example, it is clear from isotope data that Maui Petrels, previously tracked during the breeding season, likely provide different results than would tracking of any population into the

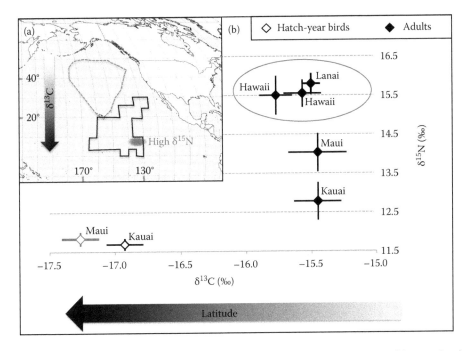

Figure 6.4. Feather data can provide insight into foraging ecology over multiple, discrete periods of the annual cycle. Here, flight feather isotope data are shown alongside at-sea locations of Hawaiian Petrels. In (a), the black line marks Hawaiian Petrel distribution from transect surveys (Spear et al. 1995). The dashed line is a typical flight path from a satellite-tracked Maui bird during the breeding season (Adams and Flora 2009). These two regions represent the predominant areas where Hawaiian Petrels occur. In (a), the blue oval denotes an approximate area where organic matter and consumers have unusually high $\delta^{15}N$ values within the Hawaiian Petrel's range (Altabet and Francois 1994, Graham et al. 2010). In (b), the blue circle surrounding Lanai and Hawaii data points identifies petrels that apparently concentrate their foraging in a region with elevated $\delta^{15}N$. In both panels, arrows emphasize the negative relationship between latitude and $\delta^{13}C$ of marine organisms (Goericke and Fry 1994, MacKenzie et al. 2011). Hatch-year birds from Maui are outlined in red to associate them with the Maui flight path. (Reprinted with permission from Wiley et al. 2013.)

nonbreeding season, when adults apparently foraged at significantly lower latitudes.

While adult Hawaiian Petrel feather data shown in Figure 6.4 represent the early nonbreeding season, feather sampling can be designed to reflect a wide range of time frames by carefully considering molt patterns and feather growth rates. For example, sampling within a feather can yield data reflecting a period of less than 24 hours (Greylag Goose, Rohwer et al. 2015), whereas sampling multiple remiges can reflect periods separated by a full year in some albatross species that take two annual molts to replace all of their flight feathers (Edwards and Rohwer 2005). Growth bars are a visual testament to a convenient truth about feathers: moving from the oldest material at the distal tip to the most recently grown material at the proximal base, each feather represents a time series (Grubb 2006). When feathers are grown in sequence, a more extended time line is recorded in plumage, and savvy isotope researchers can choose their preferred time frame to sample within this period of feather growth, whether that be days or, in some cases, months to years. For example, in adults with sequential remige molt, isotope researchers have long recognized their ability to study foraging habits throughout the period of molt by sampling multiple primaries (Thompson and Furness 1995). More recently, Rohwer and Broms (2012) used careful attention to a combination of growth bar width and feather molt overlap to estimate molt duration. They then sampled remiges at equal time intervals for stable isotope analysis, producing a very elegant isotopic time line from museum specimens.

Isotope researchers often sample a small number of whole body contour feathers from museum specimens: this approach appeals to museum curators because it typically has little to no visual impact on a study skin and leaves many similar feathers for future studies. Body contour feathers can also generate highly useful isotope data, if the timing of their growth can be constrained. For example, in Marbled Murrelets (*Brachyramphus marmoratus*), brown-tipped breast feathers represent the pre-breeding period and all-white breast feathers represent the post-breeding period (Becker and Beissinger 2006).

However, body contour feathers are not ideal samples for species in which body contour molt is protracted or poorly characterized (e.g., many petrels; Warham 1996). In such cases, isotope data from contour feathers cannot be ascribed to a particular period of the annual cycle. It can also be difficult to know the relationship between the growth period of multiple body contour feathers, outside of qualitative observations based on relative wear. For example, if a single breast contour feather is analyzed for each individual in a study, the calendrical time periods sampled for each bird may differ. If two contour feathers are analyzed for each bird, some individuals may be represented by feathers grown simultaneously, and others by feathers grown during different days, weeks, or even months. In other words, except in cases where different body contour feather types (e.g., head versus breast) are clearly molted at different periods, it may be impossible to extend a time line beyond that represented by an individual feather. A wider variety of time periods with more precise temporal constraint can generally be accessed by sampling flight feathers in birds with sequential molt. The drawback is that remiges and rectrices are considered precious to museum study skins because they have, at most, one equivalent on any given study skin, and therefore must be sampled with great care and minimal damage.

Clearly, trade-offs are inherent in the design of museum specimen sampling strategies. Researchers may navigate these trade-offs by asking two simple questions: How can feathers be sampled so as to represent the time period of interest, and how can impact to a study skin be minimized? Generally, the best way to minimize impact to a museum skin is to take only the mass required for isotope analysis and, perhaps, for a potential backup analysis. Most isotope analyses require very small amounts of material (e.g., 1 mg), and, for many feathers, this mass can be collected with very little visual alteration (see Figure 6.5). Such a sample typically represents a few days' worth of growth, when taken from one location on one feather. If a researcher desires to study a longer period of time, a whole feather can be homogenized, or homogenization can be avoided by collecting barbs from strategic locations across one or more feathers. For example, Wiley et al. (2010) described a sampling protocol where barbs taken from near the base, middle, and tip of a feather could be combined and analyzed as a single sample that accurately represented the average $\delta^{15}N$, $\delta^{13}C$, and δD values of the entire feather (see barb sampling technique, Figure 6.5a). A similar philosophy could be used to develop minimally destructive protocols that

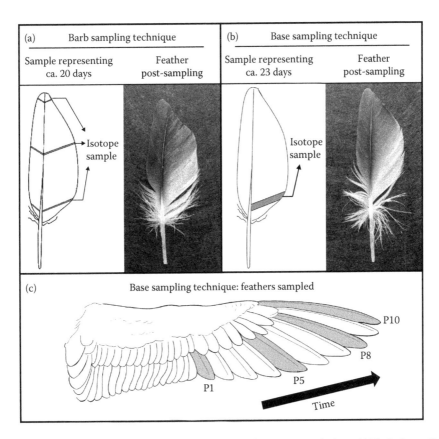

Figure 6.5. Two minimally destructive feather sampling techniques for museum study skins. (a) The barb sampling technique pictured produces isotope data reflective of the whole-feather average through strategic sampling of barbs down the length of the feather (Wiley et al. 2010). Unequal masses of barbs are taken from three locations along the feather. These subsamples are weighted so as to reflect the average distribution of mass found down the length of the feather. (b, c) The base sampling technique is less time-intensive and can easily be used for live birds. We recommend using this latter sampling technique on multiple feathers, so as to provide data from throughout the period of remige molt and data that are, together, reflective of more than very short periods of time. In (a) and (b), Hawaiian Petrel (*Pterodroma sandwichensis*) P1 is shown (length of vanes = 7 cm). Feathers are disarticulated for visualization purposes only; feathers sampled in museums and live birds are left attached to the wing. We estimated the time periods reflected by samples using growth bar width and the assumption that one growth bar is formed each day, however, we recognize the possibility that adults may form two growth bars per day, as observed in some albatross (Langston and Rohwer 1996).

represent any number of different time frames recorded in the plumage.

Alternatively, a second method of sampling remiges (Figure 6.5b,c, base sampling technique) yields isotope samples reflecting multiple, non-overlapping time periods. Here, material for a single isotope analysis is taken from the base of a primary, and multiple primaries are sampled so as to generate repeated measures from an individual bird. The resulting isotope data can be used to assess the consistency of foraging habits and the degree of individual foraging specialization throughout the period of primary molt. This protocol only compromises remiges at the base: a

region that is not visible on study skins, and is difficult to access, even with manipulation. Because the bases of remiges are often white or nearly melanin-free, this protocol limits the potentially confounding effect of melanin on carbon isotope data. It is also very fast when used on a spread wing and therefore applicable to live birds. Importantly, both the protocols described in Figure 6.5 require great patience and care when used on museum specimens, especially when wings are tightly tied, but they result in little to no visible alteration of the typical study skin.

Ideal sampling strategies will depend greatly on the species of interest and the study questions.

We note that the smaller the study species, the more destructive isotope sampling protocols will be, given the convergence of minimal isotope sample mass and feather mass. However, there are many avenues through which isotope researchers can make their sampling minimally destructive, including by taking small feathers of lesser value, by sampling small, noncritical portions of feathers, or by taking very small samples from a variety of locations. We recommend that all isotope users strive for the most minimally invasive protocols possible, when sampling museum specimens. Such a practice will increase the palatability of isotope sampling requests to museum curators and preserve valuable material for future study, including increasingly sophisticated isotope research.

CHALLENGES AND RECOMMENDATIONS FOR FUTURE MUSEUM COLLECTIONS AND ISOTOPE STUDIES

Stable isotope analysis of museum specimens can clearly be used to expand our perspective of birds' phenotypes beyond the anatomy of the specimens themselves. Although traditional study skins provide valuable isotope samples, the current state of avian museum collections presents several challenges to the stable isotope researcher. First, it can be difficult to obtain sample numbers that are large enough to accurately represent a population. Stable isotope studies typically focus on aspects of ecology that are variable within a population, such as dietary tendencies. As a result, isotope users seek large sample sizes per population per time period. When a researcher wishes to study a species across its geographic range or through a series of historical time periods, to distinguish among sexes or age classes, or to leave some portion of a museum collection series untouched by destructive sampling, the problem of small sample size can be greatly compounded. Of course, this limitation applies not only to stable isotope research, but also to many types of studies in which individuals cannot be treated as equivalents (Bolnick et al. 2003).

Museums will always be limited in the number of specimens they can collect and prepare due to finite resources and concern for the health of wild bird populations. However, increased preservation of large specimen series from select populations and species would greatly increase the knowledge discernable via stable isotope research and other ecological studies. Considering the high

value of archived museum specimens for isotope studies, we call on stable isotope researchers to contribute to growth of avian collections by collecting and preparing specimens themselves.

Supplemental, low-cost collections can also improve prospects for stable isotope studies. For example, the "feather files" maintained by the Bird Division of the National Museum of Natural History (Washington, DC) preserves bags of feathers that are normally discarded during preparation of skeletons. These feather collections take minutes to create and can be stored by the hundreds in the space of a single filing cabinet. More ambitiously, Smith et al. (2003) proposed widespread and systematic collection of feathers from migratory birds, to facilitate the myriad studies that use feathers as a sample medium. We agree that a concerted effort toward sampling live birds would provide a valuable complement to current, publically available avian collections. Indeed, one of the advantages of isotope-based methodology is that live birds can be sampled with limited impact by removing a feather or partial feather, taking a blood sample, or clipping a claw. If birds are not sacrificed, the same individuals can be sampled through time (within and between seasons or years), allowing researchers to construct ecological histories for individuals, and maybe even records of individuals over their entire lives. Such histories could be used to address understudied ecological topics, such as individual specialization in diet, and help to differentiate variation within versus between individuals.

In addition to curating samples, museums play a critical role in the curation of data. Pauli et al. (2015) have made a thoughtful call for large-scale archiving of stable isotope data into a so-called IsoBank, a stable isotope analog to GenBank. This endeavor could require many partners to help maintain and to organize metadata, and we encourage museums to help take a first step toward this important goal. At the very least, inclusion of isotope data from museum specimens into IsoBank would increase visibility and usefulness of these data.

Another limitation for stable isotope studies is the fact that many tissues of interest, such as muscle, liver, bone, and blood, are not commonly preserved in museum specimens, or else they are not preserved with the intention of supporting isotope studies. For example, many museums now preserve frozen samples of soft tissue (e.g., muscle) for genetic analyses, but curators may be reluctant to use those resources to support both isotope and genetic sampling. Similarly, preparation of bird skeletons is labor-intensive, and curators are understandably reluctant to allow cutting of bone shards from these precious specimens for isotope analysis. An obvious but potentially costly solution is to increase preservation of those tissues most commonly used for isotope analysis, especially tissues with shorter turnover times that are most likely to represent the collection locality (e.g., liver) and those most amenable to retrospective studies (e.g., bone). Just as we now preserve vouchered tissues for genetic studies, we could preserve bone, muscle, and liver samples when preparing study skins. For specific projects, stable isotope researchers can collaborate with museums to collect samples during the preparation of specimens. The preparation of skeletons and study skins typically involves discarding large volumes of material that could be gleaned for isotope analysis. Isotope researchers' sampling alongside museum preparators not only takes the burden of additional curation away from museums, it allows increased flexibility to choose a range of samples on the part of the isotope researcher. Salvaging of otherwise discarded tissues is also a means of "extending" a museum specimen (Chapter 1, this volume), as it potentially increases the value of specimens through the production of associated ecological data.

Finally, we note that, although feathers are the most accessible and commonly used tissue in isotope studies involving museum specimens, feather sampling decisions can be complex, especially when knowledge of molt is incomplete. Future molt studies will clearly benefit isotope research by clarifying the time periods reflected in plumage. Isotope studies may, themselves, contribute to this knowledge of molt (Hobson et al. 2000, Neto et al. 2006). Similarly, continued study of isotope variation within the plumage of individual birds will be important to document the effects of nontarget variables (e.g., melanin concentration) on isotope data and to encourage adoption of ideal sampling strategies.

CONCLUSIONS

Stable isotope analysis is, and will continue to be, a powerful technique for the study of avian ecology and physiology. In an era of human impact on bird populations across the globe, stable isotope analysis of museum specimens may be an increasingly important means of understanding the nature and magnitude of anthropogenic change, as well as determining how best to protect dwindling bird populations. Especially when coupled with the broad taxonomic, spatial, and temporal range represented in museum collections, stable isotope analysis can increase our knowledge of changing prey bases and shifting habitat use, as well as provide information on basic dietary and migratory patterns that is essential for conservation decisions. To encourage such research outcomes, stable isotope users and museums should work toward goals of minimally destructive sampling and increased sample availability. If met, these goals will increase the material available to future generations of isotope studies, whose particular methods and questions are as yet unknown, but whose findings may be critical to our understanding of avian biology.

ACKNOWLEDGMENTS

A large number of museum curators and collections managers have provided samples and assisted in the development of sampling techniques described in this chapter. Without their dedication, commitment, and foresight, current and future projects would not be possible. Thus, we would like to thank all institutions that house important avian collections and especially those that have been important to us, including the National Museum of Natural History, the Bernice Bishop Pauahi Bishop Museum, the Burke Museum of Natural History and Culture, and the Natural History Museum of Los Angeles County. We also thank S. Rossman, M. Webster, and an anonymous reviewer for helpful comments on our manuscript. Funding for research on the Hawaiian Petrel was provided by the National Science Foundation (Division of Environmental Biology-0745604).

LITERATURE CITED

Adams, J., and S. Flora. 2010. Correlating seabird movements with ocean winds: linking satellite telemetry with ocean scatterometry. Marine Biology 157:915–929.

Altabet, M. A., and R. Francois. 1994. Sedimentary nitrogen isotopic ratio as a recorder for surface ocean nitrate utilization. Global Biogeochemical Cycles 8:103–116.

Ambrose, S. H. 1991. Effects of diet, climate and physiology on nitrogen isotope abundances in terrestrial foodwebs. Journal of Archaeological Science 18:293–317.

Barrow, L. M., K. A. Bjorndal, and K. J. Reich. 2008. Effects of preservation method on stable carbon and nitrogen isotope values. Physiological and Biochemical Zoology 81:688–693.

Bearhop, S., W. Fiedler, R. W. Furness, S. C. Votier, S. Waldron, J. Newton, G. J. Bowen, P. Berthold, and K. Farnsworth. 2005. Assortative mating as a mechanism for rapid evolution of a migratory divide. Science 310:502–504.

Bearhop, S., R. W. Furness, G. M. Hilton, S. C. Votier, and S. Waldron. 2003. A forensic approach to understanding diet and habitat use from stable isotope analysis of (avian) claw material. Functional Ecology 17:270–275.

Bearhop, S., G. M. Hilton, S. C. Votier, and S. Waldron. 2004. Stable isotope ratios indicate that body condition in migrating passerines is influenced by winter habitat. Proceedings of the Royal Society B 271:S215–S218.

Bearhop, S., R. A. Phillips, D. R. Thompson, S. Waldron, and R. W. Furness. 2000. Variability in mercury concentrations of Great Skuas (*Catharacta Skua*): the influence of colony, diet and trophic status inferred from stable isotope signatures. Marine Ecology Progress Series 195:261–268.

Bearhop, S., S. Waldron, S. C. Votier, and R. W. Furness. 2002. Factors that influence assimilation rates and fractionation of nitrogen and carbon stable isotopes in avian blood and feathers. Physiological and Biochemical Zoology 75:451–458.

Beaudoin, C. P., E. E. Prepas, W. M. Tonn, L. I. Wassenaar, and B. G. Kotak. 2001. A stable carbon and nitrogen isotope study of lake food webs in Canada's boreal plain. Freshwater Biology 46:465–477.

Becker, B. H., and S. R. Beissinger. 2006. Centennial decline in the trophic level of an endangered seabird after fisheries decline. Conservation Biology 20:470–479.

Blight, L. K., K. A. Hobson, T. K. Kyser, and P. Arcese. 2014. Changing gull diet in a changing world: a 150-year stable isotope ($\delta^{13}C$, $\delta^{15}N$) record from feathers collected in the Pacific Northwest of North America. Global Change Biology 21:1497–1507.

Boecklen, W. J., C. T. Yarnes, B. A. Cook, and A. C. James. 2011. On the use of stable isotopes in trophic ecology. Annual Review of Ecology, Evolution, and Systematics 42:411–440.

Bolnick, D. I., R. Svanbäck, J. A. Fordyce, L. H. Yang, J. M. Davis, C. D. Hulsey, and M. L. Forister. 2003. The ecology of individuals: incidence and implications of individual specialization. American Naturalist 161:1–28.

Braune, B. M., G. M. Donaldson, and K. A. Hobson 2002. Contaminant residues in seabird eggs from the Canadian Arctic. II. Spatial trends and evidence from stable isotopes for intercolony differences. Environmental Pollution 117:133–145.

Braune, B. M., K. A. Hobson, and B. J. Malone. 2005. Regional differences in collagen stable isotope and tissue trace element profiles in populations of Long-tailed Duck breeding in the Canadian arctic. Science of the Total Environment 346:156–168.

Chamberlain, C. P., J. D. Blum, R. T. Holmes, X. Feng, T. W. Sherry, and G. R. Graves. 1997. The use of isotope tracers for identifying populations of migratory birds. Oecologia 109:132–141.

Chamberlain, C. P., J. R. Waldbauer, K. Fox-Dobbs, S. D. Newsome, P. L. Koch, D. R. Smith, M. E. Church, S. D. Chamerblain, K. J. Sorenson, and R. Risebrough. 2005. Pleistocene to recent dietary shifts in California condors. Proceedings of the National Academy of Sciences USA 102:16707–16711.

Cherel, Y., and K. Hobson. 2007. Geographical variation in carbon stable isotope signatures of marine predators: a tool to investigate their foraging areas in the Southern Ocean. Marine Ecology Progress Series 329:281–287.

Cherel, Y., K. A. Hobson, and S. Hassani. 2005. Isotopic discrimination between food and blood and feathers of captive penguins: implications for dietary studies in the wild. Physiological and Biochemical Zoology 78:106–115.

Cherel, Y., K. A. Hobson, and H. Weimerskirch. 2000. Using stable-isotope analysis of feathers to distinguish moulting and breeding origins of seabirds. Oecologia 122:155–162.

Cherel, Y., R. A. Phillips, K. A. Hobson, and R. McGill. 2006. Stable isotope evidence of diverse species-specific and individual wintering strategies in seabirds. Biology Letters 2:301–303.

Chikaraishi, Y., N. O. Ogawa, Y. Kashiyama, Y. Takano, H. Suga, A. Tomitani, H. Miyashita, H. Kitazato, and N. Ohkouchi. 2009. Determination of aquatic food-web structure based on compound-specific nitrogen isotopic composition of amino acids. Limnology and Oceanography: Methods 7:740–750.

Choy, C. A., P. C. Davison, J. C. Drazen, A. Flynn, E. J. Gier, J. C. Hoffman, J. P. McClain-Counts, T. W. Miller, B. N. Popp, S. W. Ross, and T. T. Sutton. 2012. Global trophic position comparison of two dominant mesopelagic fish families (Myctophidae, Stomiidae) using amino acid nitrogen isotopic analyses. PLoS One 7:e50133

Clarke, S. J., G. H. Miller, M. L. Fogel, A. R. Chivas, and C. V. Murray-Wallace. 2006. The amino acid and stable isotope biogeochemistry of elephant bird (*Aepyornis*) eggshells from southern Madagascar. Quaternary Science Reviews 25:2343–2356.

Coplen, T. B. 1994. Reporting of stable hydrogen, carbon, and oxygen isotopic abundances. Pure and Applied Chemistry 66:273–276.

Coplen, T. B., C. Kendall, and J. Hopple. 1983. Comparison of stable isotope reference samples. Nature 302:236–238.

Coulton, D. W., R. G. Clark, and C. E. Hebert. 2010. Determining natal origins of birds using stable isotopes (δ^{34}S, δD, δ^{15}N, δ^{13}C): model validation and spatial resolution for mid-continent Mallards. Waterbirds 33:10–21.

Dalerum, F., and A. Angerbjörn. 2005. Resolving temporal variation in vertebrate diets using naturally occurring stable isotopes. Oecologia 144:647–658.

del Rio, M. C., N. Wolf, S. A. Carleton, and L. Z. Gannes. 2009. Isotopic ecology ten years after a call for more laboratory experiments. Biological Reviews of the Cambridge Philosophical Society 84:91–111.

Devink, J., S. M. Slattery, R. G. Clark, R. T. Alisauskas, and K. A. Hobson 2011. Combining stable-isotope and body-composition analyses to assess nutrient-allocation strategies in breeding White-winged Scoters (*Melanitta Fusca*). Auk 128:166–174.

Edwards, A. E., and S. Rohwer. 2005. Large-scale patterns of molt activation in the flight feathers of two albatross species. Condor 107:835–848.

Emslie, S. D., and W. P. Patterson. 2007. Abrupt recent shift in δ^{13}C and δ^{15}N values in Adélie Penguin eggshell in Antarctica. Proceedings of the National Academy of Sciences USA 104:11666–11669.

Farmer, R. G., and M. L. Leonard. 2011. Long-term feeding ecology of Great Black-backed Gulls (*Larus marinus*) in the northwest Atlantic: 110 years of feather isotope data. Canadian Journal of Zoology 89:123–133.

Federer, R. N., T. E. Hollmén, D. Esler, M. J. Wooller, and S. W. Wang. 2010. Stable carbon and nitrogen isotope discrimination factors from diet to blood plasma, cellular blood, feathers, and adipose tissue fatty acids in Spectacled Eiders (*Somateria fischeri*). Canadian Journal of Zoology 88:866–874.

Fox, A. D., K. A. Hobson, and J. Kahlert. 2009. Isotopic evidence for endogenous protein contributions to Greylag Goose *Anser anser* flight feathers. Journal of Avian Biology 40:108–112.

Fry, B. 2006. Stable isotope ecology. Springer, New York, NY.

Gloutney, M. L., and K. A. Hobson. 1998. Field preservation techniques for the analysis of stable-carbon and nitrogen isotope ratios in eggs. Journal of Field Ornithology 69:223–227.

Goericke, R., and B. Fry. 1994. Variation of marine plankton δ^{13}C with latitude, temperature, and dissolved CO_2 in the world ocean. Global Biogeochemical Cycles 8:85–90.

González-Solís, J., M. Smyrli, T. Militão, D. Gremillet, T. Tveraa, R. A. Phillips, and T. Boulinier. 2011. Combining stable isotope analyses and geolocation to reveal kittiwake migration. Marine Ecology Progress Series 435:251–261.

Graham, B. S., P. L. Koch, S. D. Newsome, K. W. McMahon, and D. Aurioles. 2010. Using isoscapes to trace the movements and foraging behavior of top predators in oceanic ecosystems. Pp. 299–318 in J. B. West, G. J. Bowen, T. E. Dawson, and K. P. Tu (editors), Isoscapes: understanding movement, pattern, and process on Earth through isotope mapping. Springer, Dordrecht, Netherlands.

Graves, G. R., S. D. Newsome, D. E. Willard, D. A. Grosshuesch, W. W. Wurzel, and M. L. Fogel. 2012. Nutritional stress and body condition in the Great Gray Owl (*Strix nebulosa*) during winter irruptive migrations. Canadian Journal of Zoology 90: 787–797.

Grecian, W. J., R. A. McGill, R. A. Phillips, P. G. Ryan, and R. W. Furness. 2015. Quantifying variation in δ^{13}C and δ^{15}N isotopes within and between feathers and individuals: is one sample enough? Marine Biology 162:733–741.

Greenberg, R., P. P. Marra, and M. J. Wooller. 2007. Stable-isotope (C, N, H) analyses help locate the winter range of the Coastal Plain Swamp Sparrow (*Melospiza georgiana nigrescens*). Auk 124:1137–1148.

Grubb, T. C., Jr. 2006. Ptilochronology: feather time and the biology of birds. Oxford University Press, Oxford, UK.

Gunnarsson, T. G., J. A. Gill, J. Newton, P. M. Potts, and W. J. Sutherland. 2005. Seasonal matching of habitat quality and fitness in a migratory bird. Proceedings of the Royal Society B 272:2319–2323.

Hebert, C. E., J. Chao, D. Crump, T. B. Johnson, M. D. Rudy, E. Sverko, K. Williams, D. Zaruk, and M. T. Arts. 2014. Ecological tracers track changes in bird diets and possible routes of exposure to type E Botulism. Journal of Great Lakes Research 40:64–70.

Hebert, C. E., and L. I. Wassenaar. 2001. Stable nitrogen isotopes in waterfowl feathers reflect agricultural land use in western Canada. Environmental Science and Technology 35:3482–3487.

Hebert, C. E., and L. I. Wassenaar. 2005. Feather stable isotopes in western North America waterfowl: spatial patterns, underlying factors, and management applications. Wildlife Society Bulletin 33:92–102.

Hebert, C. E., D. V. Weseloh, A. Idrissi, M. T. Arts, R. O'Gorman, O. T. Gorman, B. Locke, C. P. Madenjian, and E. F. Roseman. 2008. Restoring piscivorous fish populations in the Laurentian Great Lakes causes seabird dietary change. Ecology 89:891–897.

Hedd, A., W. A. Montevecchi, H. Otley, R. A. Phillips, and D. A. Fifield. 2012. Trans-equatorial migration and habitat use by Sooty Shearwaters Puffinus griseus from the South Atlantic during the non-breeding season. Marine Ecology Progress Series 449:277–290.

Hedges, R. E., J. G. Clement, C. D. L. Thomas, and T. C. O'Connell. 2007. Collagen turnover in the adult femoral mid-shaft: modeled from anthropogenic radiocarbon tracer measurements. American Journal of Physical Anthropology 133:808–816.

Hobson, K. A. 1987. Use of stable-carbon isotope analysis to estimate marine and terrestrial protein content in gull diets. Canadian Journal of Zoology 65:1210–1213.

Hobson, K. A. 1999. Tracing origins and migration of widlife using stable isotopes: a review. Oecologia 120:314–326.

Hobson, K. A., and F. Bairlein. 2003. Isotopic fractionation and turnover in captive Garden Warblers (Sylvia borin): implications for delineating dietary and migratory associations in wild passerines. Canadian Journal of Zoology 81:1630–1635.

Hobson, K. A., G. J. Bowen, L. I. Wassenaar, Y. Ferrand, and H. Lormee. 2004. Using stable hydrogen and oxygen isotope measurements of feathers to infer geographical origins of migrating European birds. Oecologia 141:477–488.

Hobson, K. A., R. B. Brua, W. L. Hohman, and L. I. Wassenaar. 2000. Low frequency of "double molt" of remiges in Ruddy Ducks revealed by stable isotopes: implications for tracking migratory waterfowl. Auk 117:129–135.

Hobson, K. A., and R. G. Clark. 1992a. Assessing avian diets using stable isotopes I: turnover of ^{13}C in tissues. Condor 94:181–188.

Hobson, K. A., and R. G. Clark. 1992b. Assessing avian diets using stable isotopes II: factors influencing diet-tissue fractionation. Condor 94:189–197.

Hobson, K. A., A. Fisk, N. Karnovsky, M. Holst, J. Gagnon, and M. Fortier. 2002. A stable isotope ($\delta^{13}C$, $\delta^{15}N$) model for the north water food web: implications for evaluating trophodynamics and the flow of energy and contaminants. Deep Sea Research II 49:5131–5150.

Hobson, K. A., K. D. Hughes, and P. J. Ewins. 1997. Using stable-isotope analysis to identify endogenous and exogenous sources of nutrients in eggs of migratory birds: applications to Great Lakes contaminants research. Auk 114:467–478.

Hobson, K. A., K. P. McFarland, L. I. Wassenaar, C. C. Rimmer, and J. E. Goetz. 2001. Linking breeding and wintering grounds of Bicknell's Thrushes using stable isotope analyses of feathers. Auk 118:16–23.

Hobson, K. A., and W. A. Montevecchi. 1991. Stable isotopic determinations of trophic relationships of Great Auks. Oecologia 87:528–531.

Hobson, K. A., J. F. Piatt, and J. Pitocchelli. 1994. Using stable isotopes to determine seabird trophic relationships. Journal of Animal Ecology 63:786–798.

Hobson, K. A., and S. G. Sealy. 1991. Marine protein contributions to the diet of Northern Saw-Whet Owls on the Queen Charlotte Islands: a stable-isotope approach. Auk 108:437–440.

Hobson, K. A., and L. I. Wassenaar. 1997. Linking breeding and wintering grounds of neotropical migrant songbirds using stable hydrogen isotopic analysis of feathers. Oecologia 109:142–148.

Inger, R., and S. Bearhop. 2008. Applications of stable isotope analyses to avian ecology. Ibis 150:447–461.

Jackson, A. L., R. Inger, A. C. Parnell, and S. Bearhop. 2011. Comparing isotopic niche widths among and within communities: SIBER—stable isotope bayesian ellipses in R. Journal of Animal Ecology 80:595–602.

Jaeger, A., V. J. Lecomte, H. Weimerskirch, P. Richard, and Y. Cherel. 2010. Seabird satellite tracking validates the use of latitudinal isoscapes to depict predators' foraging areas in the Southern Ocean. Rapid Communications in Mass Spectrometry 24: 3456–3460.

Kaehler, S., and E. A. Pakhomov. 2001. Effects of storage and preservation on the delta-13C and delta-15N signatures of selected marine organisms. Marine Ecology Progress Series 219:299–304.

Kelly, B., J. B. Dempson, and M. Power. 2006. The effects of preservation on fish tissue stable isotope signatures. Journal of Fish Biology 69:1595–1611.

Kelly, J. F. 2000. Stable isotopes of carbon and nitrogen in the study of avian and mammalian trophic ecology. Canadian Journal of Zoology 78:1–27.

Kelly, J. F., V. Atudorei, Z. D. Sharp, and D. M. Finch. 2002. Insights into Wilson's Warbler migration from analyses of hydrogen stable-isotope ratios. Oecologia 130:216–221.

Klaassen, M., Å. Lindström, H. Meltofte, and T. Piersma. 2001. Ornithology: arctic waders are not capital breeders. Nature 413:794.

Knoche, M. J., A. N. Powell, L. T. Quakenbush, M. J. Wooller, and L. M. Phillips. 2007. Further evidence for site fidelity to wing molt locations by King Eiders: integrating stable isotope analyses and satellite telemetry. Waterbirds 30:52–57.

Kouwenberg, A., J. M. Hipfner, D. W. McKay, and A. E. Storey. 2013. Corticosterone and stable isotopes in feathers predict egg size in Atlantic puffins *Fratercula arctica*. Ibis 155:413–418.

Langston, N. E., and S. Rohwer. 1996. Molt-breeding tradeoffs in albatrosses: life history implications for big birds. Oikos 76:498–510.

Lorrain, A., B. Graham, F. Ménard, B. Popp, S. Bouillon, P. Van Breugel, and Y. Cherel. 2009. Nitrogen and carbon isotope values of individual amino acids: a tool to study foraging ecology of penguins in the Southern Ocean. Marine Ecology Progress Series 391:293–306.

Lott, C. A., T. D. Meehan, and J. A. Heath. 2003. Estimating the latitudinal origins of migratory birds using hydrogen and sulfur stable isotopes in feathers: influence of marine prey base. Oecologia 134:505–510.

MacKenzie, K. M., M. R. Palmer, A. Moore, A. T. Ibbotson, W. R. C. Beaumont, D. J. S. Poulter, and C. N. Trueman. [online]. 2011. Locations of marine animals revealed by carbon isotopes. Scientific Reports 1. doi:10.1038/srep00021. <http://www.nature.com/articles/srep00021>.

Marra, P. P., K. A. Hobson, and R. T. Holmes. 1998. Linking winter and summer events in a migratory bird by using stable-carbon isotopes. Science 282:1884–1886.

Marra, P. P., and R. T. Holmes. 2001. Consequences of dominance-mediated habitat segregation in American Redstarts during the nonbreeding season. Auk 118:92–104.

Marshall, J. D., J. R. Brooks, and K. Lajtha. 2007. Sources of variation in the stable isotopic composition of plants. Pp. 22–60 in R. Michener and K. Lajtha (editors), Stable isotopes in ecology and environmental science. Blackwell Publishing Ltd., Oxford, UK.

Mcclelland, J. W., and J. P. Montoya. 2002. Trophic relationships and the nitrogen isotopic composition of amino acids in plankton. Ecology 83:2173–2180.

McCullagh, J. S. O., J. A. Tripp, and R. E. M. Hedges. 2005. Carbon isotope analysis of bulk keratin and single amino acids from British and North American hair. Rapid Communications in Mass Spectrometry 19:3227–3231.

McKechnie, A. E., B. O. Wolf, and C. M. del Rio. 2004. Deuterium stable isotope ratios as tracers of water resource use: an experimental test with Rock Doves. Oecologia 140:191–200.

McMahon, K. W., M. J. Polito, S. Abel, M. D. McCarthy, and S. R. Thorrold. 2015. Carbon and nitrogen isotope fractionation of amino acids in an avian marine predator, the Gentoo Penguin (*Pygoscelis Papua*). Ecology and Evolution 5:1278–1290.

Meckstroth, A. M., A. K. Miles, and S. Chandra. 2007. Diets of introduced predators using stable isotopes and stomach contents. Journal of Wildlife Management 71:2387–2392.

Michalik, A., R. A. R. McGill, R. W. Furness, T. Eggers, H. J. Van Noordwijk, and P. Quillfeldt. 2010. Black and white—does melanin change the bulk carbon and nitrogen isotope values of feathers? Rapid Communications in Mass Spectrometry 24:875–878.

Michener, R. H., and L. Kaufman. 2007. Stable isotope ratios as tracers in marine foodwebs: an update. Pp. 238–282 in R. Michener and K. Lajtha (editors), Stable isotopes in ecology and environmental science. Blackwell Publishing Ltd., Oxford, UK.

Minagawa, M., and E. Wada. 1984. Stepwise enrichment of ^{15}N along food chains: further evidence and the relation between $\delta^{15}N$ and animal age. Geochimica et Cosmochimica Acta 48:1135–1140.

Moore, J. W., and B. X. Semmens. 2008. Incorporating uncertainty and prior information into stable isotope mixing models. Ecology Letters 11:470–480.

Moreno, R., L. Jover, I. Munilla, A. Velando, and C. Sanpera. 2010. A three-isotope approach to disentangling the diet of a generalist consumer: the Yellow-legged Gull in northwest Spain. Marine Biology 157:545–553.

Neto, J. M., J. Newton, A. G. Gosler, and C. M. Perrins. 2006. Using stable isotope analysis to determine the winter moult extent in migratory birds: the complex moult of Savi's warblers *Locustella luscinioides*. Journal of Avian Biology 37:117–124.

Newsome, S. D., M. T. Clementz, and P. L. Koch. 2010. Using stable isotope biogeochemistry to study marine mammal ecology. Marine Mammal Science 26:509–572.

Norris, D. R., P. Arcese, D. Preikshot, D. F. Bertram, and T. K. Kyser. 2007. Diet reconstruction and historic population dynamics in a threatened seabird. Journal of Applied Ecology 44:875–884.

Norris, D. R., and P.P. Marra. 2007. Seasonal interactions, habitat quality, and population dynamics in migratory birds. Condor 109:535–547.

Ogden, L. J. E., K. A. Hobson, and D. B. Lank. 2004. Blood isotopic ($\delta^{13}C$ and $\delta^{15}N$) turnover and diet-tissue fractionation factors in captive dunlin (*Calidris alpina pacifica*). Auk 121:170–177.

Ostrom, P. H., A. E. Wiley, S. Rossman, C. A. Stricker, and H. F. James. 2014. Unexpected hydrogen isotope variation in oceanic pelagic seabirds. Oecologia 175:1227–1235.

Parnell, A. C., R. Inger, S. Bearhop, and A. L. Jackson. 2010. Source partitioning using stable isotopes: coping with too much variation. PLoS One 5:e9672.

Parnell, A. C., D. L. Phillips, S. Bearhop, B. X. Semmens, E. J. Ward, J. W. Moore, A. L. Jackson, J. Grey, D. J. Kelly, and R. Inger. 2013. Bayesian stable isotope mixing models. Environmetrics 24:387–399.

Pauli, J. N., S. A. Steffan, and S. D. Newsome. 2015. It is time for IsoBank. BioScience 65:229–230.

Pearson, S. F., D. J. Levey, C. H. Greenberg, and C. M. Del Rio. 2003. Effects of elemental composition on the incorporation of dietary nitrogen and carbon isotopic signatures in an omnivorous songbird. Oecologia 135:516–523.

Phillips, D. L., and J. W. Gregg. 2003. Source partitioning using stable isotopes: coping with too many sources. Oecologia 136:261–269.

Powell, L. A., and K. A. Hobson. 2006. Enriched feather hydrogen isotope values for Wood Thrushes sampled in Georgia, USA, during the breeding season: implications for quantifying dispersal. Canadian Journal of Zoology 84:1331–1338.

Prevost, Y. V. E. S. 1983. The moult of the Osprey *Pandion haliaetus*. Ardea 71:199–209.

Robbins, C. T., L. A. Felicetti, and M. Sponheimer. 2005. The effect of dietary protein quality on nitrogen isotope discrimination in mammals and birds. Oecologia 144:534–540.

Robinson, W. D., M. S. Bowlin, I. Bisson, J. Shamoun-Baranes, K. Thorup, R. H. Diehl, T. H. Kunz, S. Mabey, and D. W. Winkler. 2010. Integrating concepts and technologies to advance the study of bird migration. Frontiers in Ecology and the Environment 8:354–361.

Rocque, D. A., and K. Winker. 2005. Use of bird collection in contaminant and stable-isotope studies. Auk 122:990–994.

Rohwer, S., and K. Broms. 2012. Use of feather loss intervals to estimate molt duration and to sample feather vein at equal time intervals through the primary replacement. Auk 129:653–659.

Rohwer, S., A. D. Fox, T. Daniel, and J. F. Kelly. 2015. Chronologically sampled flight feathers permits recognition of individual molt-migrants due to varying protein sources. PeerJ 3:e743.

Rohwer, S., R. E. Ricklefs, V. G. Rohwer, and M. M. Copple. 2009. Allometry of the duration of flight feather molt in birds. PLoS Biology 7:e1000132.

Rubenstein, D. R., and K. A. Hobson. 2004. From birds to butterflies: animal movement patterns and stable isotopes. Trends in Ecology and Evolution 19:256–263.

Sarakinos, H. C., M. L. Johnson, and M. J. Vander Zanden. 2002. A synthesis of tissue-preservation effects on carbon and nitrogen stable isotope signatures. Canadian Journal of Zoology 80:381–387.

Schaffner, F. C., and P. K. Swart. 1991. Influence of diet and environmental water on the carbon and oxygen isotopic signatures of seabird eggshell carbonate. Bulletin of Marine Science 48:23–38.

Schoeninger, M. J., and M. J. DeNiro. 1984. Nitrogen and carbon isotopic composition of bone collagen from marine and terrestrial animals. Geochimica et Cosmochimica Acta 48:625–639.

Sealy, J. C., N. J. Van Der Merwe, J. A. Lee Thorp, and J. L. Lanham. 1987. Nitrogen isotopic ecology in southern Africa: implications for environmental and dietary tracing. Geochimica et Cosmochimica Acta 51:2707–2717.

Smith, T. B., P. P. Marra, M. S. Webster, I. Lovette, H. L. Gibbs, R. T. Holmes, K. A. Hobson, and S. Rohwer. 2003. A call for feather sampling. Auk 120:218–221.

Spear, L. B., D. G. Ainley, N. Nur, and S. N. G. Howell. 1995. Population size and factors affecting at-sea distributions of four endangered procellariids in the tropical Pacific. Condor 97:613–638.

Stapp, P., G. A. Polis, and F. S. Pinero. 1999. Stable isotopes reveal strong marine and El Niño effects on island food webs. Nature 401:467–469.

Steffan, S. A., Y. Chikaraishi, D. R. Horton, N. Ohkouchi, M. E. Singleton, E. Miliczky, D. B. Hogg, and V. P. Jones. 2013. Trophic hierarchies illuminated via amino acid isotopic analysis. PLoS One 8:21–23.

Sweeting, C. J., N. V. C. Polunin, and S. Jennings. 2004. Tissue and fixative dependent shifts of $\delta^{13}C$ and $\delta^{15}N$ in preserved ecological material. Rapid Communications in Mass Spectrometry 18:2587–2592.

Sydeman, W. J., K. A. Hobson, S. N. Ox, P. Pyle, and E. B. Mclaren. 1997. Trophic relationships among seabirds in central California: combined stable isotope and conventional dietary approach. Condor 99:327–336.

Thode, H. G. 1991. Sulphur isotopes in nature and the environment: an overview. Pp. 1–21 in H. R. Krouse and V. A. Grineko (editors), Stable isotopes in the assessment of natural and anthropogenic sulphur in the environment. John Wiley & Sons, New York, NY.

Thompson, D. R., and R. W. Furness. 1995. Stable-isotope ratios of carbon and nitrogen in feathers indicate seasonal dietary shifts in Northern Fulmars. Auk 112:493–498.

Thompson, D. R., R. W. Furness, and S. A. Lewis. 1995. Diets and long-term changes in $\delta^{15}N$ and $\delta^{13}C$ values in Northern Fulmars *Fulmarus glacialis* from two northeast Atlantic colonies. Marine Ecology Progress Series. Oldendorf 125:3–11.

Thompson, D. R., R. A. Phillips, F. M. Stewart, and S. Waldron. 2000. Low $\delta^{13}C$ signatures in pelagic seabirds: lipid ingestion as a potential source of ^{13}C-depleted carbon in the procellariiformes. Marine Ecology Progress Series 208:265–271.

Tietje, W. D., and J. G. Teer. 1988. Winter body condition of northern shovelers on freshwater and saline habitats. Pp. 353–377 in W. W. Wellter (editor), Waterfowl in winter. University of Minnesota Press, Minneapolis, MN.

Votier, S. C., S. Bearhop, M. J. Witt, R. Inger, D. Thompson, and J. Newton. 2010. Individual responses of seabirds to commercial fisheries revealed using GPS tracking, stable isotopes and vessel monitoring systems. Journal of Applied Ecology 47:487–497.

Wainright, S. C., J. C. Haney, C. Kerr, A. N. Golovkin, and M. V. Flint. 1998. Utilization of nitrogen derived from seabird guano by terrestrial and marine plants at St. Paul, Pribilof Islands, Bering Sea, Alaska. Marine Biology 131:63–71.

Warham, J. 1996. The behaviour, population biology and physiology of the Petrels. Academic Press, London, UK.

Webster, M. S., P. P. Marra, S. M. Haig, S. Bensch, and R. T. Holmes. 2002. Links between worlds: unraveling migratory connectivity. Trends in Ecology and Evolution 17:76–83.

West, J. B., G. J. Bowen, T. E. Cerling, and J. R. Ehleringer. 2006. Stable isotopes as one of nature's ecological recorders. Trends in Ecology and Evolution 21:408–414.

Wiley, A. E., P. H. Ostrom, C. A. Stricker, H. F. James, and H. Gandhi. 2010. Isotopic characterization of flight feathers in two pelagic seabirds: sampling strategies for ecological studies. Condor 112:337–346.

Wiley, A. E., P. H. Ostrom, A. J. Welch, R. C. Fleischer, H. Gandhi, J. R. Southon, T. W. Stafford, Jr., J. F. Penniman, D. Hu, F. P. Duvall, and H. F. James. 2013. Millennial-scale isotope records from a wide-ranging predator show evidence of recent human impact to oceanic food webs. Proceedings of the National Academy of Sciences USA 110:8972–8977.

Wiley, A. E., A. J. Welch, P. H. Ostrom, H. F. James, C. A. Stricker, R. C. Fleischer, H. Gandhi, J. Adams, D. G. Ainley, F. Duvall, and N. Holmes. 2012. Foraging segregation and genetic divergence between geographically proximate colonies of a highly mobile seabird. Oecologia 168:119–130.

Wunder, M. B., J. R. Jehl, and C. A. Stricker. 2012. The early bird gets the shrimp: confronting assumptions of isotopic equilibrium and homogeneity in a wild bird population. Journal of Animal Ecology 81:1223–1232.

Wunder, M. B., C. L. Kester, F. L. Knopf, and R. O. Rye. 2005. A test of geographic assignment using isotope tracers in feathers of known origin. Oecologia 144:607–617.

CHAPTER SEVEN

What Bird Specimens Can Reveal
about Species-Level Distributional Ecology*

A. Townsend Peterson and Adolfo G. Navarro-Sigüenza

Abstract. This chapter aims to provide an overview of the emerging field of species-level distributional ecology in birds. Although such studies have been developed for decades, recent advances in making large quantities of occurrence data and environmental data openly available have fostered many advances in the field, showing that distributional data extend the utility of specimens in scientific collections far beyond the specimen itself. We review insights that are available via such analyses, and major challenges that still constrain advances in the field.

Key Words: Digital Accessible Knowledge, ecological niche, geographic distribution, primary biodiversity data, specimen locality.

Several billion scientific specimens, including many millions of bird specimens, have been accumulated by biologists over the past several centuries, and deposited for long-term care in scientific collections worldwide (Ariño 2010). These preserved biological materials provide important documentation of biological diversity, constituting the basic link among organisms, their phenotypes, their genotypes, and biological nomenclature. However, beginning with type localities, and continuing with series of additional (nontype) specimens with associated localities from which they were collected, scientific collections also provide a rich store of distributional information about species; indeed, in many cases, they offer the only information about the distribution and ecology of the species that is available. These distributional data extends the utility of specimens in scientific collections far beyond the physical specimen itself.

These rich sources of distributional information have been tapped for distributional summaries of regional avifaunas and single species. Among the older examples, for Mesoamerica, Godman and Salvin (1879–1915) provided a massive summary of distributional information, based exclusively on specimens, for much of the biota of Mexico and Central America. Later compilations treated many regions of the globe. For example, in the American tropics, where we are most familiar with the literature, many examples accumulated in the middle 20th century (Griscom 1932, Hellmayr and Conover 1942, Friedmann et al. 1950, Miller et al. 1957, Russell 1964, Meyer de Schauensee 1966), which documented distributions of species, at least in a textual sense. The

* Peterson, A. T., and A. G. Navarro-Sigüenza. 2017. What bird specimens can reveal about species-level distributional ecology. Pp. 111–125 in M. S. Webster (editor), The Extended Specimen: Emerging Frontiers in Collections-based Ornithological Research. Studies in Avian Biology (no. 50), CRC Press, Boca Raton, FL.

atlases of speciation in African birds (Hall and Moreau 1970, Snow 1978) perhaps represent the pinnacle of this genre, offering a broad compilation of specimen records from across the continent, with detailed maps of distributions of species and species complexes. The ornithological gazetteers of South America (e.g., Paynter and Traylor 1991, Paynter 1995) were apparently a step in the same direction for the Neotropics, but the species-level compilation was never completed.

Along parallel lines, other monographic treatments summarized distributions of individual species and clades. Here, systematic revisions generally treated genera or species complexes, and provided lists of localities and later maps. Of particular note is the lengthy series of detailed monographs that was produced by ornithologists at the Museum of Vertebrate Zoology (Berkeley, California) (Miller 1941; Pitelka 1950, 1951; Johnson 1963, 1980; Selander and Giller 1963; Selander 1964; Cicero 1996). These monographs have particular value as distributional summaries,

as the specimens cited generally have been reviewed and verified by an expert in the group, such that data quality can be much higher than for less-well-refereed records. An example is shown in Figure 7.1, drawn from a monograph of the *Centurus* (*Melanerpes*) woodpeckers, showing specimen localities and inferred range (from Selander and Giller 1963).

These mapping efforts, however, generally depended on hand compilation, and typically were lengthy, drawn-out processes, often extending over decades. The challenges regarding extracting data, and particularly in assigning geographic references to textual locality descriptions, were significant and often made such compilations very static entities that were never (or rarely) updated. It was not until digital data became numerous that new solutions could be explored, which led to deeper insights into distributional ecology of species.

This set of new insights and new possibilities is the subject of this overview. Specifically,

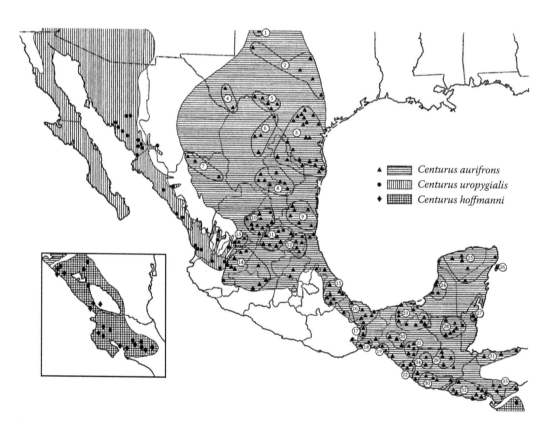

Figure 7.1. Summary of the geographic distribution of *Melanerpes aurifrons*, M. *uropygialis*, and M. *hoffmanni*. Symbols indicate localities associated with specimens examined in the original study (from Selander and Giller 1963).

we outline how specimen-based distributional ecological studies of birds can be developed and updated quickly and efficiently, based on large quantities of readily available, openly accessible primary biodiversity data. The result has been deep, rangewide insights into why species are found where they are found, why they are not found where they are not found, and how species' distributions are evolving in the face of changing environmental conditions.

JOSEPH GRINNELL: A CENTRAL FIGURE IN DISTRIBUTIONAL ECOLOGY

Joseph Grinnell (1877–1939) represents an amazing figure in this field of distributional ecology, with significant contributions spanning more than a century. He not only provided seminal thinking and ideas that laid the foundation for much of the present advance in the field of distributional ecology, but he also assembled key datasets that have made possible fundamental empirical studies in recent years. As such, in view of Grinnell's massive and multifaceted contributions, here we provide an overview of his thinking.

Grinnell's contributions are impressively broad (see obituary in H. W. Grinnell 1940). As examples, in the introduction to an early "biota of" publication, he anticipated many of the ideas of island biogeography (Grinnell and Swarth 1913). He provided incisive early thinking about speciation and phylogeny, and their interrelations (Grinnell 1924), particularly in his essay on geographic variation in the Chestnut-backed Chickadee (*Parus rufescens*; Grinnell 1904). He also provided useful, multiscalar thinking about the role of barriers to dispersal in species-level biogeography (Grinnell 1914).

Clearly, though, Grinnell's thinking about ecological niches and geographic distributions must be viewed among his most fundamental contributions. In a series of papers, he analyzed (verbally) the distributional patterns of diverse species, in each case pointing out the abiotic, biotic, and movement-related factors that structure the species' range (Grinnell 1917a,b, 1932). These papers directly influenced development of the BAM diagram (see explanation later) as a thinking framework about species' distributions (Soberón and Peterson 2005), but also listed factors important in distributional ecology that are still being explored and understood: vegetation; food supply, kind, and quantity; rainfall; humidity (relative or absolute); wetness or dryness of the soil; barometric pressure; atmospheric density; safety of breeding places; availability of temporary refuges; water (to land species); land (to aquatic species); availability of cover; nature of the ground; insolation; cloudiness; temperature (in general, or with seasonal variations); competition; parasitism; and individual preferences. This long list has occupied the attention of ecologists for the century that has elapsed since Grinnell wrote it down for the first time.

Perhaps most interesting about Grinnell, however, is the quality and organization of his empirical data. Grinnell developed an impressive precomputer relational database—a field note format—that preserved an impressive amount of information in consistent formats (Remsen 1977, Herman 1986). What is more, he directed a systematic survey of the biotas of California, which created impressive collections and resources with unprecedented detail of data at the Museum of Vertebrate Zoology. The combination of these large-scale collections with detailed data made possible impressive advances decades after Grinnell's passing, detailed next.

THE GRINNELL PROJECT

Grinnell's empirical dataset—accumulated by a large team associated with the Museum of Vertebrate Zoology over the course of decades—has paid off in major rewards of insight and understanding in recent years. This body of work has demonstrated consistent range shifts upward in elevation of ~500 m in about half of bird species in response to an average minimum daily temperature increase of ~3°C (Moritz et al. 2008). On a broader spatial scale, the Grinnell Project team demonstrated that 48 bird species (90.6% of those examined) tracked their climatic niche closely (Tingley et al. 2009), which is associated with an overall decline in bird species richness (Tingley and Beissinger 2013). Parallel studies of changes in small mammal species distributions and community composition (Rowe et al. 2014) showed that 25 of 34 species analyzed shifted their ranges upslope or downslope in at least one region, but species shifted their elevational limits differently in the three regions analyzed.

In sum, the Grinnell Project is an impressive demonstration of the scientific power that derives from careful planning and documentation of specimen collecting efforts. Deep and exciting scientific results range from the distributional and ecological results cited earlier to evidence of evolutionary change in morphology in small mammal species (Eastman et al. 2012). Although detailed and careful statistical steps are necessary to control for incomplete detectability of species (Tingley and Beissinger 2009), the power of such detailed and controlled before-and-after comparisons is impressive. Grinnell's early but detailed preservation of distributional information in the form of specimens extends a particularly rich documentation that makes many of the finest-scale comparisons possible.

PRESENT-DAY STUDIES IN DISTRIBUTIONAL ECOLOGY

Over the past three decades, the massive increases in availability of detailed distributional data in digital formats (Soberón 1999), referred to as Digital Accessible Knowledge and exemplified in Figure 7.2 (Sousa-Baena et al. 2013), have led to many interesting steps forward in understanding the geography and ecology of species' distributions. A first, important advance has been that of development of a detailed terminology and conceptual framework for discussing species' geographic distributions and ecological niches (Pulliam 2000, Soberón and Peterson 2005, Peterson et al. 2011). This area of inquiry had been fraught with confusing, overlapping, and conflicting terminology, thanks largely to the complexity of the linked geographic and environmental spaces, and how species relate to each of those spaces, which left considerable room for confusion (see, for example, the title of Godsoe 2010).

Soberón and Peterson (2005) presented a framework that united ecological and biogeographic considerations regarding species' distributions, including the suitability of biotic and abiotic environments, as well as mobility over present and past landscapes, which they termed the BAM diagram. Under BAM viewpoints, species are able to maintain populations in the long term only at sites that are suitable in both biotic (**B**) and abiotic (**A**) terms, but that also are accessible to the species via dispersal over relevant time periods (mobility = **M**); this "occupied distributional area" (**G**$_O$), refers to sets of sites, $G_O = B \cap A \cap M$. This simple heuristic has served to facilitate clear discussions of suites of factors important in shaping distributions of species, but it also can help to understand the limitations of the methods that have become popular in distributional ecology (Saupe et al. 2012).

Several approaches are available for quantitative study of such distributional ecological systems (Helaouët and Beaugrand 2009), including detailed physiological measurements (Barve et al. 2014); first-principles models (Kearney and Porter 2004); and joint simulations of niche, environment, and dispersal (Lira-Noriega et al. 2013). A simpler alternative has been to characterize distribution–environment associations via correlative approaches that take advantage of large-scale primary biodiversity occurrence datasets, frequently derived from the specimen record. Using such approaches, if one is willing to make certain assumptions about the role of biotic interactions in shaping geographic distributions of species (Soberón 2007), and if one manages assumptions about mobility and dispersal correctly (Barve et al. 2011), reasonably accurate and predictive estimates of the abiotically suitable area can be made (Peterson et al. 2011).

An important next step in such inferences, however, is to translate habitable areas and correlative estimates of habitable conditions into estimates of fundamental ecological niches of species. Such "niche models," if executed rigorously, permit researchers to address a wealth of questions that depend on predictions regarding distributions of species: where were species distributed under past conditions (Nogués-Bravo 2009, Peterson and Ammann 2013), how will species' distributions change with warming climates into the future (Peterson et al. 2002, Gibson et al. 2010, Meier et al. 2012), what is the geographic potential of an invading species (Peterson 2003, Petitpierre et al. 2012), and many others. In this sense, and particularly with the advent of massive-scale biodiversity databases (now on the order of 6×10^8 records) that are globally and openly available for use (Gaiji et al. 2013), correlative approaches to niche estimation have become enormously popular.

Correlative approaches are not without perils and problems, however. One crucial distinction is that between the fundamental niche and the existing niche (Soberón and Peterson 2011). A species' fundamental ecological niche is best conceived of as a coarse-resolution translation of its physiological requirements with respect

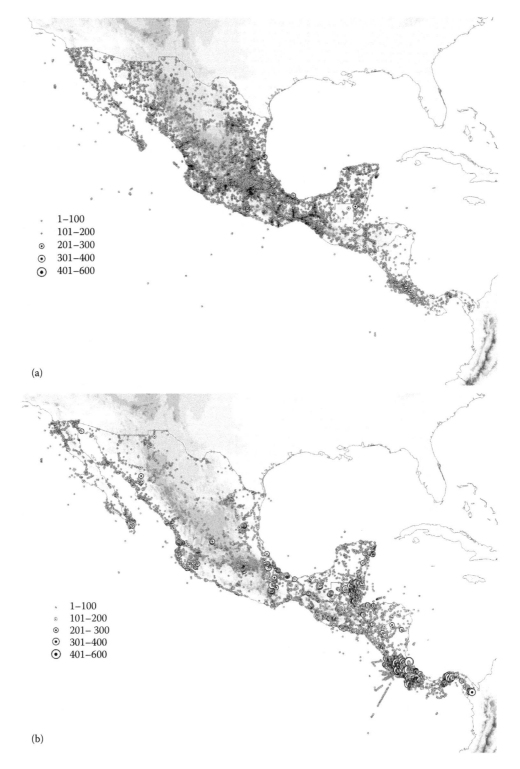

Figure 7.2. Digital Accessible Knowledge (DAK, from Sousa-Baena et al. 2013) for the birds of Mexico and Central America. Species richness at each individual locality and geographic distribution of sampling effort are illustrated by data from (a) specimen records (from GBIF, http://www.gbif.org/; VertNet, http://vertnet.org/), and (b) observational records (from AVerAves, EBird http://ebird.org/).

to key environmental parameters. These tolerable, or suitable, conditions are not necessarily all represented on the face of the Earth, however; frequently, or perhaps even universally, the full range of habitable conditions for a species is not observable across its geographic distribution or even anywhere on Earth. At another level, biogeographic restriction, or the idea that the species only ever "explores" the conditions represented across **M**, translates into further subrepresentation of the full dimensions of the ecological niche in correlative models. As such, the only set of conditions against which one is able to observe the species' response, for example, establishing populations or not, is that which is represented within the accessible area, **M**.

Another dimension of this point is the difference between potential and occupied geographic distributions. Although the niche is defined in environmental dimensions, species are subject to real-world biogeographic restrictions: a particular species evolved in one region and has not necessarily ever had access to other such regions. For example, the Resplendent Quetzal (*Pharomachrus mocinno*) evolved in the humid montane forests of the Americas and has never had access to similar—and probably habitable—forests in the Afrotropics. What is more, this particular species of *Pharomachrus* is endemic to Mesoamerica between the Isthmus of Tehuantepec, in southern Mexico, and the Isthmus of Panama, and is replaced by other *Pharomachrus* species in the Andes of South America. Again, the point is that a given species does not have, and has never had, access to its entire *potential* geographic distribution: the potential distribution can be represented as **A∩B**, but the occupied distributional area is only **A∩B∩M**. As a consequence, even if ecological niche models are successful in estimating the fundamental ecological niche and the associated abiotically suitable area, additional assumptions about history and dispersal ability, in effect, a hypothesis of **M**, are necessary before one can anticipate the species' occupied distributional area.

No technique in science is without its assumptions, caveats, and analytical pitfalls; indeed, complicated analyses represent virtual minefields of potential problems and inferential errors. Alternative approaches to understanding species' distributions include the "extent of occurrence" maps that provide only broad, simple range outlines (see comprehensive world dataset developed by BirdLife International and NatureServe 2015). Other approaches include maps derived from physiological measurements (Barve et al. 2014), first-principles models (Kearney and Porter 2004), or occupancy models of population presence (Dorazio 2012). The correlative models, however, have some advantages: (1) they are broadly applicable to most species on Earth and do not depend on availability of expensive or difficult-to-derive additional information such as physiological measurements or repeated population surveys, (2) they link explicitly to biological properties (i.e., the ecological niche) in their inferences, (3) they consider real-world geographic complexity explicitly in development of distributional hypotheses, and (4) they use as main sources for developing models the massive storehouse of primary biodiversity occurrence data found in biological research collections and databases. As a consequence, as correlative techniques have developed over the past two decades, they have gained enormous popularity and have seen considerable investment and buy-in from the broader communities of ecology and biogeography (e.g., Parra et al. 2004, Tingley et al. 2009).

WHAT CAN BE LEARNED?

Dynamics of Ecological Niches through Time

An early focus of the field of distributional ecology was on the relative stability (conservatism) of ecological niches through time, on timescales ranging from immediate (Fitzpatrick et al. 2007) to evolutionary (Wiens and Graham 2005). This focus derived from earlier debates about the process of speciation, in which significant voices argued that speciation frequently involves differentiation in ecological niche characteristics, to the point even of permitting differentiation in parapatry or sympatry (Endler 1977). Hence, an important question was (and still is) the frequency with which recently speciated lineages share the same ecological niche versus how often they may have differentiated in niche characteristics as part of the speciation process.

An early study served to offer a first step toward an answer to this question (Peterson et al. 1999): sister species pairs distributed across the Isthmus of Tehuantepec, in southern Mexico, universally shared the same ecological niche, suggesting that speciation had not involved differentiation in ecological niche characteristics. Although another study came to different conclusions regarding parallel systems (Graham et al. 2004), a detailed reanalysis of both of these early papers using much-improved methodologies (Warren et al. 2008) supported the original conclusions: sister species retain ecological niche characteristics that are significantly more similar than expected at random through time, even if they may not be identical. Many other studies purporting to document niche shifts (Broennimann et al. 2007, Fitzpatrick et al. 2007, Medley 2010) were the product of either methodological (Peterson and Nakazawa 2008) or inferential (Soberón and Peterson 2011) errors, such that the vast majority of situations on ecological and shorter evolutionary timescales are now seen to be characterized by niche conservatism (Peterson 2011, Petitpierre et al. 2012).

An issue of considerable interest, then, has been the phylogenetic pattern of when ecological niches *do* change, if they have changed demonstrably. After early methodological fumbles—compare, for example, the conclusions of Rice et al. (2003) with the more refined results of McCormack et al. (2010)—interest in these approaches has grown markedly. As an example of one research group alone, see Kozak et al. (2009), Buckley et al. (2010), Kozak and Wiens (2010a,b), and Wiens et al. (2013). In many senses, then, phylogenetic analysis of niche dimensions appears to be becoming a useful paradigm in understanding linkages between ecology and evolution.

This wealth of analyses has nonetheless relied on inappropriate methodologies, because most researchers have used environmental characteristics of known occurrences to characterize ecological niches—that is, the average of environmental values associated with the known occurrence points. This approach has significant problems: sampling intensity and bias affect niche estimates, and the niches estimated will be existing niches rather than fundamental niches. Existing niches will frequently differ but not because the fundamental niche has changed—rather, because the existing niche, $\mathbf{N}_F^*$, is defined as the reduction of the fundamental niche $\mathbf{N}_F$ by the environments represented across $\mathbf{M}$. $\mathbf{N}_F^*$ may change wildly even when $\mathbf{N}_F$ has been stable. Representation of that niche in the area that is accessible to the species is what changes, particularly in a speciating

lineage in which $\mathbf{M}$'s can differ quite markedly in extent and characteristics. As a consequence, conclusions of many studies have been compromised, and methodologies better founded in a detailed conceptual framework for distributional ecology should be used instead (Saupe et al., in prep.).

Dynamics of Distributional Areas through Time

Once the idea of estimating ecological niches was conceived, an immediate and obvious potential was that of estimating geographic distributions of species. Under single sets of present-day environmental conditions, this function allowed researchers to explore and estimate the full potential distributions of species, even when sampling had been incomplete and inadequate. In more radical examples, researchers have been able to anticipate the existence of previously unknown populations of species (Guisan et al. 2006, Menon et al. 2010, Menon et al. 2012) and even previously unknown species (Raxworthy et al. 2003).

Providing that niches have been estimated appropriately and robustly, a further possibility is that of using niche estimates to classify landscapes from other time periods, past or future, as suitable or unsuitable. A limited number and diversity of paleoclimate model outputs now exists, thanks to several large-scale organized paleoclimate model intercomparison projects (Pinot et al. 1999, Haywood et al. 2015), and many of these data are now openly and conveniently available (Lima-Ribeiro et al. 2015) via a website called ecoClimate (http://ecoclimate.org/). Transferring models calibrated in the present day to these paleoclimate conditions has yielded numerous interesting results, including hypotheses of Pleistocene refugia to complement phylogeographic results (Smith et al. 2011, Arbeláez-Cortés et al. 2014), tests of the Pleistocene refugium hypothesis in the Amazon (Bonaccorso et al. 2006, Peterson and Nyári 2007) and elsewhere (Jakob et al. 2009), exploration of factors causing extinctions (Nogués-Bravo et al. 2008), and examinations of past distributional connectivity (Waltari and Guralnick 2009). The robustness of these paleoprojections has been tested in several studies, with most indicating quite-positive results (Martínez-Meyer et al. 2004, Martínez-Meyer and Peterson 2006, Waltari et al. 2007), and only one asserting lack of intertemporal predictive ability (Varela et al. 2010). An illustration of a Pleistocene model transfer—identifying potential climatic refugia for the Thrush-like Mourner (*Schiffornis turdinus*)—that resulted from an early study (Peterson and Nyári 2007) is shown in Figure 7.3.

The same idea of intertemporal model transfer can be applied to future scenarios of climate, thanks to large stores of climate model outputs made available by the Intergovernmental Panel on Climate Change (http://www.ipcc.ch/). Beginning with early explorations for Mexican faunas (Peterson et al. 2001), this paradigm has gained enormous popularity to the point that it is now seen as the go-to methodology for anticipating

distributional shifts of birds in response to ongoing climate-change processes (Prieto-Torres et al. 2016).

Model transfers for past and present climates as just described can be, nonetheless, rather tenuous and risky endeavors. Several studies have documented broad variation in intertemporal model transfer results among niche modeling algorithms (Pearson et al. 2006), and a recent detailed analysis of risk of extrapolation indicated considerable potential for inappropriate conclusions from such exercises (Owens et al. 2013). The point is that developing quick-and-dirty intertemporal

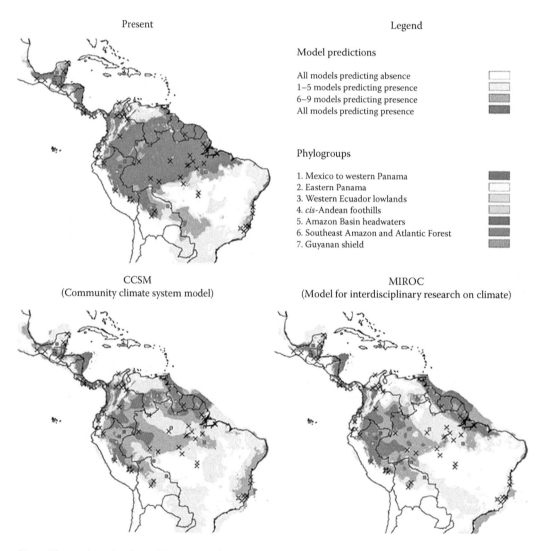

Figure 7.3. Ecological niche model estimates of present-day and Last Glacial Maximum potential distribution areas, the latter under two different general circulation model estimates. Also shown are sample points used in niche model development (Xs), and sampling localities for molecular sequence data (squares). (From Peterson and Nyári 2007.)

projections of niche models is now quite easy, but executing this process appropriately and rigorously is quite a bit more challenging (Anderson 2013), requiring careful attention to hypotheses of accessible areas and BAM configurations (Barve et al. 2011, Saupe et al. 2012), as well as detailed assessments of model complexity and spatial autocorrelation effects (Diniz-Filho et al. 2003, Segurado et al. 2006, Warren and Seifert 2011, Shcheglovitova and Anderson 2013, Boria et al. 2014, de Oliveira et al. 2014). Quick-and-dirty model results are unlikely to yield robust results, particularly in the case of intertemporal model transfers.

Links among Niche, Geography, and Population Biology

Niche models were originally conceived as a means of estimating coarse-resolution environmental parameters associated with suitability of landscapes for species. However, to the extent that the niches estimated are coarse-resolution translations of physiological responses (Barve et al. 2014), a link to population density or abundance might be expected to exist. Several studies have sought an association between suitability indices produced by niche modeling algorithms but have failed to find a close relationship (e.g., Nagaraju et al. 2013).

Martínez-Meyer et al. (2012), however, appealed to the biological sense of an ecological niche estimate and sought a different sort of relationship. That is, instead of comparing suitability estimates with abundance, they built a secondary index of niche centrality versus peripherality, under the idea that central populations ought to be able to reach higher populations than peripheral ones, an idea that had been proposed previously in terms of geographic range centrality (Brown 1984, Hengeveld 1992). They compared these relationships in terms of both geographic (spatial) and environmental (niche) centrality, and found strong associations only in the latter case (Martínez-Meyer et al. 2012). Lira-Noriega and Manthey (2014) showed a further relationship between niche centrality and standing genetic variation in populations. These possibilities of anticipating population processes from niche estimates derived from occurrence data are still being explored and tested (Holt et al. 2002, Manthey et al. 2014), and offer exciting possibilities for further, and deeper, inferences about population ecology of species.

WHAT COMES NEXT?

The emerging field of distributional ecology has indeed emerged in the course of just a couple of decades. As such, the massive progress and exciting insights reviewed briefly earlier have been explored and derived in a very short period of time. Indeed, methods used just a few years ago are now seen as dated, and the field has become fast-moving, multimodal, and complex.

A first major next step will be estimation of **M** in quantitative, robust, repeatable ways. A paper by Barve et al. (2011) made a clear case that (1) hypotheses about **M** drive essentially all results and conclusions using ecological niche modeling approaches, and (2) that **M** is the area over which models should be calibrated, if such calibration is to be appropriate and rigorous. Barve et al. (2011) presented a "cartoon" of one approach to how **M** might be estimated quantitatively, which consists of the simultaneous simulation of niche, distribution, dispersal, and environmental change (see later); if the entire area to which the species is able to disperse over a relevant time interval can be tracked, then an appropriate hypothesis of **M** could be assembled. Until such a methodology is implemented rigorously, however, **M** hypotheses remain provisional and improvised, such that many models are likely to prove suboptimal.

A second major step forward will be that of developing and testing algorithms that allow fitting biologically realistic response shapes, rather than the overfit, highly complex shapes that most current algorithms yield (Peterson et al. 2011). That is, both theoretical expectations and empirical results (e.g., Birch 1953) in ecophysiology suggest that responses to environmental variables should be unimodal, and therefore that multivariate niches should be convex. Response surfaces fit by popular niche modeling algorithms, however, are frequently quite complex and multimodal (Peterson et al. 2011). The result is a niche estimate that does not match current understanding about niche shape and properties, and will not have optimal predictive power during model transfer procedures. Such highly complex niche estimates will also not provide a robust basis for estimating qualities of niches, such as breadth, position, and volume.

Finally, and perhaps most challenging, will be joint modeling of niche, dispersal, and distribution. At present, essentially all approaches to modeling ecological niches neglect the geometry and topology of the individual pixels being used as geographic and environmental units in the modeling process. Two pixels may be next to one another or may be quite distant, but this spatial relationship is not taken into account in the environment-only models that are the norm at present. Rigorous **M** hypotheses are perhaps a Band-Aid on the problem: the nature of **M** will evolve through time: an early invader may have an **M** limited by individual dispersal distances, whereas an established species may be limited more by major biogeographic barriers. Although several initial explorations of such niche-dispersal-environment simulations have been developed (e.g., Lira-Noriega et al. 2013), full implementation remains rather a profound challenge.

In sum, in the many arguments about the value of specimens in modern organismal biology (Remsen 1995, Peterson et al. 2007), a repeated theme has been that specimens are rich and permanent records of the occurrence of a species in a place at a point in time, and that they will be put to many diverse uses well into the future (see the other novel uses explored in Hobson 1999, Chakraborty et al. 2006, and the other chapters in this volume). Many of these extended uses of specimens were never envisioned or imagined when the specimen was originally collected. The role of specimen data as the backbone central to this emerging field of distributional ecology represents still another example of this evolving utility of specimens.

ACKNOWLEDGMENTS

We thank E. Martínez-Meyer, A. Gordillo, colleagues at Museo de Zoología (Facultad de Ciencias, UNAM), and the KU Ecological Niche Modeling Group for input and inspiration over many past years. Comments from M. Webster and K. Huyvaert improved the manuscript.

LITERATURE CITED

Anderson, R. P. 2013. A framework for using niche models to estimate impacts of climate change on species distributions. Annals of the New York Academy of Sciences 1297:8–28.

Arbeláez-Cortés, E., B. Milá, and A. G. Navarro-Sigüenza. 2014. Multilocus analysis of intraspecific differentiation in three endemic bird species from the dry forest of the northern Neotropics. Molecular Phylogenetics and Evolution 70:362–377.

Ariño, A. H. 2010. Approaches to estimating the universe of natural history collections data. Biodiversity Informatics 7:81–92.

Barve, N., V. Barve, A. Jiménez-Valverde, A. Lira-Noriega, S. P. Maher, A. T. Peterson, J. Soberón, and F. Villalobos. 2011. The crucial role of the accessible area in ecological niche modeling and species distribution modeling. Ecological Modelling 222:1810–1819.

Barve, N., C. E. Martin, and A. T. Peterson. 2014. Role of physiological optima in shaping the geographic distribution of Spanish moss. Global Ecology and Biogeography 23:633–645.

Birch, L. C. 1953. Experimental background to the study of the distribution and abundance of insects: I. The influence of temperature, moisture and food on the innate capacity for increase of three grain beetles. Ecology 34:698–711.

BirdLife International, and NatureServe. 2015. Bird species distribution maps of the world. BirdLife International, Cambridge, UK.

Bonaccorso, E., I. Koch, and A. T. Peterson. 2006. Pleistocene fragmentation of Amazon species' ranges. Diversity and Distributions 12:157–164.

Boria, R. A., L. E. Olson, S. M. Goodman, and R. P. Anderson. 2014. Spatial filtering to reduce sampling bias can improve the performance of ecological niche models. Ecological Modelling 275:73–77.

Broennimann, O., U. A. Treier, H. Müller-Schärer, W. Thuiller, A. T. Peterson, and A. Guisan. 2007. Evidence of climatic niche shift during biological invasion. Ecology Letters 10:701–709.

Brown, J. H. 1984. On the relationship between distribution and abundance. American Naturalist 124:255–279.

Buckley, L. B., T. J. Davies, D. D. Ackerly, N. J. B. Kraft, S. P. Harrison, B. L. Anacker, H. V. Cornell, E. I. Damschen, J.-A. Grytnes, B. A. Hawkins, C. M. McCain, P. R. Stephens, and J. J. Wiens. 2010. Phylogeny, niche conservatism and the latitudinal diversity gradient in mammals. Proceedings of the Royal Society B 277:2131–2138.

Chakraborty, A., M. Sakai, and Y. Iwatsuki. 2006. Museum fish specimens and molecular taxonomy: a comparative study on DNA extraction protocols and preservation techniques. Journal of Applied Ichthyology 22:160–166.

Cicero, C. A. 1996. Sibling species of titmice in the *Parus inornatus* complex (Aves: Paridae). University of California Publications in Zoology 128:1–217.

de Oliveira, G., T. F. Rangel, M. S. Lima-Ribeiro, L. C. Terribile, and J. A. F. Diniz-Filho. 2014. Evaluating, partitioning, and mapping the spatial autocorrelation component in ecological niche modeling: a new approach based on environmentally equidistant records. Ecography 37:637–647.

Diniz-Filho, J. A. F., L. M. Bini, and B. A. Hawkins. 2003. Spatial autocorrelation and red herrings in geographical ecology. Global Ecology and Biogeography 12:53–64.

Dorazio, R. M. 2012. Predicting the geographic distribution of a species from presence-only data subject to detection errors. Biometrics 68:1303–1312.

Eastman, L. M., T. L. Morelli, K. C. Rowe, C. J. Conroy, and C. Moritz. 2012. Size increase in high elevation ground squirrels over the last century. Global Change Biology 18:1499–1508.

Endler, J. A. 1977. Geographic variation, speciation and clines. Princeton University Press, Princeton, NJ.

Fitzpatrick, M. C., J. F. Weltzin, N. J. Sanders, and R. R. Dunn. 2007. The biogeography of prediction error: why does the introduced range of the fire ant over-predict its native range? Global Ecology and Biogeography 16:24–33.

Friedmann, H., L. Griscom, and R. T. Moore. 1950. Distributional check-list of the birds of Mexico. Part I. Pacific Coast Avifauna 29:1–202.

Gaiji, S., V. Chavan, A. H. Ariño, J. Otegui, D. Hobern, R. Sood, and E. Robles. 2013. Content assessment of the primary biodiversity data published through GBIF network: status, challenges and potentials. Biodiversity Informatics 8:94–172.

Gibson, L., A. McNeill, P. de Tores, A. Wayne, and C. Yates. 2010. Will future climate change threaten a range restricted endemic species, the quokka (*Setonix brachyurus*), in south west Australia? Biological Conservation 143:2453–2461.

Godman, F. D., and O. Salvin. 1879–1915. Biologia Centrali-Americana [published in 215 parts by various authors]. Taylor & Francis, London, UK.

Godsoe, W. 2010. I can't define the niche but I know it when I see it: a formal link between statistical theory and the ecological niche. Oikos 119:53–60.

Graham, C., S. Ron, J. Santos, C. Schneider, and C. Moritz. 2004. Integrating phylogenetics and environmental niche models to explore speciation mechanisms in dendrobatid frogs. Evolution 58:1781–1793.

Grinnell, H. W. 1940. Joseph Grinnell: 1877–1939. Condor 42:3–34.

Grinnell, J. 1904. The origin and distribution of the Chestnut-backed Chickadee. Auk 21:364–382.

Grinnell, J. 1914. Barriers to distribution as regards birds and mammals. American Naturalist 48:248–254.

Grinnell, J. 1917a. Field tests of theories concerning distributional control. American Naturalist 51:115–128.

Grinnell, J. 1917b. The niche-relationships of the California Thrasher. Auk 34:427–433.

Grinnell, J. 1924. Geography and evolution. Ecology 5:225–229.

Grinnell, J. 1932. Habitat relations of the Giant Kangaroo Rat. Journal of Mammalogy 13:305–320.

Grinnell, J., and H. S. Swarth. 1913. An account of the birds and mammals of the San Jacinto area of southern California with remarks upon the behavior of geographic races on the margins of their habitats. University of California Publications in Zoology 10:197–406.

Griscom, L. 1932. The distribution of bird life in Guatemala: a contribution to a study of the origin of Central American bird-life. Bulletin of the American Museum of Natural History 9:1–439.

Guisan, A., O. Broennimann, R. Engler, M. Vust, N. G. Yoccoz, A. Lehmann, and N. E. Zimmermann. 2006. Using niche-based models to improve the sampling of rare species. Conservation Biology 20:501–511.

Hall, B. P., and R. E. Moreau. 1970. An atlas of speciation in African passerine birds. Trustees of the British Museum (Natural History), London, UK.

Haywood, A. M., H. J. Dowsett, A. M. Dolan, D. Rowley, A. Abe-Ouchi, B. Otto-Bliesner, M. A. Chandler, S. J. Hunter, D. J. Lunt, M. Pound, and U. Salzmann. 2015. Pliocene Model Intercomparison (PlioMIP) Phase 2: scientific objectives and experimental design. Climate of the Past 11:4003–4038.

Helaouët, P., and G. Beaugrand. 2009. Physiology, ecological niches and species distribution. Ecosystems 12:1235–1245.

Hellmayr, C. E., and B. Conover. 1942. Catalogue of birds of the Americas and the adjacent islands in Field Museum of Natural History. Field Museum of Natural History Zoology Series 6:1–636.

Hengeveld, R. 1992. Dynamic biogeography. Cambridge University Press, Cambridge, UK.

Herman, S. G. 1986. The naturalist's field journal: a manual of instruction based on a system established by Joseph Grinnell. Buteo Books, Vermillion, SD.

Hobson, K. A. 1999. Tracing origins and migration of wildlife using stable isotopes: a review. Oecologia 120:314–326.

Holt, A. R., K. J. Gaston, and F. He. 2002. Occupancy-abundance relationships and spatial distribution: a review. Basic and Applied Ecology 3:1–13.

Jakob, S. S., E. Martinez-Meyer, and F. R. Blattner. 2009. Phylogeographic analyses and paleodistribution modeling indicate Pleistocene in situ survival of *Hordeum* species (Poaceae) in southern Patagonia without genetic or spatial restriction. Molecular Biology and Evolution 26:907–923.

Johnson, N. K. 1963. Biosystematics of sibling species of flycatchers in the *Empidonax hammondii-oberholseri-wrightii* complex. University of California Publications in Zoology 66:79–238.

Johnson, N. K. 1980. Character variation and evolution of sibling species in the *Empidonax difficilis-flavescens* complex (Aves: Tyrannidae). University of California Publications in Zoology 112:1–151.

Kearney, M., and W. P. Porter. 2004. Mapping the fundamental niche: physiology, climate, and the distribution of a nocturnal lizard. Ecology 85:3119–3131.

Kozak, K. H., and J. J. Wiens. 2010a. Accelerated rates of climatic-niche evolution underlie rapid species diversification. Ecology Letters 13:1378–1389.

Kozak, K. H., and J. J. Wiens. 2010b. Niche conservatism drives elevational diversity patterns in Appalachian salamanders. American Naturalist 176:40–54.

Kozak, K. H., J. J. Wiens, and D. Pfennig. 2009. Does niche conservatism promote speciation? a case study in North American salamanders. Evolution 60:2604–2621.

Lima-Ribeiro, M. S., S. Varela, J. González-Hernández, G. D. Oliveira, J. A. F. Diniz-Filho, and L. C. Terribile. 2015. EcoClimate: a database of climate data from multiple models for past, present, and future for macroecologists and biogeographers. Biodiversity Informatics 10:1–21.

Lira-Noriega, A., and J. D. Manthey. 2014. Relationship of genetic diversity and niche centrality: a survey and analysis. Evolution 68:1082–1093.

Lira-Noriega, A., J. Soberón, and C. P. Miller. 2013. Process-based and correlative modeling of desert mistletoe distribution: a multiscalar approach. Ecosphere 4:1–23.

Manthey, J. D., L. P. Campbell, E. E. Saupe, J. Soberón, C. M. Hensz, C. E. Myers, H. L. Owens, K. Ingenloff, A. T. Peterson, N. Barve, A. Lira-Noriega, and V. Barve. 2014. A test of niche centrality as a determinant of population trends and conservation status in threatened and endangered North American birds. Endangered Species Research 26:201–208.

Martínez-Meyer, E., D. Díaz-Porras, A. T. Peterson, and C. Yáñez-Arenas. 2012. Ecological niche structure determines rangewide abundance patterns of species. Biology Letters 9:20120637.

Martínez-Meyer, E., and A. T. Peterson. 2006. Conservatism of ecological niche characteristics in North American plant species over the Pleistocene-to-Recent transition. Journal of Biogeography 33:1779–1789.

Martínez-Meyer, E., A. T. Peterson, and W. W. Hargrove. 2004. Ecological niches as stable distributional constraints on mammal species, with implications for Pleistocene extinctions and climate change projections for biodiversity. Global Ecology and Biogeography 13:305–314.

McCormack, J. E., A. J. Zellmer, and L. L. Knowles. 2010. Does niche divergence accompany allopatric divergence in *Aphelocoma* jays as predicted under ecological speciation? Insights from tests with niche models. Evolution 64:1231–1244.

Medley, K. A. 2010. Niche shifts during the global invasion of the Asian tiger mosquito, *Aedes albopictus* Skuse (Culicidae), revealed by reciprocal distribution models. Global Ecology and Biogeography 19:122–133.

Meier, E. S., H. Lischke, D. R. Schmatz, and N. E. Zimmermann. 2012. Climate, competition and connectivity affect future migration and ranges of European trees. Global Ecology and Biogeography 21:164–178.

Menon, S., B. I. Choudhury, M. L. Khan, and A. T. Peterson. 2010. Ecological niche modeling predicts new populations of *Gymnocladus assamicus*, a critically endangered tree species. Endangered Species Research 11:175–181.

Menon, S., M. L. Khan, A. Paul, and A. T. Peterson. 2012. *Rhododendron* species in the Indian Eastern Himalayas: new approaches to understanding rare plant species' distributions. Journal of the American Rhododendron Society Spring 2012:78–84.

Meyer de Schauensee, R. 1966. The species of birds of South America and their distribution. Academy of Natural Sciences, Philadelphia, PA.

Miller, A. H. 1941. Speciation in the avian genus *Junco*. University of California Publications in Zoology 44:173–434.

Miller, A. H., H. Friedmann, L. Griscom, and R. T. Moore. 1957. Distributional check-list of the Birds of Mexico. Part 2. Pacific Coast Avifauna 33:1–436.

Moritz, C., J. L. Patton, C. J. Conroy, J. L. Parra, G. C. White, and S. R. Beissinger. 2008. Impact of a century of climate change on small-mammal communities in Yosemite National Park, USA. Science 322:261–264.

Nagaraju, S. K., R. Gudasalamani, N. Barve, J. Ghazoul, G. K. Narayanagowda, and U. S. Ramanan. 2013. Do ecological niche model predictions reflect the adaptive landscape of species? A test using *Myristica malabarica* Lam., an endemic tree in the Western Ghats, India. PLoS One 8:e82066.

Nogués-Bravo, D. 2009. Predicting the past distribution of species climatic niches. Global Ecology and Biogeography 18:521–531.

Nogués-Bravo, D., J. Rodríguez, J. Hortal, P. Batra, and M. B. Araújo. 2008. Climate change, humans, and the extinction of the Woolly Mammoth. PLoS Biology 6:e79.

Owens, H. L., L. P. Campbell, L. Dornak, E. E. Saupe, N. Barve, J. Soberón, K. Ingenloff, A. Lira-Noriega, C. M. Hensz, C. E. Myers, and A. T. Peterson. 2013. Constraints on interpretation of ecological niche models by limited environmental ranges on calibration areas. Ecological Modelling 263:10–18.

Parra, J. L., C. C. Graham, and J. F. Freile. 2004. Evaluating alternative data sets for ecological niche models of birds in the Andes. Ecography 27:350–360.

Paynter, R. A., Jr. 1995. Ornithological gazetteer of Argentina (2nd ed.). Museum of Comparative Zoology, Cambridge, MA.

Paynter, R. A., Jr., and M. A. Traylor, Jr. 1991. Ornithological gazetteer of Brazil. Museum of Comparative Zoology, Cambridge, MA.

Pearson, R. G., W. Thuiller, M. B. Araújo, E. Martínez-Meyer, L. Brotons, C. McClean, L. Miles, P. Segurado, T. P. Dawson, and D. C. Lees. 2006. Model-based uncertainty in species range prediction. Journal of Biogeography 33:1704–1711.

Peterson, A. T. 2003. Predicting the geography of species' invasions via ecological niche modeling. Quarterly Review of Biology 78:419–433.

Peterson, A. T. 2011. Ecological niche conservatism: a time-structured review of evidence. Journal of Biogeography 38:817–827.

Peterson, A. T., and C. M. Ammann. 2013. Global patterns of connectivity and isolation of populations of forest bird species in the late Pleistocene. Global Ecology and Biogeography 22:596–606.

Peterson, A. T., R. T. Brumfield, R. G. Moyle, Á. Nyári, J. V. Remsen, Jr., and M. B. Robbins. 2007. The need for proper vouchering in phylogenetic studies of birds. Molecular Phylogenetics and Evolution 45:1042–1044.

Peterson, A. T., and Y. Nakazawa. 2008. Environmental data sets matter in ecological niche modeling: an example with *Solenopsis invicta* and *Solenopsis richteri*. Global Ecology and Biogeography 17:135–144.

Peterson, A. T., and Á. Nyári. 2007. Ecological niche conservatism and Pleistocene refugia in the Thrush-like Mourner, *Schiffornis* sp., in the Neotropics. Evolution 62:173–183.

Peterson, A. T., M. A. Ortega-Huerta, J. Bartley, V. Sánchez-Cordero, J. Soberón, R. H. Buddemeier, and D. R. B. Stockwell. 2002. Future projections for Mexican faunas under global climate change scenarios. Nature 416:626–629.

Peterson, A. T., V. Sánchez-Cordero, J. Soberón, J. Bartley, R. H. Buddemeier, and A. G. Navarro-Sigüenza. 2001. Effects of global climate change on geographic distributions of Mexican Cracidae. Ecological Modelling 144:21–30.

Peterson, A. T., J. Soberón, R. G. Pearson, R. P. Anderson, E. Martínez-Meyer, M. Nakamura, and M. B. Araújo. 2011. Ecological niches and geographic distributions. Princeton University Press, Princeton, NJ.

Peterson, A. T., J. Soberón, and V. Sánchez-Cordero. 1999. Conservatism of ecological niches in evolutionary time. Science 285:1265–1267.

Petitpierre, B., C. Kueffer, O. Broennimann, C. Randin, C. Daehler, and A. Guisan. 2012. Climatic niche shifts are rare among terrestrial plant invaders. Science 335:1344–1348.

Pinot, S., G. Ramstein, S. Harrison, I. Prentice, J. Guiot, M. Stute, and S. Joussaume. 1999. Tropical paleoclimates at the Last Glacial Maximum: comparison of Paleoclimate Modeling Intercomparison Project (PMIP) simulations and paleodata. Climate Dynamics 15:857–874.

Pitelka, F. A. 1950. Geographic variation and the species problem in the shore bird genus *Limnodromus*. University of California Publications in Zoology 50:1–108.

Pitelka, F. A. 1951. Speciation and ecologic distribution in American Jays of the genus *Aphelocoma*. University of California Publications in Zoology 50:195–464.

Prieto-Torres, D. A., A. G. Navarro-Sigüenza, D. Santiago-Alarcón, and O. R. Rojas-Soto. 2016. Response of the endangered tropical dry forests to climate change and the role of Mexican protected areas for their conservation. Global Change Biology 22:364–379.

Pulliam, H. R. 2000. On the relationship between niche and distribution. Ecology Letters 3:349–361.

Raxworthy, C. J., E. Martínez-Meyer, N. Horning, R. A. Nussbaum, G. E. Schneider, M. A. Ortega-Huerta, and A. T. Peterson. 2003. Predicting distributions of known and unknown reptile species in Madagascar. Nature 426:837–841.

Remsen, J., Jr. 1977. On taking field notes. American Birds 31:946–953.

Remsen, J. V. 1995. The importance of continued collecting of bird specimens to ornithology and bird conservation. Bird Conservation International 5:145–180.

Rice, N. H., E. Martínez-Meyer, and A. T. Peterson. 2003. Ecological niche differentiation in the *Aphelocoma* jays: a phylogenetic perspective. Biological Journal of the Linnaean Society 80:369–383.

Rowe, K. C., K. M. C. Rowe, M. W. Tingley, M. S. Koo, J. L. Patton, C. J. Conroy, J. D. Perrine, S. R. Beissinger, and C. Moritz. 2014. Spatially heterogeneous impact of climate change on small mammals of montane California. Proceedings of the Royal Society B 282:20141857.

Russell, S. M. 1964. A distributional study of the birds of British Honduras. Ornithological Monographs, no. 1. American Ornithologists' Union, Washington DC.

Saupe, E. E., V. Barve, C. E. Myers, J. Soberón, N. Barve, C. M. Hensz, A. T. Peterson, H. Owens, and A. Lira-Noriega. 2012. Variation in niche and distribution model performance: the need for a priori assessment of key causal factors. Ecological Modelling 237:11–22.

Segurado, P., M. B. Araújo, and W. E. Kunin. 2006. Consequences of spatial autocorrelation for niche-based models. Journal of Applied Ecology 43:433–444.

Şekercioğlu, Ç. H., R. B. Primack, and J. Wormworth. 2012. The effects of climate change on tropical birds. Biological Conservation 148:1–18.

Selander, R. K. 1964. Speciation in wrens of the genus *Campylorhynchus*. University of California Publications in Zoology 74:1–305.

Selander, R. K., and D. R. Giller. 1963. Species limits in the woodpecker genus *Centurus* (Aves). Bulletin of the American Museum of Natural History 124:217–263.

Shcheglovitova, M., and R. P. Anderson. 2013. Estimating optimal complexity for ecological niche models: a jackknife approach for species with small sample sizes. Ecological Modelling 269:9–17.

Smith, B. T., P. Escalante, B. E. H. Baños, A. G. Navarro-Sigüenza, S. Rohwer, and J. Klicka. 2011. The role of historical and contemporary processes on phylogeographic structure and genetic diversity in the Northern Cardinal, *Cardinalis cardinalis*. BMC Evolutionary Biology 11:136.

Snow, D. W. 1978. An atlas of speciation in African non-passerine birds. British Museum (Natural History), London, UK.

Soberón, J. 1999. Linking biodiversity information sources. Trends in Ecology and Evolution 14:291.

Soberón, J. 2007. Grinnellian and Eltonian niches and geographic distributions of species. Ecology Letters 10:1115–1123.

Soberón, J., and A. Peterson. 2005. Interpretation of models of fundamental ecological niches and species' distributional areas. Biodiversity Informatics 2:1–10.

Soberón, J., and A. T. Peterson. 2011. Ecological niche shifts and environmental space anisotropy: a cautionary note. Revista Mexicana de Biodiversidad 82:1348–1353.

Sousa-Baena, M. S., L. C. Garcia, and A. T. Peterson. 2013. Completeness of Digital Accessible Knowledge of the plants of Brazil and priorities for survey and inventory. Diversity and Distributions 20:369–381.

Tingley, M. W., and S. R. Beissinger. 2009. Detecting range shifts from historical species occurrences: new perspectives on old data. Trends in Ecology and Evolution 24:625–633.

Tingley, M. W., and S. R. Beissinger. 2013. Cryptic loss of montane avian richness and high community turnover over 100 years. Ecology 94:598–609.

Tingley, M. W., W. B. Monahan, S. R. Beissinger, and C. Moritz. 2009. Birds track their Grinnellian niche through a century of climate change. Proceedings of the National Academy of Sciences USA 106:19637–19643.

Varela, S., J. M. Lobo, J. Rodriguez, and P. Batra. 2010. Were the Late Pleistocene climatic changes responsible for the disappearance of the European spotted hyena populations? hindcasting a species geographic distribution across time. Quaternary Science Reviews 29:2027–2035.

Waltari, E., and R. Guralnick. 2009. Ecological niche modelling of montane mammals in the Great Basin, North America: examining past and present connectivity of species across basins and ranges. Journal of Biogeography 36:148–161.

Waltari, E., S. Perkins, R. Hijmans, A. T. Peterson, Á. Nyári, and R. Guralnick. 2007. Locating Pleistocene refugia: comparing phylogeographic and ecological niche model predictions. PLoS One 2:e563.

Warren, D. L., R. E. Glor, and M. Turelli. 2008. Environmental niche equivalency versus conservatism: quantitative approaches to niche evolution. Evolution 62:2868–2883.

Warren, D. L., and S. N. Seifert. 2011. Ecological niche modeling in Maxent: the importance of model complexity and the performance of model selection criteria. Ecological Applications 21:335–342.

Wiens, J. J., and C. H. Graham. 2005. Niche conservatism: integrating evolution, ecology, and conservation biology. Annual Review of Ecology, Evolution and Systematics 36:519–539.

Wiens, J. J., K. H. Kozak, and N. Silva. 2013. Diversity and niche evolution along aridity gradients in North American lizards (Phrynosomatidae). Evolution 67: 1715–1728.

CHAPTER EIGHT

Using Museum Specimens to Study Flight and Dispersal*

Santiago Claramunt and Natalie A. Wright

Abstract. Flight performance is an important aspect of avian biology, potentially affecting the energetics, life history, ecology, dispersal, and evolution of birds. Yet measuring flight performance directly can be challenging and time intensive. Here we discuss morphological proxies of flight performance that can be measured from museum specimens. These include wing shape, wing size, and flight muscle size. We provide details on how to obtain indices of these characters from traditional round study skin specimens as well as from new specimen types, such as spread wings and skeletons. We also offer suggestions for specimen preparators on how to maximize the usefulness of specimens for studies of flight performance. Finally, we discuss recent applications of the use of museum specimens in studies of flight performance and dispersal, including how dispersal ability relates to diversification and how flight constrains genome size.

Key Words: aspect ratio, dispersal, flight, flight muscles, wing area, wing shape.

The ability to fly is a prominent characteristic of birds that determines many aspects of their biology. Birds use a variety of flight styles in their locomotion, including gliding and soaring, powered flapping flight, and hovering (Rayner 1988, Norberg 1990, Videler 2005, Pennycuick 2008). Some species are capable of great long-distance migrations, whereas others struggle to fly more than a few meters at a time. These vastly different flight strategies affect and are affected by aspects of birds' physiology, morphology, life history, and evolution (Norberg 1990, Hedenström 2002, Dial 2003, Pennycuick 2008). Although directly measuring flight performance across many different species or lineages can elucidate how flight relates to other aspects of bird biology (e.g., Altshuler et al. 2004a, Altshuler 2006), doing so is extremely time consuming and difficult. Fortunately, many aspects of flight performance and strategy are tied directly to morphological traits (Rayner 1988, Norberg 1990, Pennycuick 2008), many of which can be assessed in museum specimens.

Flight directly affects dispersal, from individual movements to macroecological and macroevolutionary scales. Because most birds disperse via flight, and flight performance is determined by the characteristics of the flight apparatus, birds are a promising model for comparative studies of dispersal and its influence on ecological and

* Claramunt, S., and N. A. Wright. 2017. Using museum specimens to study flight and dispersal. Pp. 127–141 in M. S. Webster (editor), The Extended Specimen: Emerging Frontiers in Collections-based Ornithological Research. Studies in Avian Biology (no. 50), CRC Press, Boca Raton, FL.

evolutionary processes. Traditionally, it has been assumed that dispersal is determined by behavioral choices about when, where, and how far to disperse. In particular, limited dispersal in tropical birds was attributed to "behavioral flightlessness," or fear of flying over inhospitable habitat (Diamond 1981, Laurance et al. 2004). However, recent evidence suggests that flight ability may have an important influence on the dispersal process. "Dispersal challenge" experiments demonstrated that some tropical birds cannot maintain sustained flight over 300 meters (Moore et al. 2008) and the mean distance flown has been found to be tightly correlated with wing morphology (Claramunt et al. 2012), suggesting that morphology of the flight apparatus influences dispersal capabilities of these species. Morphological constraints on dispersal are likely correlated with behavioral ones, as birds with a strong flight apparatus are more likely to engage in long distance flights or flights over inhospitable habitats, whereas birds with a weak flight apparatus are less likely to do so (Rayner 1988, Videler 2005, Pennycuick 2008). The advantage of using morphology over behavior for the study of dispersal is that the former is relatively easy to measure using museum specimens, thus also extending the value of the specimen beyond traditional specimen-based studies.

Here, we review methods that can be used to assess both flight performance and dispersal ability from phenotypic information measurable from museum specimens. Some of these methods can be used to estimate the relative dispersal ability of bird species and shed light on the role of dispersal in multiple ecological and evolutionary phenomena. Other methods are useful in estimating the metabolic demands of flight, predator escape ability, or endurance flight capability. At the same time, basic research on bird flight is sometimes limited by the availability of reliable anatomical data, and will benefit from an effort from research collections to expand and preserve certain specimen types from which wing area and wingspan can be obtained. We start with general notions about the relationship between morphology and flight performance, followed by description of several methods that can be used for assessing various aspects of flight performance from research specimens. We illustrate some of the properties of these methods with our own published and unpublished data

from specimens housed at the Louisiana State University Museum of Natural Science (LSUMZ, Baton Rouge, LA), the University of Washington Burke Museum of Natural History and Culture (UWBM, Seattle, WA), the American Museum of Natural History (AMNH) (New York, NY), the Florida Museum of Natural History (Gainesville, FL), and the Museum of Southwestern Biology (Albuquerque, NM). We end with examples from the literature that have used these techniques for studies of bird biology and evolution. Together, these methods extend museum specimens for use in large-scale comparative studies on the ecological and evolutionary consequences of flight.

MORPHOLOGICAL TRAITS AND FLIGHT

A suite of morphological characters influences bird flight performance, including wing size, wing shape, flight muscle size, and heart size. A common measure of relative wing size is wing loading, which is calculated as the mass or weight of the bird divided by its wing area. For a given body size, larger values of wing loading indicate smaller wings and lower values indicate larger wings. However, because wing loading scales positively with body size, comparing birds of different sizes becomes tricky, and some authors have used wing loading indices that are "corrected" for body size (Rayner 1988, Norberg 1990). For a given body size, birds with higher wing loading require higher wingbeat frequencies and/or faster airspeeds to stay aloft, and, therefore, flight requires more energy per unit time, or power, for these species than for those with low wing loading (Rayner 1988, Norberg 1990). Relative wing size may be an indicator of the "intensity" of flight, or its energetic cost per unit time. However, relative wing size is not related to the efficiency of long-distance flights measured as the cost of transport, which is the work done in transporting a unit of body weight over a unit of distance (Viedeler 2005, Pennycuick 2008). The reason behind this phenomenon is that birds with relatively small wings such as ducks, loons, and auks, fly fast, and, although their flights are energetically costly per unit time, they travel long distances in short periods of time, resulting in a low cost of transport (Rayner 1988, Pennycuick 2008).

The morphological characteristic that best reflects the economy of long-distance flight is the aspect ratio (Rayner 1988, Norberg 1990, Pennycuick 2008). Aspect ratio is the relationship between the wingspan and the wing width; higher values indicate longer and narrower wings, which are typical of long-distance migrants, seabirds, and aerial foragers. According to aerodynamic models of bird flight, high aspect ratio wings produce more lift and less drag compared to low aspect ratio wings, reducing the cost of transport (Rayner 1988, Norberg 1990, Pennycuick 2008). Therefore, the aspect ratio is the most relevant morphological proxy for long-distance flight performance, with multiple potential applications in the study of dispersal in ecology, evolution, and biogeography.

Less widely used, but nonetheless extremely informative, are characters directly related to providing the power for flight, such as the relative sizes of the pectoral flight muscles. The pectoral flight muscles include the pectoralis major, which draws the wing down/forward, and the smaller supracoracoideus, which pulls the wing up/backward. These pectoral flight muscles vary greatly in size across volant birds, ranging from ~5% of the body mass in rails and ground cuckoos to over 30% in some hummingbirds and pigeons (Hartman 1961; Wright et al. 2016; N. A. Wright, unpubl. data). Together, these muscles provide the vast majority of the power for flapping flight and are, therefore, most important during takeoffs and powered bursts of speed in flight when power requirements are greatest (Tobalske et al. 2003). Large flight muscles likely indicate selection for rapid takeoff ability or powered acceleration during flapping flight. Accordingly, relative flight muscle size may reflect the importance of predation pressure selecting for rapid takeoffs (Wright and Steadman 2012, Wright et al. 2016).

The sizes of both the pectoral flight muscles and heart are directly related to the metabolic demands of flight. Use of the pectoral flight muscles is energetically costly, and the size of these muscles is a primary determinant of exercise-induced maximum metabolic rate and thermogenic capacity (Hohtola 1982, Hohtola and Stevens 1986, Chappell et al. 1999, Altshuler et al. 2004b, Swanson et al. 2013). Whereas variation in relative flight muscle size indicates the degree of importance of burst power, heart size is an indicator of endurance flight performance (Bishop 1997). The size of the heart constrains its stroke volume, thereby limiting maximum cardiac output, aerobic power, exercise-induced maximum metabolic rate, and aerobic scope (Bishop and Butler 1995; Bishop 1997, 1999). As routine powered flight incurs a 10- to 20-fold sustained increase in metabolic rate (Ward et al. 2002), the sizes of the heart and flight muscles should reflect a bird's aerobic power and metabolic requirements, respectively, for sustained flight.

METHODS FOR ESTIMATING FLIGHT PERFORMANCE FROM SPECIMENS

Estimating Flight Performance from Traditional Study Skins

The traditional study skin is prepared with closed wings, making direct measurements of important aspects of wing size and shape impossible. Nevertheless, several methods have been developed to assess wing shape and size from closed wings (e.g., Kipp 1959, Lockwood et al. 1998). Instead of providing an extensive list of indices, here we review the indices of wing shape and wing area that we consider the most useful for studies of flight performance because they have clear theoretical and empirical support.

Hand-Wing Indices

The hand-wing index was first proposed by Kipp (1959) as a simple index of aspect ratio and therefore an indicator of flight performance. The hand-wing index is calculated as

$$\text{Hand-wing index } 1 = 100 \frac{D_K}{L_w}$$

where L_w is the length of the closed wing, that is, the traditional measurement of wing length, from the carpal joint to the tip of the longest primary (Baldwin et al. 1931, Stiles and Altshuler 2004); and D_K is the distance between the tip of the longest primary and the tip of the first secondary (Figure 8.1), also known as "Kipp's distance." Note that Kipp's distance is not the same as the "primary projection" used in bird identification; the latter is the distance from the tip of the longest primary to the tip of the longest secondary, which is not the first secondary but a proximal secondary

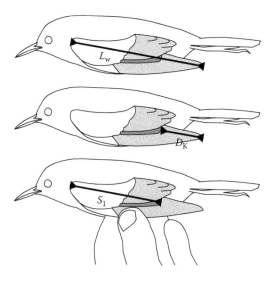

Figure 8.1. Linear measurements required for calculating hand-wing and wing area indices taken on the wing of a standard study skin specimen. L_w: Length of closed wing, from the carpal joint to the tip of the longest primary. D_K: Kipp's distance, from the tip of the first secondary (dark gray) to the tip of the longest primary. S_1: From the carpal joint to the tip of the first secondary (dark gray), usually visible only after gently spreading out primary and secondary feathers. Measuring L_w and S_1 is sufficient for most purposes. Measurements are easily taken with calipers in small- to medium-size birds, and with rulers or tree calipers for large birds.

or a tertial. Alternatively, a hand-wing index can be calculated as

$$\text{Hand-wing index 2} = 100 \frac{L_w - S_1}{L_w}$$

where L_w is wing length as before, and S_1 the distance between the carpal joint and the tip of the first secondary (Figure 8.1; Claramunt et al. 2012).

Measurements required for calculating both versions of the hand-wing index are easily taken from traditional study skins (Figure 8.1) but finding the tip of the first secondary feather is not a trivial task in some cases. The first secondary is sandwiched between primary and secondary feathers and accessing its tip requires gently opening the other remiges. Identification of the first secondary feather is facilitated by the fact that it appears to be the shortest remige in the folded wing and has a distinctive inward curvature compared to the primaries. If in doubt, counting may

help because in most birds the first secondary is the 11th feather counting from the outermost primary inward; exceptions to this include nine-primaried oscines, in which the outermost primary is reduced and concealed (Hall 2005), and some non-passerines that have 11 or 12 primaries.

Although not exactly equivalent, both versions of the hand-wing index are very similar. How similar they are depends on the wing's cross-section curvature. Only for wings that are completely flat, $L_w - S_1$ equals D_K, and both versions should give exactly the same number. In practice, both versions of the index are highly correlated. We compared hand-wing indices from Kipp (1959), who used the first version, with those computed using the second version. The strong correlation (Pearson's correlation coefficient $r = 0.996$; Figure 8.2) demonstrates the equivalence of the two indices, at least for this diverse interspecific dataset. This high correlation also suggests that species means are highly repeatable, even using measurements from different specimens made by different investigators. However, when comparing closely related species or individuals within species, for species with curved wings, the equivalence between the

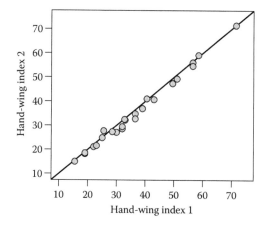

Figure 8.2. Correlation between the two versions of the hand-wing index for 24 European bird species. Values of the hand-wing index 1 were taken from Kipp (1959), which are values of single individuals or minimum and maximum values based on two to six individuals; in the latter case, the average of the two values was used for the plot. Values of the hand-wing index 2 are species averages based on measurements taken by S. Claramunt at the LSUMZ and the UWBM (two to six specimens per species, including males and females).

two versions and the high repeatability are not guaranteed.

The second formulation of the index makes the relationship between the hand-wing index and the aspect ratio more apparent due to the relationship between L_w and S_1 and the extent and the width of the wing, respectively (Claramunt et al. 2012). It is also more robust to variations in wing posture and preparation because it does not require the feathers to be aligned, as is the case for measuring Kipp's distance. In fact, L_w and S_1 can be measured independently, in different wings, or in wings that are extended or disarticulated. This becomes useful, for example, for estimating the hand-wing index of fossil specimens.

The hand-wing index is correlated with the aspect ratio of the wing but not in a linear manner (Figure 8.3). This occurs because the highest aspect ratios in birds are attained by elongation of the arm-wing, including the arm and forearm, rather than the hand, and thus are underestimated by the hand-wing index. As a result, birds with the highest aspect ratios such as albatrosses do not show the highest hand-wing indices (Figure 8.3). Other birds with relatively long arm-wings, such as those in the Procellariidae, Fregatidae, Gaviidae, Rallidae, Accipitridae, Strigidae, and Alcedinidae, also rank lower on the hand-wing index scale compared to the aspect ratio scale.

Some studies have used Kipp's distance rather than the hand-wing index as an index of wing shape or aspect ratio. However, this is only appropriate for comparisons involving birds of similar sizes and similar wing lengths. Kipp's distance confounds differences in wing shape with differences in wing size that could be misleading. For example, Kipp's distances in swallows (Hirundinidae) and chickens (*Gallus gallus*) are similar (60–70 mm), despite great differences in wing morphology and flight performance.

Wing Area Indices

Indices of wing area can be estimated from folded wings on study skins. Considering that extended bird wings are generally shaped as one-quarter of an ellipse, the wing area can be estimated as the area of an ellipse divided by four using the same measurements used for calculating hand-wing indices (Wright et al. 2014):

$$\text{Wing area index} = L_w S_1 \frac{\pi}{4}$$

where L_w is the length of the closed wing and S_1 is the distance between the carpal joint and the tip of the first secondary, which approximate the semiaxes of the ellipse. The same formula applies also to wings that are shaped like half an ellipse. This estimate of wing area ignores slotting, variations in wing shape that deviate from rounded (e.g., pointed wings), and the area contributed by the secondaries. The lengths of the manus and the longest primary feather comprise L_w, and both lengths are correlated with overall wing length and scale isometrically with other wing components such as the ulna (Nudds 2007, Nudds et al. 2011, Wang et al. 2011). Changes in the lengths of the ulna and the unmeasured area of the secondaries are, therefore, generally reflected in isometric changes in the lengths of the manus and the longest primary, which are measured, and thus are reflected in the wing area index.

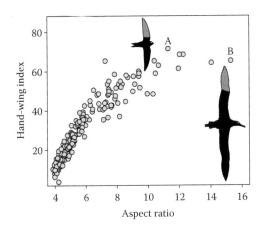

Figure 8.3. Comparison of the hand-wing index and the aspect ratio among 181 species. Bird silhouettes illustrate the species with the highest hand-wing index (A, *Apus apus*, Apodidae), and the species with the highest aspect ratio (B, *Diomedea exulans*, Diomedeidae). Because the hand-wing index considers only the hand portion of the wing (gray shading), it misses the contribution of the arm to the aspect ratio. Aspect ratios from the Wings database, part of the Flight program (Pennycuick 2008), Claramunt et al. (2012), and estimates based on specimens at LSUMZ and UWBM (S. Claramunt, unpubl. data). Hand-wing indices from Kipp (1959), Claramunt et al. (2012), and specimens at LSUMZ and AMNH (S. Claramunt, unpubl. data).

In aerodynamic models, the relevant wing area includes not only the area of both wings but also the area of the body between the wings (Pennycuick 2008). An index that approximates this area is

$$\text{Total wing area index} = 3L_wS_1$$

The total wing area index is highly correlated with the actual total wing area estimated using extended wings in a wide sample of birds (Pearson's $r = 0.98$; Figure 8.4). This high correlation may not be surprising because both quantities scale tightly with body size in this diverse dataset. We also evaluated the effect of using the total wing area index to estimate wing loadings and found a high correlation with estimates of wing loadings using actual wing areas (Pearson's $r = 0.97$). Therefore, at least for datasets encompassing a wide range of bird sizes, wing area indices can be used as surrogates for wing area in studies of flight performance.

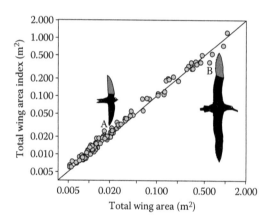

Figure 8.4. Relationship between the total wing area (the area of both wings plus the area of the body between the wings) and the total wing area index ($3L_wS_1$) among 163 bird species. Bird silhouettes illustrate a case in which the index overestimates the actual area (A, *Apus apus*, Apodidae) and a case in which the index underestimates the actual area (B, *Diomedea exulans*, Diomedeidae). In both cases, deviations are caused by the proportion of the area corresponding to the hand wing (gray shading, from which L_w and S_1 are measured) in relation to the arm. Total wing area data from the wings database (Pennycuick 2008), Claramunt et al. (2012), and spread wing specimens from LSUMZ and UWBM (S. Claramunt, unpubl. data). The total wing area index was calculated using L_w and S_1 measurements from Claramunt et al. (2012) and specimens from LSUMZ and AMNH (S. Claramunt, unpubl. data).

Estimating Flight Performance from Extended Specimens

When direct estimates of wingspan and wing area are available, it is preferable to use indicators of flight performance that are more directly linked to aerodynamic theory. Estimation of wingspan and wing area requires data or preparation types not typically preserved in bird collections, but which are relatively easy to collect during specimen preparation in the field or the laboratory. Skeletal specimens and muscle and heart masses provide additional information on flight performance. In the following sections we describe, first, how the necessary data can be collected during specimen preparation, and then how this information is used to estimate more advanced indicators of flight performance.

Preparation of Specimens Useful for Aerodynamic Calculations

Specimens become an important source of information on flight performance if they include a measurement of the wingspan and a voucher (physical or digital) of one spread wing. Neither of these is novel to museum practices (e.g., Chapin 1929, Baldwin et al. 1931) and can be easily incorporated into specimen preparation, storage, and curation. The wingspan is the distance between the tips of the wings in the fully extended position. The wingspan is measured by placing the bird on its back over a ruler and stretching its wings to their full natural extent, without forcefully overstretching them, and the distance between the most distal feather tips is recorded (Figure 8.5; Baldwin et al. 1931). In birds that have concave wings, the tips of the wing will not touch the ruler; in such cases, respect the curvature of the wing, that is, do not flatten the feathers. Instead, take the measurement by looking immediately over each tip, perpendicular to the ruler. Alternatively, the bird could be placed on its belly so the wing tips touch the surface of the ruler, facilitating the reading (Pennycuick 2008). For large birds or when one wing is damaged, the semispan can be measured instead, which is the distance between the wing tip and the center of the body, and then double the obtained value (Pennycuick 2008).

Measurement of the wingspan is easiest in fresh, dead specimens before or after *rigor mortis*.

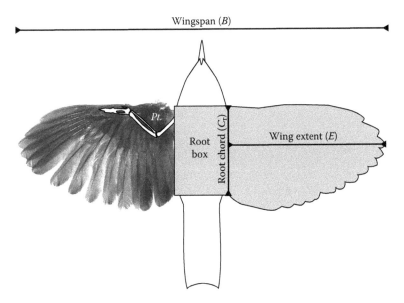

Figure 8.5. Measurements required for estimating flight performance. The left wing is a spread wing specimen of *Synallaxis scutata* (LSUMZ 169297) on which the approximate position of wing bones and the propatagium (Pt.) is indicated. The right wing is based on the same spread wing specimen after digitization, editing, and automated outline recognition for estimation of the area. Wing extent and wing root chord, measured in the extended wing specimen, are used in combination with the wingspan to estimate the area of the root box (see text for details).

Measurement in living individuals is potentially problematic because birds may actively resist fully stretching the wings and may become stressed or injured. Most preservation styles prevent measuring wingspan after specimen preparation. When wings are disarticulated, as in the case of study skins or spread wings, the original distance between the wings is lost. Also, full stretching of wings is impossible for both study skins and fluid-preserved specimens. Therefore, the measurement of the wingspan should be taken before specimen preparation and can be part of the standard information recorded on the specimen label and the catalog.

The wing area relevant to aerodynamic models is the area of both wings plus the area of the body between the wings (i.e., the "root box"; Figure 8.5). Although the total wing area could be estimated directly from an entire bird with the wings fully stretched, it is more convenient to estimate it based on a single wing (Pennycuick 2008). Here we propose that a voucher of a spread wing should be preserved for this purpose. This voucher can be a tracing of the wing's outline on archival-quality paper, a digital photograph of the wing including a scale (Stiles and Altshuler 2004), or the actual wing as a spread wing specimen.

Regardless of voucher type, because wing area changes depending on the flying style and during the flapping cycle, it is critical to standardize the posture of the wing for preservation. Following Stiles and Altshuler (2004), we recommend that the base of the rachis of the outermost primary feather should continue the leading edge of the wing, perpendicular to the axis of the body (Figure 8.5).

A spread wing can be traced on the back of the collector's field catalog (Figure 8.6) or on a separate sheet if it is too large. Although high quality permanent ink is generally preferred for recording data in catalogs, an initial tracing with a graphite pencil is recommended to avoid staining the feathers with ink. At least the specimen catalog number should be recorded in or immediately next to the tracing, but we also recommend recording the species name, body mass, and wingspan, which become convenient for later usage. A digital photograph may be a convenient way of recording the spread wing for area estimation (Stiles and Altshuler 2004). The photograph should be taken with a normal or telephoto lens, never with a wide-angle lens, and perpendicular to the wing surface to minimize distortions. A tripod or photocopy stand can help, and using a

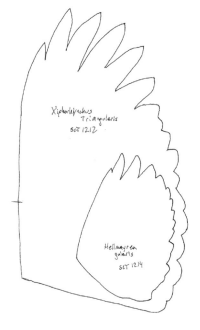

Figure 8.6. Wing tracings on the back of S. Claramunt's field catalog, available at the LSU Museum of Natural Science, including *Heliangelus micraster* (SCT 1215, Trochilidae), *Xiphorhychus triangularis* (SCT 1212, Furnariidae), and *Hellmayrea gularis* (SCT 1214, Furnariidae).

contrasting background facilitates automatic outline recognition by image-processing software. The image should include at least the specimen number (but preferably the whole specimen label or catalog page as part of the image) and a reference ruler to properly scale the image before area estimation. The image can then be stored as an associated file in the collection's database.

Finally, the most valuable and informative voucher is the wing itself. Spread wing preparations are becoming more common in bird collections as their usefulness for examining wing morphology, coloration, and molt is becoming more widely appreciated. A spread wing preparation is obtained by clipping the wing at its base, stripping off the muscles, and then pinning the wing for drying. The clipping usually requires cutting the humerus where it emerges from the pectoral muscles, but it can alternatively be performed through the elbow joint so an intact

humerus can be preserved for other purposes. These alternatives, as well as slight differences on how the proximal feathers are arranged, may result in different estimates for the area of the single wing, but they do not much affect estimates of the total wing area, as explained later. The next step is to remove the muscles and tendons, for which large wings often require an additional incision to remove forearm and hand muscles (Chapin 1929, Blake 1949). Secondary feathers should never be detached from the ulna, as this may distort feather position. The wing is then pinned on cardboard or Styrofoam board for drying. The wing should be fully extended, with a straight leading edge, including the proximal rachis of the outermost primary feather (Figure 8.5), and with the proximal edge arranged so as to form a smooth border perpendicular to the leading edge, that is, parallel to the main body axis and direction of flight (Figure 8.5). A smooth proximal edge is attained by stretching the propatagium back in its extended position, as it usually collapses after wing clipping, and arranging properly the proximal feathers including secondaries, tertials, and wing coverts. To facilitate this latter step and to obtain a clean and straight proximal border, we suggest that scapular feathers should not be preserved as part of the spread wing preparation. Use as many pins as needed to attain the desired position of the wing and a neat arrangement of feathers. A note on the label of the main specimen voucher should indicate that a wing tracing, a wing photograph, or a spread wing has been preserved. All data associated with the specimen should be recorded on the spread wing tag, of which body mass and wingspan should not be missing so that the wing is useful for studies of flight.

Estimation of Total Wing Area

As explained earlier, aerodynamic models consider the total wing area, which is the area of both wings plus the area of the body between the wings (the root box; Figure 8.5). Estimation of the total wing area begins with the estimation of the area of a single wing. Several methods exist to estimate the area of the vouchered single wing. The traditional method of estimating areas of irregular forms using grid paper is straightforward to use on wing tracings or spread wing specimens (Pennycuick 2008). It is also straightforward to digitize the wings and use image analysis software to obtain the area of the wing. For wing tracing and spread wings, the first step is digitization, which can be done with a digital camera, as described earlier for the image voucher, or with a regular CCD document scanner, which are robust to scale and perspective distortions. Once in digital format, the image is imported into image analysis software such as ImageJ (U.S. National Institutes of Health, Bethesda, MD). If the background of the image contrasts with the line tracing or the wing, image analysis software can automatically recognize the outline of the wing. At this stage, small corrections can be performed such as erasing undesired or disarranged wing coverts, or cloning a remige to substitute a missing or growing one. Once the scale is set properly using the scale ruler included in the photograph, the software can be used to calculate the area of the single wing.

In addition to doubling the area of the single wing, the total wing area includes the root box, a rectangle representing the area between the wings (Figure 8.5). The width of the root box is the portion of the wingspan that does not include the extent of the two wings (Figure 8.5); thus, the extent of the single wing needs to be measured on the vouchered specimen or image, from the proximal edge to the tip of the most distal primary feather tip, parallel to the leading edge of the wing, and perpendicular to the main axis of the body and flight direction (Figure 8.5). The length of the root box equals the width of the wing at its base (the root chord, Figure 8.5), which also needs to be measured from the voucher or image. Then, the total wing area is

$$\text{Total wing area} = 2A_w + (B - 2E)C_r$$

where A_w is the area of a single wing, B is the wingspan, E is the extent of the single wing, and C_r is the root chord. From these formulas, it is evident that the wingspan is crucial for estimating the total wing area, and we urge collectors and preparators to measure and record the wingspan for every specimen for which a spread wing voucher is to be preserved.

Aspect Ratio

The aspect ratio is the relationship between the extent and the width of the wing, or more precisely the relationship between the wingspan and the "chord" of the wing, where the chord is the straight-line distance between the anterior and posterior borders of the wing. However, because avian wings have irregular shapes, their chords change from proximal to distal sections of the wing. For that reason, aspect ratios are calculated using the average chord of a wing, which can be estimated as the area of the wing divided by the wingspan, leading to the usual way of computing avian aspect ratios:

$$\text{Aspect ratio} = \frac{B^2}{A_{tot}}$$

where B is the wingspan and A_{tot} is the total wing area.

Lift-to-Drag Ratio

The lift-to-drag ratio is a key determinant of horizontal flight economy and maximum range in aerodynamic models of bird flight (Pennycuick 1989, 2008; Norberg 1990). Estimating lift-to-drag ratios requires estimating the amount of energy consumed per unit time, or power, required for horizontal flight, which requires information on body mass, gravitational acceleration, air density, air velocity, and drag coefficients, in addition to several morphological traits such as wingspan and wing area. The lift-to-drag ratio can incorporate aspects of flight aerodynamics that may be important for some applications. For example, when comparing flight performance across altitudinal gradients, the potential effects of differences in air density can be taken into account. Here we outline a simple version of a basic model developed by Pennycuick (1989, 2008) for forward flapping flight, which has been used for studies of migratory birds.

In uniform horizontal flight, lift forces counteract the weight of the bird, and the thrust forces counteract the drag produced by the friction of the air on the body's surfaces. The lift-to-drag ratio is calculated as

$$\text{Lift:drag ratio} = \frac{mgV}{P}$$

where P is the total power required, m is the bird's mass, g is the gravitational acceleration (i.e., mg is the weight in force units), and V is the forward speed. The lift-to-drag ratio is a nondimensional quantity suitable for comparing birds of different sizes.

Using blade-element and momentum jet theories it is possible to calculate the total aerodynamic power required for horizontal flight using a set of morphological measurements, mass, speed, and physical parameters (Pennycuick 1989, 2008; Norberg 1990). The lift-to-drag ratio depends on velocity, not only because it appears in the numerator but also because the total power required depends on speed in a nonlinear manner: in a plot of power as a function of velocity, the relationship between power and speed is U-shaped (Tobalske et al. 2003). This relationship implies that there is a velocity that minimizes the total power required for flight, called the minimum power velocity (V_{mp}), and a different velocity that minimizes the power required per unit distance, called the maximum range velocity (V_{mr}).

The total power required for steady horizontal flight can be divided into three components: induced power (P_{ind}), parasite power (P_{par}), and profile power (P_{pro}). Induced power is the power required to support the weight of the bird. For a bird in horizontal flapping flight, this quantity can be calculated as

$$\text{Induced power} = 2\frac{(mg)^2}{V\pi B^2 \rho}$$

in which mg is the body weight, V is the air speed, π is the constant pi, B is the wingspan, and ρ is the air density. Departures from the idealized conditions of the aerodynamic model result in biased estimates of the actual induced power, and an induced power factor (k) has been used as a correction factor (Pennycuick 2008). However, the precise value of k is not known and likely depends on velocity, flight style, and wing morphology (Pennycuick et al. 2013).

Parasite power is the power required for counteracting the drag of the body and is calculated as

$$\text{Parasite power} = \frac{V^3 A_b c_b \rho}{2}$$

where A_b is the body's frontal area and c_b is the body's drag coefficient. In the absence of direct measurements, body frontal area can be estimated as $A_b = 0.01m^{2/3}$, in which the coefficient 0.01 is a combination of several estimates and the exponent 2/3 represents the isometric relationship between mass and frontal area (Hedenström and Rosén 2003, Nudds and Rayner 2006). The body drag coefficient is poorly known. Early estimates of $c_b = 0.4$ based on frozen birds in wind tunnels may be too high and it has been argued that a streamlined bird should have a drag coefficient closer to 0.05 (Pennycuick et al. 1996). Estimates based on living birds, either in wind tunnels or during migration, suggest that values around 0.10 are possible (Hedenström and Rosén 2003, Pennycuick et al. 2013).

Profile power is the power required to overcome the drag of the wings that is not accounted for already as induced drag associated with lift. This is the least known component of power and likely depends not only on wingspan and wing area but also on other details of wing anatomy and kinematics (Pennycuick 2008). Contrary to the previous components, no formula has been derived from first principles for estimating profile power. It is not even clear how the parasite power changes with velocity. In Pennycuick's (2008) model, profile power is estimated as

$$\text{Profile power} = \frac{P_{am}c_p}{R_a}$$

where P_{am} is the absolute minimum power required, which is the sum of the induced and parasite powers ($P_{ind} + P_{par}$) when flying at the minimum power velocity (V_{mp}); c_p is a constant (8.4 in Pennycuick's model); and R_a is the aspect ratio. The minimum power velocity is obtained using the first derivative of $P_{ind} + P_{par}$ as a function of V, which results in

$$V_{mp} = \frac{0.807(mg)^{1/2}}{(B\rho)^{1/2}(A_b c_b)^{1/4}}$$

In the absence of empirical flight speed data, we propose the use of V_{mp} to estimate all power components and the lift-to-drag ratio. Empirical data suggest that daily commuting flight speeds in birds are closer to the estimated minimum power velocity than to the maximum range velocity (Pennycuick 1997), and large birds may be incapable of flying at maximum range velocity because of physiological constraints (Pennycuick et al. 2013). Therefore, the minimum power velocity can be used as a standard speed for comparative purposes. Then, the lift-to-drag ratio can be calculated using the minimum power speed and the power components described earlier:

$$\text{Lift:drag ratio} = \frac{mgV_{mp}}{P_{ind} + P_{par} + P_{pro}}$$

We implemented these calculations in functions in the R language (R Core Team 2016), which are available at the GitHub repository (https://github .com/evolucionario/birdflight).

Flight Muscles and Hearts

Flight muscle size provides an estimate of the power available for flight and should be a particularly good indicator of takeoff performance, as power requirements are often greatest during takeoff (Tobalske et al. 2003). Flight muscle and heart sizes are also highly correlated with metabolic rate (Hohtola 1982, Chappell et al. 1999, Altshuler et al. 2004b, Swanson et al. 2013), and, as such, can be used as indicators of the metabolic demands of flight. Masses of flight muscles and hearts can be quickly and easily recorded during specimen preparation (Wright and Steadman 2012, Wright et al. 2014). In many cases, they are already being extracted for use as tissue samples, and recording their masses requires very little extra time of the specimen preparator. To extract the heart, grasp the heart anteriorly with forceps and tug gently. Trim any excess blood vessels and blot to remove blood before placing on the scale. A simple way to remove the pectoralis major is to run scissors or a scalpel along one side of the sternal keel, cutting through the thin layer of fascia connecting the muscle to the keel. The pectoralis can then be peeled back, sometimes with additional cutting of fascia needed, until it is only attached via the tendon connecting it to the humerus. Cut this tendon without leaving any of the pectoralis remaining at the shoulder. The supracoracoideus sometimes comes off the carcass in conjunction with the pectoralis. If not, it is easily peeled off the ventral surface of the sternum. Because the flight muscles are symmetrical, only one side need be measured; total pectoral flight muscle mass can be calculated by multiplying the mass of one side by 2 (Wright and Steadman 2012, Wright et al. 2014). For an experienced preparator, this extraction process only takes about 30 to 60 seconds, and if tissues or skeletons are being saved, these muscles would likely be removed from the carcass anyway. The only additional time costs are in recording the data.

If flight muscle size was not recorded during specimen preparation, it can be estimated from skeletal specimens. The flight muscles attach to the sternal keel, and keel size is closely related to flight muscle size. A single measurement, the diagonal length of the keel, which encompasses both length and depth, is an excellent predictor of flight muscle size, correlating strongly with flight muscle mass both within and among species (Wright and Steadman 2012, Wright et al. 2016). This correlation between flight muscle size and keel length holds for a wide variety of taxa, including songbirds (e.g., *Rhipidura* fantails: $P < 0.001$, $r^2 = 0.90$), pigeons (e.g., *Macropygia mackinlayi*: $P = 0.0029$, $r^2 = 0.61$; *Ptilinopus* fruit doves: $P < 0.001$, $r^2 = 0.85$), and hummingbirds (e.g., *Amazilia tobaci*: $P < 0.001$, $r^2 = 0.80$). Partial skeletons that include the sternum are quick and easy to preserve while preparing skin and/or spread wing specimens, and are becoming more common in collections.

Using skeletal measurements as indices of flight muscle size may be preferable over actual masses of pectoral flight muscles, depending on the research question. Flight muscle size often fluctuates within an individual due to season (Swanson and Merkord 2012), migration stage (Battley et al. 2000), breeding (Veasey et al. 2000), and molt (Gaunt et al. 1990). These fluctuations are not usually reflected in keel size. Thus, if the goal is to study interspecific or even population-level differences in flight muscle size (e.g., Wright and Steadman 2012, Wright et al. 2014, Wright et al. 2016), skeletal measurements may be better suited to address the question of interest than masses of flight muscle sizes themselves, as seasonal fluctuations in flight muscle size would simply add unnecessary noise to the dataset. On the other hand, sternal keel measurements would obviously not be an appropriate choice for a study on seasonal changes in flight muscle size.

Flight muscle size may also be estimated using ultrasound on live birds (Swanson and Merkord 2012). A strength of this method is that it allows for repeated measures on living birds, but may be less useful for museum collectors and preparators. Ultrasonography or computed tomography methods could be particularly powerful, however, for allowing estimates of flight muscle size from alcohol-preserved specimens without the need for damaging dissections.

THE USE OF MUSEUM SPECIMENS TO ASSESS FLIGHT PERFORMANCE AND ITS ECOLOGICAL AND EVOLUTIONARY CONSEQUENCES

Diversification Rates

Using research specimens to quantify the relative flight capabilities of different species has great potential for revealing the effects of dispersal on ecological and evolutionary processes. For example, Claramunt et al. (2012) used the hand-wing index to study the relationship between dispersal ability and diversification rates in the family Furnariidae. Few studies have addressed this relationship (e.g., Mayr and Diamond 2001, Phillimore et al. 2006), and most were limited by the use of crude categories of dispersal propensity. The use of the hand-wing index allows for exploration of a larger universe of potential models beyond simple positive or negative associations. Claramunt et al. (2012) found that the relationship between dispersal ability and diversification rates in Furnariidae was negative, with the most dispersive lineages having the lowest diversification rates. This finding is consistent with the idea that speciation is initiated by vicariance events, with the rise of geographic barriers severing the distributions of lineages with low dispersal ability more often than those of lineages with high dispersal ability. They speculated that this result might apply mostly to continental clades and that in geographical settings characterized by isolation such as oceanic archipelagoes a positive or at least a hump-shaped relationship could be expected. To test this idea, Weeks and Claramunt (2014) analyzed the relationship between dispersal ability and diversification rates among bird clades from Australasian archipelagoes. Contrary to the prediction, they found that monotonically negative models described the pattern better than either positive or unimodal models. This finding runs against the idea that dispersal has a positive role in diversification on oceanic archipelagoes. Indeed, high dispersal ability may have the "positive"

role of allowing some lineages to reach oceanic islands, but at the expense of reducing their diversification rates (Weeks and Claramunt 2014). Alternatively, dispersal ability may evolve rapidly, with lineages experiencing phases of geographic expansion, during which they colonize oceanic islands, and phases of isolation, during which speciation and diversification occur (Diamond et al. 1976, Ricklefs and Birmingham 2002). The use of morphological proxies with high phylogenetic inertia such as the hand-wing index will allow the study of how dispersal ability changes through evolutionary time using phylogenetic comparative methods (Weeks and Claramunt 2014). Much more research is needed in this area, and improvements in the use of specimen-based proxies have the potential to lead to great advances. For example, the use of the hand-wing index as a proxy for dispersal ability is providing new insights into the macroecology of community assembly, co-occurrence, and range expansion (Pigot and Tobias 2015, Pigot et al. 2016, White 2016).

Genome Size Evolution

Estimates of flight performance from museum specimens have shed light on patterns of genome size evolution. For example, the observation that all three lineages of volant vertebrates—bats, birds, and pterosaurs—have smaller genomes than their closest nonflying relatives, has led to the hypothesis that the metabolic demands of flight select for small cells with reduced DNA content (Hughes and Hughes 1995, Organ and Shedlock 2009, Zhang and Edwards 2012). This metabolic rate hypothesis holds that the size of the genome constrains the minimum size of cells, and smaller cells' greater surface areas allow for higher metabolic rates (Gregory 2001). Powered flight requires sustained high metabolic output (Ward et al. 2002), resulting in selection for smaller genomes to reduce cell size (Gregory 2001). Alternative hypotheses include the ideas that reduced genomes allow for body mass reductions (Gregory 2002) or that reduced genome sizes enhance the efficiency of neural function, an adaptation for maneuverability in flight (Gregory 2002, Andrews and Gregory 2009).

Wright et al. (2014) used museum specimens to test the hypothesized link between flight and small genomes in birds, and to differentiate among hypothesized mechanisms. To do so, they needed multiple estimates of flight ability, some directly tied to the metabolic demands of flight and others independent of metabolism. They examined flight muscle size, heart size, wing shape, wing size, and body mass as metrics of flight ability. During specimen collection, personnel at the Museum of Southwestern Biology (MSB; Albuquerque, NM) regularly save blood smear slides from which genome size can be estimated (Hardie et al. 2002, Gregory et al. 2009), and also record the fresh masses of pectoral flight muscles and hearts (see earlier section "Estimating Flight Performance from Extended Specimens"). The majority of MSB specimens are prepared as study skins with closed wings and partial skeletons, without spread wings. Thus wing shape and size were estimated using hand-wing and wing area indices (see earlier section "Estimating Flight Performance from Study Skins"). Across a wide swath of the avian tree, genome size was negatively correlated with estimates of flight performance. The best predictors of genome size were the relative sizes of the flight muscles and heart, traits linked to the metabolic demands of flight (Wright et al. 2014). These results support the hypothesis that genome size reduction in volant vertebrates is due to the high metabolic demands of flight selecting for more energetically efficient cells and, thus, smaller genomes.

FUTURE DIRECTIONS

Museum collections regularly become useful to scientists in novel ways as new technologies and disciplines develop new ways of using old specimens, extending the specimen. We have presented examples of how current topics in ornithology benefit from estimates of flight performance and dispersal from museum specimens. Future applications of these measurements are likely to be more varied and interesting than we can currently predict. Preserving as much data as possible from specimens, including spread wings, skeletons, and muscles/hearts, means making possible an ever-widening variety of future comparative research.

ACKNOWLEDGMENTS

We thank all those who contribute to museum collections via specimen collection, preparation, and management; without their hard work our research would not be possible. We thank D. Dittman, A. Cuervo, R. Gibbons, J. Maley, and L. Naka, who contributed with wingspan measurements, extended wing preparations, and wing outlines. We thank A. Bravo for help digitizing and measuring extended wings at LSUMZ. We thank C. Witt for his insightful advice and discussions, and A. Johnson, A. Kratter, and D. Steadman for access to collections and assistance collecting flight muscle and heart masses. S. Claramunt's work was supported by the LSU Department of Biological Sciences, the LSU Museum of Natural Science, and the AMNH Frank M. Chapman Memorial Fund. N. A. Wright is supported by a Drollinger-Dial Postdoctoral Fellowship.

LITERATURE CITED

Altshuler, D. L. 2006. Flight performance and competitive displacement of hummingbirds across elevational gradients. American Naturalist 167:216–229.

Altshuler, D. L., R. Dudley, J. A. McGuire, and D. B. Wake. 2004a. Resolution of a paradox: hummingbird flight at high elevation does not come without a cost. Proceedings of the National Academy of Sciences USA 101:17731–17736.

Altshuler, D. L., F. G. Stiles, and R. Dudley. 2004b. Of hummingbirds and helicopters: hovering costs, competitive ability, and foraging strategies. American Naturalist 163:16–25.

Andrews, C. B., and T. R. Gregory. 2009. Genome size is inversely correlated with relative brain size in parrots and cockatoos. Genome 52:261–267.

Baldwin, S. P., H. C. Oberholser, and L. G. Worley. 1931. Measurements of birds. Scientific Publications of the Cleveland Museum of Natural History 2:1–165.

Battley, P. F., T. Piersma, M. W. Dietz, S. Tang, A. Dekinga, and K. Hulsman. 2000. Empirical evidence for differential organ reductions during trans-oceanic bird flight. Proceedings of the Royal Society B 67:191–195.

Bishop, C. M. 1997. Heart mass and the maximum cardiac output of birds and mammals: implications for estimating the maximum aerobic power input of flying animals. Philosophical Transactions of the Royal Society B 352:447–456.

Bishop, C. M. 1999. The maximum oxygen consumption and aerobic scope of birds and mammals: getting to the heart of the matter. Proceedings of the Royal Society B 266:2275–2281.

Bishop, C. M., and P. J. Butler. 1995. Physiological modeling of oxygen consumption in birds during flight. Journal of Experimental Biology 198:2153–2163.

Blake, E. R. 1949. Preserving birds for study. Fieldiana Technique 7:1–38.

Chapin, J. P. 1929. The preparation of birds for study. American Museum of Natural History, New York, NY.

Chappell, M. A., C. Bech, and W. A. Buttemer. 1999. The relationship of central and peripheral organ masses to aerobic performance variation in house sparrows. Journal of Experimental Biology 202:2269–2279.

Claramunt, S., E. P. Derryberry, J. V. Remsen, Jr., and R. T. Brumfield. 2012. High dispersal ability inhibits speciation in a continental radiation of passerine birds. Proceedings of the Royal Society B 279:1567–1574.

Dial, K. P. 2003. Evolution of avian locomotion: correlates of flight style, locomotor modules, nesting biology, body size, development, and the origin of flapping flight. Auk 120:941–952.

Diamond J, M. E. Gilpin, E. Mayr. 1976. Species-distance relation for birds of the Solomon Archipelago, and the paradox of the great speciators. Proceedings of the National Academy of Sciences USA 73: 2160–2164.

Gaunt, A. S., R. S. Hikida, J. R. Jehl, and L. Fenbert. 1990. Rapid atrophy and hypertrophy of an avian flight muscle. Auk 107:649–659.

Gregory, T. R. 2001. The bigger the C-value, the larger the cell: genome size and red blood cell size in vertebrates. Blood Cells, Molecules and Diseases 27:830–843.

Gregory, T. R. 2002. A bird's-eye view of the c-value enigma: genome size, cell size, and metabolic rate in the class Aves. Evolution 56:121–130.

Gregory, T. R., C. B. Andrews, J. A. McGuire, and C. C. Witt. 2009. The smallest avian genomes are found in hummingbirds. Proceedings of the Royal Society B 276:3753–3757.

Hall, K. S. S. 2005. Do nine-primaried passerines have nine or ten primary feathers? The evolution of a concept. Journal of Ornithology 146:121–126.

Hardie, D. C., T. R. Gregory, and P. D. N. Hebert. 2002. From pixels to picograms: a beginners' guide to genome quantification by Feulgen image analysis densitometry. Journal of Histochemistry and Cytochemistry 50:735–749.

Hartman, F. A. 1961. Locomotor mechanisms of birds. Smithsonian Miscellaneous Collections, vol. 143. Smithsonian Insitution, Washington, DC.

Hedenström, A. 2002. Aerodynamics, evolution and ecology of avian flight. Trends in Ecology and Evolution 17:415–422.

Hedenström, A., and F. Liechti. 2001. Field estimates of body drag coefficient on the basis of dives in passerine birds. Journal of Experimental Biology 204:1167–1175.

Hedenström, A., and M. Rosén. 2003. Body frontal area in passerine birds. Journal of Avian Biology 34:159–162.

Heers, A. M., and K. P. Dial. 2015. Wings versus legs in the avian bauplan: development and evolution of alternative locomotor strategies. Evolution 69:305–320.

Hohtola, E. 1982. Thermal and electromyographic correlates of shivering thermogenesis in the pigeon. Comparative Biochemistry and Physiology. Part A, Molecular and Integrative Physiology 73:159–166.

Hohtola, E., and E. D. Stevens. 1986. The relationship of muscle electrical activity, tremor and heat production to shivering thermogenesis in the Japanese quail. Journal of Experimental Biology 125:119–135.

Hughes, A. L., and M. K. Hughes. 1995. Small genomes for better flyers. Nature 377:391.

Kipp, F. A. 1959. Der Handflügel-Index als flugbiologisches MaB. Die Vogelwarte 20:77–86.

Laurance, S. G., P. C. Stouffer, and W. F. Laurance. 2004. Effects of road clearings on movement patterns of understory rainforest birds in central Amazonia. Conservation Biology 18:1099–1109.

Lockwood, R., J. P. Swaddle, and J. M. V. Rayner. 1998. Avian wingtip shape reconsidered: wingtip shape indices and morphological adaptations to migration. Journal of Avian Biology 29:273–292.

Mayr, E., and J. M. Diamond. 2001. The birds of Northern Melanesia. Oxford University Press, New York, NY.

Moore, R. P., W. D. Robinson, I. J. Lovette, and T. R. Robinson. 2008. Experimental evidence for extreme dispersal limitation in tropical forest birds. Ecology Letters 11:960–968.

Norberg, U. M. 1990. Vertebrate flight. Springer, Berlin, Germany.

Nudds, R. L. 2007. Wing-bone length allometry in birds. Journal of Avian Biology 38:515–519.

Nudds, R. L., and J. M. V. Rayner. 2006. Scaling of body frontal area and body width in birds. Journal of Morphology 267:341–346.

Nudds, R. L., G. W. Kaiser, and G. J. Dyke. 2011. Scaling of avian primary feather length. PLoS One 6:e15665.

Organ, C. L., and A. M. Shedlock. 2009. Palaeogenomics of pterosaurs and the evolution of small genome size in flying vertebrates. Biology Letters 5:47–50.

Paradis, E., S. R. Baillie, W. J. Sutherland, and R. D. Gregory. 1998. Patterns of natal and breeding dispersal in birds. Journal of Animal Ecology 67:518–536.

Pennycuick, C. J. 1975. Mechanics of flight. Avian Biology 5:1–75.

Pennycuick, C. J. 1989. Bird flight performance: a practical calculation manual. Oxford University Press, Oxford, UK.

Pennycuick, C. J. 1997. Actual and "optimum" flight speeds: field data reassessed. Journal of Experimental Biology 200:2355–2361.

Pennycuick, C. J. 2008. Modeling the flying bird. Academic Press, Cambridge, MA.

Pennycuick, C. J., S. Åkesson, and A. Hedenström. 2013. Air speeds of migrating birds observed by ornithodolite and compared with predictions from flight theory. Journal of the Royal Society Interface 10:20130419.

Pennycuick, C. J., M. Klaassen, A. Kvist, and Å. Lindström. 1996. Wingbeat frequency and the body drag anomaly: wind-tunnel observations on a thrush nightingale (Luscinia luscinia) and a teal (Anas crecca). Journal of Experimental Biology 199:2757–2765.

Phillimore, A. B., R. P. Freckleton, C. D. Orme, and I. P. F. Owens. 2006. Ecology predicts large-scale patterns of phylogenetic diversification in birds. American Naturalist 168:220–229.

Pigot, A. L., and J. A. Tobias. 2015. Dispersal and the transition to sympatry in vertebrates. Proceedings of the Royal Society B 282:20141929.

Pigot, A. L., J. A. Tobias, and W. Jetz. 2016. Energetic constraints on species coexistence in birds. PLoS Biology 14:e1002407.

R Core Team. [online]. 2016. R: A language and environment for statistical computing. R Foundation for Statistical Computing, Vienna, Austria. <https://www.R-project.org/> (20 March 2016).

Rayner, J. M. V. 1979. A new approach to animal flight mechanics. Journal of Experimental Biology 80:17–54.

Rayner, J. M. V. 1988. Form and function in avian flight. Current Ornithology 5:1–66.

Ricklefs, R. E., and E. Bermingham. 2002. The concept of the taxon cycle in biogeography. Global Ecology and Biogeography 11:353–361.

Stiles, F. G., and D. L. Altshuler. 2012. Conflicting terminology for wing measurements in ornithology and aerodynamics. Auk 121:973–976.

Swanson, D. L., and C. Merkord. 2012. Seasonal phenotypic flexibility of flight muscle size in small birds: a comparison of ultrasonography and tissue mass measurements. Journal of Ornithology 154:119–127.

Swanson, D. L., Y. Zhang, and M. O. King. 2013. Individual variation in thermogenic capacity is correlated with flight muscle size but not cellular metabolic capacity in American goldfinches (Spinus tristis). Physiological and Biochemical Zoology 86:421–431.

Tobalske, B. W., T. L. Hedrick, K. P. Dial, and A. A. Biewener. 2003. Comparative power curves in bird flight. Nature 421:363–366.

Veasey, J. S., D. C. Houston, and N. B. Metcalfe. 2000. Flight muscle atrophy and predation risk in breeding birds. Functional Ecology 14:115–121.

Videler, J. J. 2005. Avian flight. Oxford University Press, Oxford, UK.

Wang, X., A. J. McGowan, and G. J. Dyke. 2011. Avian wing proportions and flight styles: first step towards predicting the flight modes of Mesozoic birds. PLoS One 6:e28672.

Ward, S., C. M. Bishop, A. J. Woakes, and P. J. Butler. 2002. Heart rate and the rate of oxygen consumption of flying and walking barnacle geese (Branta leucopsis) and bar-headed geese (Anser indicus). Journal of Experimental Biology 205:3347–3356.

Weeks, B. C., and S. Claramunt. 2014. Dispersal has inhibited avian diversification in Australasian archipelagoes. Proceedings of the Royal Society B 281:20141257.

White, A. E. 2016. Geographical barriers and dispersal propensity interact to limit range expansions of Himalayan birds. American Naturalist 188: 99–112.

Wright, N. A., T. R. Gregory, and C. C. Witt. 2014. Metabolic "engines" of flight drive genome size reduction in birds. Proceedings of the Royal Society B 281:20132780.

Wright, N. A., and D. W. Steadman. 2012. Insular avian adaptations on two Neotropical continental islands. Journal of Biogeography 39:1891–1899.

Wright, N. A., D. W. Steadman, and C. C. Witt. 2016. Predictable evolution toward flightlessness in volant island birds. Proceedings of the National Academy of Sciences USA 113:4765–4770.

Zhang, Q., and S. V. Edwards. 2012. The evolution of intron size in amniotes: a role for powered flight? Genome Biology and Evolution 4:1033–1043.

CHAPTER NINE

Transforming Museum Specimens into Genomic Resources*

John E. McCormack, Flor Rodríguez-Gómez, Whitney L. E. Tsai, and Brant C. Faircloth

Abstract. Scientists have been studying ancient DNA for three decades, but next-generation sequencing (NGS) has made the process of sequencing ancient DNA much easier. This technological leap has huge potential for genomic analysis of the very museum specimens that already form the foundation of prior knowledge about biodiversity. We are still determining which methods of preparing ancient DNA for NGS work best for various questions, especially with regard to whether whole genomes need to be sequenced or whether a subset of the genome is more desirable and computationally tractable. In this chapter, we discuss different methods for preparing ancient DNA for NGS, and various analytical issues specific to NGS data output from ancient DNA sources. We focus mainly on birds and discuss several case studies where NGS has been applied to very old museum specimens, like subfossils, as well as younger specimens, like those from museum study skins collected in the last century. Although few recently published studies using NGS are about ancient DNA from bird museum specimens, the number is expected to grow rapidly. The case studies demonstrate that systematics and

taxonomy are important applications of NGS to museum specimens, and that plenty remains to be learned from specimens about population-level processes such as the genetics of changes in population size, some of which have led to extinction. Better methods of extracting ancient DNA from museum specimens are badly needed, as well as careful consideration of how the research community archives DNA extractions and the billions of DNA sequencing reads now being produced from museum specimens on NGS platforms. Methods for correcting errors that occur during NGS, as well as those introduced during the process of DNA degradation in the specimen itself (e.g., deamination), are in development, but easy-to-use pipelines are still lacking. In sum, although methods are still in development, the field of next-generation museum genomics is burgeoning, with high potential to extend the utility of museum specimens in bird systematics, historical demography, and conservation.

Key Words: birds, evolution, genomes, natural history collections, next-generation sequencing, phylogenetics.

* McCormack, J. E., F. Rodríguez-Gómez, W. L. E. Tsai, and B. C. Faircloth. 2017. Transforming museum specimens into genomic resources. Pp. 143–156 in M. S. Webster (editor), The Extended Specimen: Emerging Frontiers in Collections-based Ornithological Research. Studies in Avian Biology (no. 50), CRC Press, Boca Raton, FL.

BACKGROUND: FIRST FORAYS INTO THE GENOMES OF MUSEUM SPECIMENS

The first museum specimen to reveal anything about its genomic content to the world was a salted pelt of a quagga (*Equus quagga quagga*), a handsome extinct subspecies of the plains zebra stored at the Natural History Museum in Mainz, Germany (Higuchi et al. 1984). Lacking a method to increase trace amounts of DNA to high concentration, the researchers of 1984 required large amounts of starting material, over 10 ng of DNA for a single reaction in the case of the quagga study, to clone into a bacterial vector and sequence with a relatively new method at the time, dideoxy sequencing, also called Sanger sequencing (Sanger et al. 1977). Recovered DNA sequences allowed the researchers to place an extinct animal in the molecular tree of life for the first time.

Progress in ancient DNA research has always been tightly linked to technological advancements. Not long after the quagga success, the polymerase chain reaction (PCR) was invented (Saiki et al. 1985), which allowed small amounts of DNA from ancient samples to be amplified hundreds of thousands of times to the levels needed for Sanger sequencing (Pääbo 1989). The subsequent discovery of highly conserved regions in mitochondrial DNA (mtDNA) provided convenient priming sites for PCR amplification of variable DNA across major branches of the tree of life (Kocher et al. 1989). With this plethora of new genomic sites to target, ancient DNA studies began to proliferate.

The first published use of ancient DNA from a bird museum specimen was an attempt to describe a new shrike by comparing its DNA to the DNA of other bird species, some of which was derived from study skins (Smith et al. 1991). A deeper time study of ancient avian DNA, from birds more than 3,000 years old, occurred when researchers extracted DNA from skin, bone, and muscle tissue of four specimens of an extinct order of birds, the moas of New Zealand (Cooper et al. 1992). The 390 DNA bases that Cooper and colleagues painstakingly stitched together from many smaller pieces, each targeted by individual primer pairs, suggested that moas were not the closest relative of another endemic New Zealand bird order, the kiwis. New Zealand, according to these data, had been colonized twice by a largely flightless group of birds known as the paleognaths, which also includes birds like emus and ostriches. This study marked a major achievement for the use of DNA sequencing with ornithological museum specimens.

Every new technological advance comes with its problems, and this was also true for PCR and DNA sequencing in their application to museum samples, particularly at their inception. Amplifying tiny amounts of genetic material made DNA sequencing more accessible, but it also greatly increased the risk of amplifying contaminant DNA, either introduced to the specimen through years of handling or by exposure to laboratories that were now, thanks to the advent of PCR, awash in ultrahigh concentration DNA. Several early successes in ancient DNA were found to be the result of contaminating DNA (e.g., DeSalle et al. 1992). Researchers soon called for ultraclean laboratory conditions (Cooper et al. 2001, Ho and Gilbert 2010) and exacting standards for replication of results (Handt et al. 1994, Cooper and Poinar 2000).

Sequencing DNA from museum specimens, even when it was the right DNA, was not an easy process. DNA degrades and fragments over time through a variety of biochemical processes (Willerslev and Cooper 2005). A recent study using extracted DNA from a time series of museum specimens (McCormack et al. 2016) demonstrates this fragmentation process through time (Figure 9.1). Recent specimens have DNA quality similar to fresh tissue. Specimens up to 20 years old might still contain high molecular-weight DNA. But specimens 30 years old and older are increasingly fragmented, with most fragments eventually being less than 500 base pairs (i.e., low quality). Thus, a researcher starting from high-quality DNA—extracted, say, from frozen tissue—can use universal primers to target a long span of variable DNA (0.5 to 10 kilobases), but this is rarely possible with ancient DNA. Instead, the degraded fragments of ancient DNA require many sets of primers, often designed from scratch, that span variable sections of DNA; design of such primers is a time-consuming and challenging task. In addition to being difficult to develop, the resulting primers, because they are placed in regions of variable DNA specific to the organism under study, often lose their universality, which is one of their principal benefits.

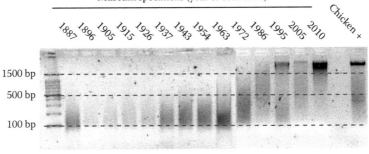

Figure 9.1. DNA extracted from a time series of museum specimens, as compared to DNA extracted from fresh, frozen tissue from a chicken.

THE GAME-CHANGER: NEXT-GENERATION SEQUENCING

Most of the truly remarkable feats in ancient DNA sequencing in recent years were made possible by what is colloquially referred to as "next-generation sequencing," or NGS. Developed during the late 1990s, NGS was a radical departure from previous DNA sequencing methods (Shendure and Ji 2008). Although NGS platforms differ in their exact approaches and chemistries (Glenn 2011), all NGS methods clonally produce millions of DNA sequencing reads from a single run in "massively parallel" fashion compared to the achingly serial nature of Sanger sequencing.

NGS appeared just as ancient DNA studies were becoming more numerous. If an aura of exuberance surrounds the potential of NGS to transform our view of biocollections (Nachman 2013, Burrell et al. 2015, Wood and De Pietri 2015, Linderholm 2016), it is because NGS—at least in concept—appears to solve many of the problems that plagued prior ancient DNA studies (Knapp and Hofreiter 2010). First, the dramatically increased throughput of NGS, combined with the resulting decrease in the cost of sequencing each DNA base, turned the problem of contaminating DNA into less of an existential concern. The issue became less whether *any* target DNA would be sequenced at all, and more about how to separate target DNA from the inevitable contaminating DNA. Second, NGS platforms typically output short DNA reads, which seemed well suited to the degraded DNA input of ancient DNA studies. It is thus not surprising that many of the first applications of NGS involved museum specimens (Noonan et al. 2005, Poinar et al. 2006, Mason et al. 2011), with the first published study on

birds focusing on the development of microsatellite loci from "shotgun" NGS reads of the extinct moa genome (Allentoft et al. 2009).

CHOOSING A NEXT-GENERATION SEQUENCING TECHNIQUE

The term "next-generation sequencing" belies a uniformity of method that does not exist. The applications of NGS to studies of museum specimens are as varied as the specific protocols used to prepare the samples for sequencing and the specific platforms used for sequencing (e.g., see Buerki and Baker 2015 and references within). One of the early decisions a researcher must make is whether to target the whole genome of the study organism (Figure 9.2a) or to focus on a subsample of the genome (Figure 9.2b). This decision rests in large part on the specific research questions. Guidelines can be found in other reviews (Lerner and Fleischer 2010, Ekblom and Galindo 2011, McCormack et al. 2013b, Toews et al. 2015). For whole genome sequencing, the method is relatively straightforward. One of the first steps of standard genome sequencing is to shear genomic DNA into smaller fragments because most NGS platforms sequence short reads between 100 and 250 base pairs. Ancient DNA is already fragmented, so this step is often unnecessary although more recent samples often need to be sheared (McCormack et al. 2016). Degraded ancient DNA can often be input directly onto the sequencer after some preparation steps.

If whole genomes are not desired, then the researcher faces another series of choices (Figure 9.2c–f). Here, the specific attributes of ancient DNA may play a larger role in method selection than

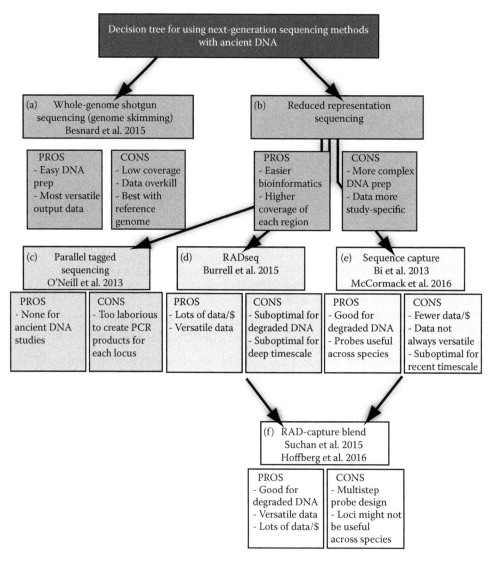

Figure 9.2. Decision tree for using NGS with ancient DNA from museum specimens. Methods denoted by letters are referenced in text.

the research question itself because certain ways of subsampling the genome might not be as effective, or even possible, when using degraded DNA. For example, parallel tagged sequencing (Figure 9.2c), a method of pooling PCR products for NGS (Meyer et al. 2008), suffers from many of the same inefficiencies of pre-NGS methods. Although the sequencing of NGS products occurs in parallel, which makes it suitable for medium-sized projects from cryopreserved tissue (O'Neill et al. 2013), the process is still time intensive when applied to ancient DNA because it requires primer design, PCR optimization, and serial generation of PCR products.

Restriction-site associated DNA sequencing (RADseq; Figure 9.2d) is a popular method that operates by creating a genomic subsample using restriction enzymes to cut up DNA in a systematic fashion, resulting in DNA fragments of similar sizes that are sequenced en masse on an NGS platform (Baird et al. 2008). When starting from a high-quality DNA source, RADseq produces many thousands of single nucleotide polymorphisms (SNPs). A major benefit of RADseq is that a reference genome is not required to identify variant sites, making it especially useful when applied to the many species found in biocollections that lack existing genomic resources.

The uses of RADseq data for studies of birds range from conservation genetics (Oyler-McCance et al. 2015) and phylogeography (Harvey and Brumfield 2015) to migratory connectivity (Ruegg et al. 2014b) and speciation genomics (Ruegg et al. 2014a). However, when applied to older or ancient samples present in biocollections, RADseq can be problematic. For instance, as shown in Figure 9.3b, DNA degradation can produce very small DNA fragments, most of which lack the needed restriction sites (Burrell et al. 2015), especially when using double-digest RADseq (ddRADseq; Peterson et al. 2012) that requires two different restriction sites on the same fragment. Also, the lack of restriction sites on some fragments can produce null alleles, which

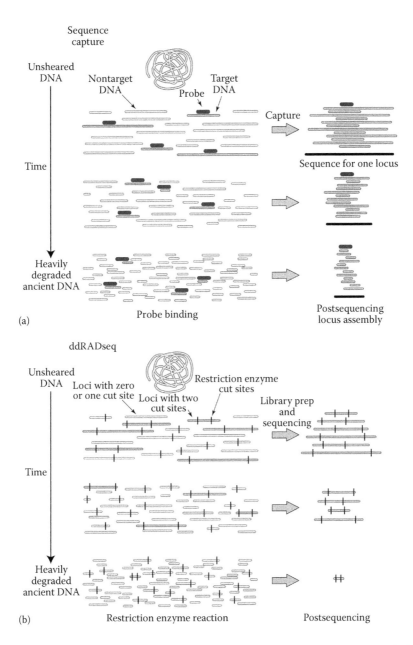

Figure 9.3. A comparison of how degraded DNA affects both (a) sequence capture and (b) ddRADseq methods. Sequence capture on ancient DNA leads to shorter assembled loci after sequencing. Ancient DNA in conjunction with ddRADseq leads to fewer loci sequenced because fewer fragments contain both restriction-digest cut sites.

may mislead downstream analyses (Graham et al. 2015). RADseq also generally requires a larger amount of high-quality starting DNA than other methods (especially ddRADseq; Puritz et al. 2014), which is not always accessible from ancient DNA.

Target enrichment—also called sequence capture, hybrid enrichment, or bait capture (Figure 9.2e)—offers another efficient alternative to whole genome sequencing that is well suited, at least in concept, to short, degraded DNA regions (Jones and Good 2016). Here, DNA or RNA probes that target a reduced subset of the genome are added to sequencing libraries prepared from extracted DNA (Gnirke et al. 2009). The probes are hybridized to their targets, and magnetic beads separate target and nontarget DNA (Mamanova et al. 2010). Following a round of PCR amplification, the targeted DNA is sequenced all at once using an NGS platform. One benefit of target enrichment is that it does not rely on systematic fragmentation of DNA inputs. Thus, in theory, the vagaries of how and where DNA degrades are less of a concern. After sequencing the captured products, the resulting DNA data for a given locus are assembled from the variously sized DNA sequences that overlap a given probe (Figure 9.3a). Although some preexisting genomic information is required to design the probe regions, once designed, probe sets are often broadly applicable across the tree of life, especially when they are designed from conserved genomic regions like exons (Bi et al. 2012, Ilves and López Fernández 2014, Prum et al. 2015) or ultraconserved elements (UCEs; Faircloth et al. 2012, Faircloth et al. 2013). Probes can even be created from PCR products (Peñalba et al. 2014).

A concern of using highly conserved genomic regions with ancient DNA is that assembled ancient DNA loci tend to be shorter than loci assembled from fresh tissue samples (Figure 9.3a; McCormack et al. 2016). If sequence capture probes use a highly conserved central core, then the worry of short loci is that insufficient variable DNA sites will be captured from the flanking regions. Another concern is that sequence capture usually targets a particular locus type (e.g., UCEs or exons), which can limit the versatility of the resulting data for addressing many types of questions. For example, DNA flanking UCEs is thought to be largely noncoding compared to the core UCE region, which is thought to be under strong stabilizing selection (Katzman et al. 2007). UCE flanking DNA, being largely noncoding, is

therefore likely not useful for making associations between genotype and phenotype. It is, however, quite useful for questions of phylogenetics and demographic history that prefer "neutrally evolving" DNA (Crawford et al. 2012, McCormack et al. 2013a, Smith et al. 2013). Meanwhile, RADseq data typically include both coding and noncoding sites, and are therefore more versatile in their application to different research questions, especially given the recent advent of phylogenetic methods that use SNP data (e.g., Bryant et al. 2012). One drawback of RADseq data, however, is that they appear to be better suited to questions at more recent timescales because increasing genetic divergence over time mutates the cut sites, leading to fewer and fewer homologous fragments among more distantly related species (Rubin et al. 2012).

New methods are now being developed that blend the best attributes of RADseq and sequence capture (Ali et al. 2016, Hoffberg et al. 2016, Suchan et al. 2016). In these approaches, sequence-capture probes are designed from RADseq markers originally detected from data generated from fresh tissue samples (Figure 9.2f). This minimizes the concern over where the RADseq digest cut sites will occur in degraded museum specimen DNA, while also allowing for the collection of a large number of genomic loci that are maximally versatile for addressing research questions involving both neutral and nonneutral processes. A remaining question is whether these approaches will be useful across species or whether new probes will need to be repeatedly designed as divergence increases among the targeted species.

PROBLEMS WITH SEQUENCE DATA SPECIFIC TO ANCIENT DNA

Thanks to NGS, DNA data production now outstrips, by a wide margin, our ability to render judgment on the quality and value of those data. To restate a prior point: the problem now is not whether we are going to sequence any of the target organism's genome, but how we are going to assess the quality of the resulting sequence data. For instance, how will we separate the DNA of museum specimens from contaminant DNA? And once we have done that, what are the specific problems with ancient DNA that we must account for?

After an organism dies, there are relatively rapid biochemical processes that break down tissues as well as slower-acting biochemical processes that degrade and damage DNA. Because storage conditions for museum specimens are not optimized for molecular stability, these degradation processes affect the DNA in museum specimens, which is one major factor making molecular work with older museum specimens difficult (Wandeler et al. 2007). The processes of DNA degradation are incompletely understood, but include both enzymatic and biochemical effects that alter DNA bases (mainly conversions from cytosine [C] to thymine [T]) through deamination (Dabney et al. 2013) and inhibit the effectiveness of DNA polymerase when synthesizing new copies of ancient DNA using PCR. Although each of these effects is troublesome, fragmentation of ancient DNA is perhaps the most problematic. Beyond a certain point, DNA sequences become too short to capture and sequence with current technology. Empirical studies showing an excess of purines (adenine and guanine bases) in the genomic positions directly adjacent to the ends of ancient DNA fragments suggest that depurination is the major cause of fragmentation (Briggs et al. 2007, Orlando et al. 2011).

The multifarious processes that are responsible for degrading DNA in museum specimens also differ in how they operate through time. Deamination events, primarily C to T transitions, appear to increase at a steady rate with specimen age (Sawyer et al. 2012), a pattern that is evident in DNA from mammal skins (Bi et al. 2013) and in plant herbarium tissue (Staats et al. 2013) collected during the last 100 years, as well as in DNA extracted from more ancient samples (Hofreiter et al. 2001, Briggs et al. 2009). As far as fragmentation is concerned, a study on specimens ranging in age from 18 to 60,000 years old did not find that DNA became more fragmented with age (Sawyer et al. 2012). However, this contrasts with a recent study that focused exclusively on DNA extracted from avian toe pads collected during the previous 120 years (McCormack et al. 2016). This study found that the length of assembled DNA loci was shorter when using DNA from older specimens, even when older specimens had large numbers of NGS reads associated with them. In this latter study, the rate of DNA fragmentation was faster during the first 30 years of specimen storage, after which time the DNA was heavily fragmented (also see Figure 9.1). It is possible that Sawyer et al. (2012) did not observe this trend because

their study examined a much longer window of time that included few specimens younger than 30 years. Supporting the relationship between fragmentation and age, another recent study looking at herbarium specimens also found shorter fragments in older specimens preserved over the last 300 years (Weiß et al. 2015).

While the processes affecting DNA degradation and timing of ancient DNA damage are still being investigated, it is clear that these issues are a concern for those working on samples thousands of years old. But it is important to realize that these same processes also affect DNA of historical specimens collected during the last 100 years. A simple way to identify deamination in samples is to plot the occurrence of all possible DNA base changes (e.g., A to G, C to T, etc.) against distance from the end of sequenced DNA fragments (Briggs et al. 2007, Bi et al. 2013). Because deamination is known to occur in greater frequency toward the ends of fragments, a signature of deamination is an elevated signal of C to T transitions close to the 5′ end of the sequenced fragment. In fact, this pattern is so ubiquitous that it has been proposed as a way of validating that DNA is truly derived from an ancient source and not the result of more modern contamination (Dabney et al. 2013). Other, more sophisticated analytical methods for estimating deamination, like mapDamage2.0 (Jónsson et al. 2013), and contamination, like PMDtools (Skoglund et al. 2014), build on the models originally described in Briggs et al. (2007). At the moment, these techniques are better suited for human data where high-quality reference sequences are available, but they may become more suitable for nonhuman genetic studies, including those of birds, as additional avian reference sequences are developed (Zhang et al. 2014). In addition, kits for correcting deamination prior to sequencing are commercially available.

CASE STUDIES IN ORNITHOLOGY

Still relatively few studies apply NGS to bird museum specimens. Early studies, however, provide tantalizing hints of the potential for NGS to extend the mission of ornithological biocollections, especially collections that contain older or extremely rare specimens from which modern sources of cryopreserved DNA are lacking. Many smaller, regional collections at state and county natural history museums contain these kinds of rare specimens. Genomic applications for museum specimens might therefore have their biggest impact on these smaller collections, which hold important specimens, but often struggle to obtain needed space and funding from administrators and funding bodies (Snow 2005).

The study of extinct or highly endangered species is an obvious and important application of NGS to museum specimens because no high-quality DNA will likely ever be available for many of these species. In fact, most published studies to date using NGS with museum specimens focus on extinct or endangered species. The Passenger Pigeon (*Ectopistes migratorius*) is a recent favorite subject for genomic study as a result of the 100-year anniversary of its demise (Greenberg 2014) and ongoing efforts to revive the species through de-extinction approaches (Seddon et al. 2014). Recent NGS studies of Passenger Pigeons sequenced two complete mtDNA genomes (Hung et al. 2013) and roughly half the nuclear genome of four specimens from different parts of their geographic range (Hung et al. 2014). By combining DNA extracted from museum specimens, broad sampling of the genome, and new analytical methods for assessing effective population size from limited population sampling, the nuclear genomic study revealed novel conclusions about Passenger Pigeons that could not have been reached in any other way. In particular, these analyses suggest that Passenger Pigeons frequently went through dramatic population fluctuations, which left them vulnerable to extinction, and were in a decline phase that was exacerbated by human exploitation (Hung et al. 2014).

Another use of whole genome shotgun sequencing using NGS, where many random parts of the genome are targeted, is a recent study of crowned pigeons in the genus *Goura*. The three living members of this genus are threatened, which makes the acquisition of fresh tissue impossible (Besnard et al. 2015). Here, the authors used "genome skimming" on DNA extracted from museum specimens. Genome skimming involves filtering through millions of shotgun NGS sequence reads to find the few reads associated with particular DNA regions of interest. While perhaps not the most efficient use of sequencing effort, this method allowed the authors to assemble a small phylogenetic dataset consisting of both mtDNA and nuclear genes that placed crowned pigeons with high confidence into the existing pigeon

tree of life, closely related to several extinct or highly endangered island species like the Dodo (*Raphus cucullatus*). The Dodo itself could not have been included in this genetic analysis if DNA had not been sequenced previously from the cortical bone of a museum specimen (Shapiro et al. 2002). Perhaps more than anything else, this study demonstrates the kind of opportunistic genome skimming studies that will arise in greater frequency as more and more NGS data are produced for species across the bird tree of life, provided such data are archived (see later).

Other studies used target enrichment to subsample genomic DNA from museum specimens. One such study used an mtDNA probe set to assemble whole mtDNA genomes from subfossils of elephant birds (family Aepyornithidae), showing that they are most closely related to the New Zealand Kiwi. This surprising result implicates dispersal, not vicariance, as a major diversifying force in ratite birds (Mitchell et al. 2014). Another study enriched thousands of UCE loci from a time series of California Scrub-Jay (*Aphelocoma californica*) and Woodhouse's Scrub-Jay (*Aphelocoma woodhouseii*) study skins and identified variable loci and SNPs that allowed both phylogeographic and population genetic analyses (McCormack et al. 2016). This study demonstrated that sequence capture of conserved regions could produce phylogenetic data from degraded nuclear DNA in museum specimens. But along with this success, it also showed that age and starting DNA quality and quantity mattered for data matrix completeness and locus lengths, both important mileposts for producing high-quality phylogenetic datasets. Compared to whole genome shotgun sequencing, the targeted capture approach allowed for many individuals to be queried at thousands of loci, with high efficiency and little missing data.

FUTURE DIRECTIONS

Returning to Fundamentals: DNA Extraction

Underlying the previous discussion is the preeminent importance of starting DNA quality. It is ironic, given the pace of technological advancements in DNA sequencing, that researchers are more frequently returning to retool antiquated protocols for retrieving DNA from museum specimens. Unlike the case with DNA sequencing, we are still awaiting paradigm-shifting advances in

DNA extraction, and they are badly needed. A few studies have tested different ancient DNA extraction protocols (e.g., Rohland and Hofreiter 2007), without a clear consensus emerging on universal best practices. At least one study claims that the low yields of DNA from ancient material is more a product of inefficient extraction methods than low DNA content of the samples themselves (Barta et al. 2014), which serves to highlight the need for improved protocols.

New studies should not only test different protocols, but different sources of starting material from bird study skins. For example, toe pads are currently the most commonly used source material (Mundy et al. 1997), but other options include skin punches from featherless skin tracts or from the feathers themselves (Sefc et al. 2003, Rawlence et al. 2009). Ancient DNA studies on mammals suggest that finely ground bone produces the highest DNA yields (Pruvost et al. 2007, Hawkins et al. 2016). It would be interesting to test whether the same is true in birds, perhaps by arthroscopic retrieval of small bone pieces from inside bird study skins, which would have the added benefit of leaving no trace of the sampling on the outside of the specimen. Of course, differences in bone structure between mammals and birds might have important implications for the retrieval of DNA between these two groups. Until new DNA extraction methods are developed, we must work with what we have.

Archiving Data: One Researcher's Trash Is Another's Treasure

Subsampling the genome with NGS approaches like sequence capture and RADseq appears to be taking hold in phylogeography and phylogenetics as methods of reducing the overall complexity of datasets that still contain a large and representative genomic sample (McCormack et al. 2013b). However, this is not universally true, and many high-profile studies have featured whole-genome sequencing, with subsets of the genome later being extracted in silico and analyzed independently in a manner similar to genome skimming (e.g., Jarvis et al. 2014). The utopian situation time and money are maximized by sequencing whole genomes, with different teams later tackling different questions with different parts of the genome and different analyses, is becoming a reality for living organisms with the rapidly

advancing genomes initiatives for multiple animal groups (Haussler et al. 2009, i5K Consortium 2013). Making sure these initiatives sequence genomes linked to museum vouchers whenever possible is one important step toward increasing the scientific value of individual specimens by linking a genotype to a vouchered phenotype.

By the same token, independent researchers conducting *ad hoc* low-coverage, shotgun sequencing of museum specimens should make all their data open to the broader scientific community, using, for example, the National Center for Biotechnology Information's Short Read Archive (http://www .ncbi.nlm.nih.gov/sra). It is probably also worth archiving all sequencing reads generated from genomic subsampling approaches like sequence capture, because these methods produce millions of off-target reads from the nuclear and organellar genomes, including off-target mtDNA reads generated during sequence capture, that might be of interest to other researchers (e.g., off-target mtDNA reads generated during sequence capture; Meiklejohn et al. 2014, do Amaral et al. 2015).

Analytical Programs for Detecting DNA Damage and Contamination

A number of analytical issues that are specific to ancient DNA exist for which automated pipelines are currently lacking. For instance, analysis to detect sources of contamination should be carried out as a matter of course for ancient DNA studies, but few tools are available for automating this process. One could, for instance, imagine a program that keeps a database of all DNA sequences that have passed through a lab. When a new study is carried out, the resulting sequence data are screened against the database as the most likely source of contamination. Deamination is another issue particular to ancient DNA studies whose detection and correction would benefit from more study and automated pipelines to detect deamination events. Similar to other analytical issues associated with NGS, methods of analysis and data processing pipelines lag behind our ability to generate incredible amounts of data.

CONCLUSIONS

Although currently few studies use NGS on bird museum specimens, this number is expected to grow rapidly in the coming years, as protocols are developed and computational pipelines become more user friendly. This will undoubtedly mirror the growth in use of NGS on nondegraded samples, which multiplied rapidly in the last 5 years as researchers started to speculate about possible applications (Lerner and Fleischer 2010) and later tested various methods and described those that seemed most successful (Ekblom and Galindo 2011, McCormack et al. 2013b). Similarly, as NGS methods using degraded DNA from museum specimens become more established, we will undoubtedly see a shift from studies that focus almost exclusively on systematics and phylogenetics toward those that require deeper sampling from populations. This will open the door to large-scale study of genetic change through time (Holmes et al. 2016), both natural and human-mediated, already hinted at in the existing case study of the Passenger Pigeon (Hung et al. 2014). The field of paleornithology is poised to benefit greatly from NGS methods (Wood and De Pietri 2015), as ancient DNA has been successfully extracted from eggshells (Oskam et al. 2010), ancient feathers (Rawlence et al. 2009), and even coprolites (Wood et al. 2013) and sedimentary deposits (Willerslev et al. 2003). As those in the museum community know and have long advocated, one of the truly unique features of biocollections is that they offer a snapshot of biodiversity at a particular moment in time. Accessing the genomes of the organisms captured in each successive snapshot will add to the extended specimen and will be the work of future generations.

ACKNOWLEDGMENTS

We thank M. Webster and other organizers of the symposium, as well as N. Mason and D. Toews for helpful comments on the manuscript.

LITERATURE CITED

Ali, O. A., S. M. O'Rourke, S. J. Amish, M. H. Meek, G. Luikart, C. Jeffres, and M. R. Miller. 2016. RAD capture (Rapture): flexible and efficient sequence-based genotyping. Genetics 202:389–400.

Allentoft, M. E., S. Schuster, R. Holdaway, M. Hale, E. McLay, C. L. Oskam, M. T. P. Gilbert, P. Spencer, E. Willerslev, and M. Bunce. 2009. Identification of microsatellites from an extinct moa species using high-throughput (454) sequence data. Biotechniques 46:195–200.

Baird, N. A., P. D. Etter, T. S. Atwood, M. C. Currey, A. L. Shiver, Z. A. Lewis, E. U. Selker, W. A. Cresko, and E. A. Johnson. 2008. Rapid SNP discovery and genetic mapping using sequenced RAD markers. PLoS One 3:e3376.

Barta, J. L., C. Monroe, J. E. Teisberg, M. Winters, K. Flanigan, and B. M. Kemp. 2014. One of the key characteristics of ancient DNA, low copy number, may be a product of its extraction. Journal of Archaeological Science 46:281–289.

Besnard, G., J. A. M. Bertrand, B. Delahaie, Y. X. C. Bourgeois, E. Lhuillier, and C. Thébaud. 2015. Valuing museum specimens: high-throughput DNA sequencing on historical collections of New Guinea crowned pigeons (*Goura*). Biological Journal of the Linnean Society 117:71–82.

Bi, K., T. Linderoth, D. Vanderpool, J. M. Good, R. Nielsen, and C. Moritz. 2013. Unlocking the vault: next-generation museum population genomics. Molecular Ecology 22:6018–6032.

Bi, K., D. Vanderpool, S. Singhal, T. Linderoth, C. Moritz, and J. M. Good. 2012. Transcriptome-based exon capture enables highly cost-effective comparative genomic data collection at moderate evolutionary scales. BMC Genomics 13:403.

Briggs, A. W., U. Stenzel, P. L. F. Johnson, R. E. Green, J. Kelso, K. Prüfer, M. Meyer, J. Krause, M. T. Ronan, and M. Lachmann. 2007. Patterns of damage in genomic DNA sequences from a Neandertal. Proceedings of the National Academy of Sciences USA 104:14616–14621.

Briggs, A. W., U. Stenzel, M. Meyer, J. Krause, M. Kircher, and S. Pääbo. 2009. Removal of deaminated cytosines and detection of in vivo methylation in ancient DNA. Nucleic Acids Research 38:e87.

Bryant, D., R. Bouckaert, J. Felsenstein, N. A. Rosenberg, and A. RoyChoudhury. 2012. Inferring species trees directly from biallelic genetic markers: bypassing gene trees in a full coalescent analysis. Molecular Biology and Evolution 29:1917–1932.

Buerki, S., and W. J. Baker. 2015. Collections-based research in the genomic era. Biological Journal of the Linnean Society 117:5–10.

Burrell, A. S., T. R. Disotell, and C. M. Bergey. 2015. The use of museum specimens with high-throughput DNA sequencers. Journal of Human Evolution 79:35–44.

Cooper, A., C. Lalueza-Fox, S. Anderson, A. Rambaut, J. Austin, and R. Ward. 2001. Complete mitochondrial genome sequences of two extinct moas clarify ratite evolution. Nature 409:704–707.

Cooper, A., C. Mourer-Chauviré, G. K. Chambers, A. von Haeseler, A. C. Wilson, and S. Pääbo. 1992. Independent origins of New Zealand moas and kiwis. Proceedings of the National Academy of Sciences USA 89:8741–8744.

Cooper, A., and H. N. Poinar. 2000. Ancient DNA: do it right or not at all. Science 289:1139.

Crawford, N. G., B. C. Faircloth, J. E. McCormack, R. T. Brumfield, K. Winker, and T. C. Glenn. 2012. More than 1000 ultraconserved elements provide evidence that turtles are the sister group of archosaurs. Biology Letters 8:783–786.

Dabney, J., M. Meyer, and S. Pääbo. 2013. Ancient DNA damage. Cold Spring Harbor Perspectives in Biology 5:a012567.

DeSalle, R., J. Gatesy, W. Wheeler, and D. Grimaldi. 1992. DNA sequences from a fossil termite in Oligo-Miocene amber and their phylogenetic implications. Science 257:1933–1936.

do Amaral, F. R., L. G. Neves, M. F. Resende Jr., F. Mobili, C. Y. Miyaki, K. C. Pellegrino, and C. Biondo. 2015. Ultraconserved elements sequencing as a low-cost source of complete mitochondrial genomes and microsatellite markers in non-model amniotes. PLoS One 10:e0138446.

Ekblom, R., and J. Galindo. 2011. Applications of next generation sequencing in molecular ecology of non-model organisms. Heredity 107:1–15.

Faircloth, B. C., J. E. McCormack, N. G. Crawford, M. G. Harvey, R. T. Brumfield, and T. C. Glenn. 2012. Ultraconserved elements anchor thousands of genetic markers spanning multiple evolutionary timescales. Systematic Biology 61:717–726.

Faircloth, B. C., L. Sorenson, F. Santini, and M. E. Alfaro. 2013. A phylogenomic perspective on the radiation of ray-finned fishes based upon targeted sequencing of ultraconserved elements (UCEs). PLoS One 8:e65923.

Glenn, T. C. 2011. Field guide to next-generation DNA sequencers. Molecular Ecology Resources 11:759–769.

Gnirke, A., A. Melnikov, J. Maguire, P. Rogov, E. M. LeProust, W. Brockman, T. Fennell, G. Giannoukos, S. Fisher, C. Russ, S. Gabriel, D. B. Jaffe, E. S. Lander, and C. Nusbaum. 2009. Solution hybrid selection with ultra-long oligonucleotides for massively parallel targeted sequencing. Nature Biotechnology 27:182–189.

Graham, C. F., T. C. Glenn, A. G. McArthur, D. R. Boreham, T. Kieran, S. Lance, R. G. Manzon, J. A. Martino, T. Pierson, and S. M. Rogers. 2015. Impacts of degraded DNA on restriction enzyme associated DNA sequencing (RADseq). Molecular Ecology Resources 15:1304–1315.

Greenberg, J. 2014. A feathered river across the sky: the passenger pigeon's flight to extinction. Bloomsbury Publishing USA, New York, NY.

Handt, O., M. Höss, M. Krings, and S. Pääbo. 1994. Ancient DNA: methodological challenges. Experientia 50:524–529.

Harvey, M. G., and R. T. Brumfield. 2015. Genomic variation in a widespread Neotropical bird (Xenops minutus) reveals divergence, population expansion, and gene flow. Molecular Phylogenetics and Evolution 83:305–316.

Haussler D., S. J. O'Brien, O. A. Ryder, F. K. Barker, M. Clamp, A. J. Crawford, R. Hanner, O. Hanotte, W. E. Johnson, and J. A. McGuire. 2009. Genome 10K: a proposal to obtain whole-genome sequence for 10,000 vertebrate species. Journal of Heredity 100:659–674.

Hawkins, M. T. R., C. A. Hofman, T. Callicrate, M. M. McDonough, M. T. N. Tsuchiya, E. E. Gutiérrez, K. M. Helgen, and J. E. Maldonado. 2016. In-solution hybridization for mammalian mitogenome enrichment: pros, cons, and challenges associated with multiplexing degraded DNA. Molecular Ecology Resources 16:1173–1188.

Higuchi, R., B. Bowman, M. Freiberger, O. A. Ryder, and A. C. Wilson. 1984. DNA sequences from the quagga, an extinct member of the horse family. Nature 312:282–284.

Ho, S. Y. W., and M. T. P. Gilbert. 2010. Ancient mitogenomics. Mitochondrion 10:1–11.

Hoffberg, S. L., T. J. Kieran, J. M. Catchen, A. Devault, B. C. Faircloth, R. Mauricio, and T. C. Glenn. 2016. RADcap: sequence capture of dual-digest RADseq libraries with identifiable duplicates and reduced missing data. Molecular Ecology Resources 16:1264–1278.

Hofreiter, M., V. Jaenicke, D. Serre, A. von Haeseler, and S. Pääbo. 2001. DNA sequences from multiple amplifications reveal artifacts induced by cytosine deamination in ancient DNA. Nucleic Acids Research 29:4793–4799.

Holmes, M. W., T. T. Hammond, G. O. U. Wogan, R. E. Walsh, K. LaBarbera, E. A. Wommack, F. M. Martins, J. C. Crawford, K. L. Mack, L. M. Bloch, and M. W. Nachman. 2016. Natural history collections as windows on evolutionary processes. Molecular Ecology 25:864–881.

Hung, C.-M., R.-C. Lin, J.-H. Chu, C.-F. Yeh, C.-J. Yao, and S.-H. Li. 2013. The de novo assembly of mitochondrial genomes of the extinct passenger pigeon (Ectopistes migratorius) with next generation sequencing. PLoS One 8:e56301.

Hung, C.-M., P.-J. L. Shaner, R. M. Zink, W.-C. Liu, T.-C. Chu, W.-S. Huang, and S.-H. Li. 2014. Drastic population fluctuations explain the rapid extinction of the passenger pigeon. Proceedings of the National Academy of Sciences USA 111:10636–10641.

i5K Consortium. 2013. The i5K Initiative: advancing arthropod genomics for knowledge, human health, agriculture, and the environment. Journal of Heredity 104:595–600.

Ilves, K. L., and H. López Fernández. 2014. A targeted next-generation sequencing toolkit for exon-based cichlid phylogenomics. Molecular Ecology Resources 14:802–811.

Jarvis, E. D., S. Mirarab, A. J. Aberer, B. Li, P. Houde, C. Li, S. Y. W. Ho, B. C. Faircloth, B. Nabholz, J. T. Howard, A. Suh, C. C. Weber, R. R. da Fonseca, J. Li, F. Zhang, H. Li, L. Zhou, N. Narula, L. Liu, G. Ganapathy, B. Boussau, M. S. Bayzid, V. Zavidovych, S. Subramanian, T. Gabaldón, S. Capella-Gutiérrez, J. Huerta-Cepas, B. Rekepalli, K. Munch, M. Schierup, B. Lindow, W. C. Warren, D. Ray, R. E. Green, M. W. Bruford, X. Zhan, A. Dixon, S. Li, N. Li, Y. Huang, E. P. Derryberry, M. F. Bertelsen, F. H. Sheldon, R. T. Brumfield, C. V. Mello, P. V. Lovell, M. Wirthlin, M. P. C. Schneider, F. Prosdocimi, J. A. Samaniego, A. M. V. Velazquez, A. Alfaro-Núñez, P. F. Campos, B. Petersen, T. Sicheritz-Ponten, A. Pas, T. Bailey, P. Scofield, M. Bunce, D. M. Lambert, Q. Zhou, P. Perelman, A. C. Driskell, B. Shapiro, Z. Xiong, Y. Zeng, S. Liu, Z. Li, B. Liu, K. Wu, J. Xiao, X. Yinqi, Q. Zheng, Y. Zhang, H. Yang, J. Wang, L. Smeds, F. E. Rheindt, M. Braun, J. Fjeldså, L. Orlando, F. K. Barker, K. A. Jønsson, W. Johnson, K. P. Koepfli, S. O'Brien, D. Haussler, O. A. Ryder, C. Rahbek, E. Willerslev, G. R. Graves, T. C. Glenn, J. McCormack, D. Burt, H. Ellegren, P. Alström, S. V. Edwards, A. Stamatakis, D. P. Mindell, J. Cracraft, E. L. Braun, T. Warnow, J. Wang, M. T. P. Gilbert, and G. Zhang. 2014. Whole-genome analyses resolve early branches in the tree of life of modern birds. Science 346:1320–1331.

Jones, M. R., and J. M. Good. 2016. Targeted capture in evolutionary and ecological genomics. Molecular Ecology 25:185–202.

Jónsson, H., A. Ginolhac, M. Schubert, P. L. F. Johnson, and L. Orlando. 2013. mapDamage2.0: fast approximate Bayesian estimates of ancient DNA damage parameters. Bioinformatics 29:1682–1684.

Katzman, S., A. D. Kern, G. Bejerano, G. Fewell, L. Fulton, R. K. Wilson, S. R. Salama, and D. Haussler. 2007. Human genome ultraconserved elements are ultraselected. Science 317:915.

Knapp, M., and M. Hofreiter. 2010. Next generation sequencing of ancient DNA: requirements, strategies and perspectives. Genes 1:227–243.

Kocher, T. D., W. K. Thomas, A. Meyer, S. V. Edwards, S. Pääbo, F. X. Villablanca, and A. C. Wilson. 1989. Dynamics of mitochondrial DNA evolution in animals: amplification and sequencing with conserved primers. Proceedings of the National Academy of Sciences USA 86:6196–6200.

Lerner, H. R. L., and R. C. Fleischer. 2010. Prospects for the use of next-generation sequencing methods in ornithology. Auk 127:4–15.

Linderholm, A. 2016. Ancient DNA: the next generation–chapter and verse. Biological Journal of the Linnean Society 117:150–160.

Mamanova, L., A. J. Coffey, C. E. Scott, I. Kozarewa, E. H. Turner, A. Kumar, E. Howard, J. Shendure, and D. J. Turner. 2010. Target-enrichment strategies for next-generation sequencing. Nature Methods 7:111–118.

Mason, V. C., G. Li, K. M. Helgen, and W. J. Murphy. 2011. Efficient cross-species capture hybridization and next-generation sequencing of mitochondrial genomes from noninvasively sampled museum specimens. Genome Research 21:1695–1704.

McCormack, J., W. L. E. Tsai, and B. C. Faircloth. 2016. Sequence capture of ultraconserved elements from bird museum specimens. Molecular Ecology Resources 16:1189–1203.

McCormack, J. E., M. G. Harvey, B. C. Faircloth, N. G. Crawford, T. C. Glenn, and R. T. Brumfield. 2013a. A phylogeny of birds based on over 1,500 loci collected by target enrichment and high-throughput sequencing. PLoS One 8:e54848.

McCormack, J. E., S. M. Hird, A. J. Zellmer, B. C. Carstens, and R. T. Brumfield. 2013b. Applications of next-generation sequencing to phylogeography and phylogenetics. Molecular Phylogenetics and Evolution 66:526–538.

Meiklejohn, K. A., M. J. Danielson, B. C. Faircloth, T. C. Glenn, E. L. Braun, and R. T. Kimball. 2014. Incongruence among different mitochondrial regions: a case study using complete mitogenomes. Molecular Phylogenetics and Evolution 78:314–323.

Meyer, M., U. Stenzel, and M. Hofreiter. 2008. Parallel tagged sequencing on the 454 platform. Nature Protocols 3:267–278.

Mitchell, K. J., B. Llamas, J. Soubrier, N. J. Rawlence, T. H. Worthy, J. Wood, M. S. Y. Lee, and A. Cooper. 2014. Ancient DNA reveals elephant birds and kiwi are sister taxa and clarifies ratite bird evolution. Science 344:898–900.

Mundy, N. I., P. Unitt, and D. S. Woodruff. 1997. Skin from feet of museum specimens as a non-destructive source of DNA for avian genotyping. Auk 114:126–129.

Nachman, M. W. 2013. Genomics and museum specimens. Molecular Ecology 22:5966–5968.

Noonan, J. P., M. Hofreiter, D. Smith, J. R. Priest, N. Rohland, G. Rabeder, J. Krause, J. C. Detter, S. Pääbo, and E. M. Rubin. 2005. Genomic sequencing of Pleistocene cave bears. Science 309:597–599.

O'Neill, E. M., R. Schwartz, C. T. Bullock, J. S. Williams, H. B. Shaffer, X. Aguilar-Miguel, G. Parra-Olea, and D. W. Weisrock. 2013. Parallel tagged amplicon sequencing reveals major lineages and phylogenetic structure in the North American tiger salamander (*Ambystoma tigrinum*) species complex. Molecular Ecology 22:111–129.

Orlando, L., A. Ginolhac, M. Raghavan, J. Vilstrup, M. Rasmussen, K. Magnussen, K. E. Steinmann, P. Kapranov, J. F. Thompson, and G. Zazula. 2011. True single-molecule DNA sequencing of a Pleistocene horse bone. Genome Research 21:1705–1719.

Oskam, C. L., J. Haile, E. McLay, P. Rigby, M. E. Allentoft, M. E. Olsen, C. Bengtsson, G. H. Miller, J.-L. Schwenninger, and C. Jacomb. 2010. Fossil avian eggshell preserves ancient DNA. Proceedings of the Royal Society B 277:1991–2000.

Oyler-McCance, S. J., R. S. Cornman, K. L. Jones, and J. A. Fike. 2015. Genomic single-nucleotide polymorphisms confirm that Gunnison and Greater sage-grouse are genetically well differentiated and that the Bi-State population is distinct. Condor 117:217–227.

Pääbo, S. 1989. Ancient DNA: extraction, characterization, molecular cloning, and enzymatic amplification. Proceedings of the National Academy of Sciences USA 86:1939–1943.

Peñalba, J. V., L. L. Smith, M. A. Tonione, C. Sass, S. M. Hykin, P. L. Skipwith, J. A. McGuire, R. C. Bowie, and C. Moritz. 2014. Sequence capture using PCR-generated probes: a cost-effective method of targeted high-throughput sequencing for non-model organisms. Molecular Ecology Resources 14:1000–1010.

Peterson, B. K., J. N. Weber, E. H. Kay, H. S. Fisher, and H. E. Hoekstra. 2012. Double digest RADseq: an inexpensive method for de novo SNP discovery and genotyping in model and non-model species. PLoS One 7:e37135.

Poinar, H. N., C. Schwarz, J. Qi, B. Shapiro, R. D. E. MacPhee, B. Buigues, A. Tikhonov, D. H. Huson, L. P. Tomsho, and A. Auch. 2006. Metagenomics to paleogenomics: large-scale sequencing of mammoth DNA. Science 311:392–394.

Prum, R. O., J. S. Berv, A. Dornburg, D. J. Field, J. P. Townsend, E. M. Lemmon, and A. R. Lemmon. 2015. A comprehensive phylogeny of birds (Aves) using targeted next-generation DNA sequencing. Nature 526:569–573.

Pruvost, M., R. Schwarz, V. B. Correia, S. Champlot, S. Braguier, N. Morel, Y. Fernandez-Jalvo, T. Grange, and E.-M. Geigl. 2007. Freshly excavated fossil bones are best for amplification of ancient DNA. Proceedings of the National Academy of Sciences USA 104:739–744.

Puritz, J. B., M. V. Matz, R. J. Toonen, J. N. Weber, D. I. Bolnick, and C. E. Bird. 2014. Demystifying the RAD fad. Molecular Ecology 23:5937–5942.

Rawlence, N., J. Wood, K. Armstrong, and A. Cooper. 2009. DNA content and distribution in ancient feathers and potential to reconstruct the plumage of extinct avian taxa. Proceedings of the Royal Society B 276:3395–3402.

Rohland, N., and M. Hofreiter. 2007. Comparison and optimization of ancient DNA extraction. Biotechniques 42:343–352.

Rubin, B. E., R. H. Ree, and C. S. Moreau. 2012. Inferring phylogenies from RAD sequence data. PLoS One 7:e33394.

Ruegg, K., E. C. Anderson, J. Boone, J. Pouls, and T. B. Smith. 2014a. A role for migration-linked genes and genomic islands in divergence of a songbird. Molecular Ecology 23:4757–4769.

Ruegg, K. C., E. C. Anderson, K. L. Paxton, V. Apkenas, S. Lao, R. B. Siegel, D. F. DeSante, F. Moore, and T. B. Smith. 2014b. Mapping migration in a songbird using high-resolution genetic markers. Molecular Ecology 23:5726–5739.

Saiki, R. K., S. Scharf, F. Faloona, K. B. Mullis, G. T. Horn, H. A. Erlich, and N. Arnheim. 1985. Enzymatic amplification of beta-globin genomic sequences and restriction site analysis for diagnosis of sickle cell anemia. Science 230:1350–1354.

Sanger, F., S. Nicklen, and A. R. Coulson. 1977. DNA sequencing with chain-terminating inhibitors. Proceedings of the National Academy of Sciences USA 74:5463–5467.

Sawyer, S., J. Krause, K. Guschanski, V. Savolainen, and S. Pääbo. 2012. Temporal patterns of nucleotide misincorporations and DNA fragmentation in ancient DNA. PLoS One 7:e34131.

Seddon, P. J., A. Moehrenschlager, and J. Ewen. 2014. Reintroducing resurrected species: selecting DeExtinction candidates. Trends in Ecology and Evolution 29:140–147.

Sefc, K. M., R. B. Payne, M. D. Sorenson, and R. C. Fleischer. 2003. Microsatellite amplification from museum feather samples: effects of fragment size and template concentration on genotyping errors. Auk 120:982–989.

Shapiro, B., D. Sibthorpe, A. Rambaut, J. Austin, G. M. Wragg, O. R. Bininda-Emonds, P. L. Lee, and A. Cooper. 2002. Flight of the dodo. Science 295:1683.

Shendure, J., and H. Ji. 2008. Next-generation DNA sequencing. Nature Biotechnology 26:1135–1145.

Skoglund, P., B. H. Northoff, M. V. Shunkov, A. P. Derevianko, S. Pääbo, J. Krause, and M. Jakobsson. 2014. Separating endogenous ancient DNA from modern day contamination in a Siberian Neandertal. Proceedings of the National Academy of Sciences USA 111:2229–2234.

Smith, B. T., M. G. Harvey, B. C. Faircloth, T. C. Glenn, and R. T. Brumfield. 2013. Target capture and massively parallel sequencing of ultraconserved elements for comparative studies at shallow evolutionary time scales. Systematic Biology 63:83–95.

Smith, E. F., P. Arctander, J. Fjeldså, and O. G. Amir. 1991. A new species of shrike (Laniidae: Laniarius) from Somalia, verified by DNA sequence data from the only known individual. Ibis 133:227–235.

Snow, N. 2005. Successfully curating smaller herbaria and natural history collections in academic settings. BioScience 55:771–779.

Staats, M., R. H. J. Erkens, B. van de Vossenberg, J. J. Wieringa, K. Kraaijeveld, B. Stielow, J. Geml, J. E. Richardson, and F. T. Bakker. 2013. Genomic treasure troves: complete genome sequencing of herbarium and insect museum specimens. PLoS One 8:e69189.

Suchan, T., C. Pitteloud, N. Gerasimova, A. Kostikova, S. Schmid, N. Arrigo, M. Pajkovic, M. Ronikier, and N. Alvarez. 2016. Hybridization capture using RAD probes (hyRAD), a new tool for performing genomic analyses on collection specimens. PLoS One 11(3):e0151651.

Toews, D. P., L. Campagna, S. A. Taylor, C. N. Balakrishnan, D. T. Baldassarre, P. E. Deane-Coe, M. G. Harvey, D. M. Hooper, D. E. Irwin, and C. D. Judy. 2015. Genomic approaches to understanding population divergence and speciation in birds. Auk 133:13–30.

Wandeler, P., P. E. Hoeck, and L. F. Keller. 2007. Back to the future: museum specimens in population genetics. Trends in Ecology and Evolution 22:634–642.

Weiß, C. L., V. J. Schuenemann, J. Devos, G. Shirsekar, E. Reiter, B. A. Gould, J. R. Stinchcombe, J. Krause, and H. A. Burbano. 2015. Temporal patterns of damage and decay kinetics of DNA retrieved from plant herbarium specimens. Royal Society Open Science 3:160239.

Willerslev, E., and A. Cooper. 2005. Review paper: ancient DNA. Proceedings of the Royal Society B 272:3–16.

Willerslev, E., A. J. Hansen, J. Binladen, T. B. Brand, M. T. P. Gilbert, B. Shapiro, M. Bunce, C. Wiuf, D. A. Gilichinsky, and A. Cooper. 2003. Diverse plant and animal genetic records from Holocene and Pleistocene sediments. Science 300:791–795.

Wood, J. R., and V. L. De Pietri. 2015. Next-generation paleornithology: technological and methodological advances allow new insights into the evolutionary and ecological histories of living birds. Auk 132:486–506.

Wood, J. R., J. M. Wilmshurst, S. J. Richardson, N. J. Rawlence, S. J. Wagstaff, T. H. Worthy, and A. Cooper. 2013. Resolving lost herbivore community structure using coprolites of four sympatric moa species (Aves: Dinornithiformes). Proceedings of the National Academy of Sciences USA 110:16910–16915.

Zhang, G., B. Li, C. Li, M. T. P. Gilbert, E. D. Jarvis, and J. Wang. 2014. Comparative genomic data of the Avian Phylogenomics Project. GigaScience 3:26.

CHAPTER TEN

Methods for Specimen-based Studies of Avian Symbionts*

Holly L. Lutz, Vasyl V. Tkach, and Jason D. Weckstein

Abstract. The collection of avian voucher specimens has long played an important role in studying the basic biology, ecology, and evolution of birds. However, symbionts (such as parasites and pathogens) of avian hosts have been largely neglected by ornithologists and are largely underrepresented in most major museum collections. Museum-oriented research expeditions to collect bird specimens capture a diversity of metadata, but the proper collection of symbionts for optimal use in downstream research projects remains uncommon. In this chapter, we provide methods for the comprehensive sampling of a diverse suite of symbionts from avian hosts, including blood parasites (haematozoans), microbial symbionts (bacteria and viruses), ectoparasites (arthropods), and endoparasites (helminths), while attempting to illustrate the research avenues opened by collecting such samples. Our objective is to encourage a view of birds as ecosystems in and of themselves, and to empower field ornithologists, particularly those participating in the collection of voucher specimens, to sample the plethora of micro- and macroorganisms that live in and on avian hosts. By collecting these additional specimens, ornithologists will not only unlock new aspects of avian biology, but also will expand the scientific community's ability to address ecological and evolutionary questions, while aiding in the discovery of new biodiversity and maximizing the utility of the "extended" avian specimen.

Key Words: field workflow, microbiome, museum collections, parasite, pathogen, symbiont, voucher.

Collections of avian specimens have been used to address complex ecological and evolutionary questions, and these museum specimens have served as an invaluable resource for the scientific community for centuries. As we develop new tools and methods, the scientific potential of individual bird specimens continues to expand, demanding that we take a more comprehensive approach to collecting modern whole bird specimens—considering them as ecosystems in-and-of themselves. Birds are capable of hosting a plethora of symbionts, some visible to the naked eye and others microscopic; some are ectoparasitic and some internal (Figure 10.1). Relationships between these symbionts and their avian hosts range from mutualistic to parasitic to pathogenic, and all have the potential to influence avian behavior, ecology, and evolution (Combes

* Lutz, H. L., V. V. Tkach, and J. D. Weckstein. 2017. Methods for specimen-based studies of avian symbionts. Pp. 157–183 in M. S. Webster (editor), The Extended Specimen: Emerging Frontiers in Collections-based Ornithological Research. Studies in Avian Biology (no. 50), CRC Press, Boca Raton, FL.

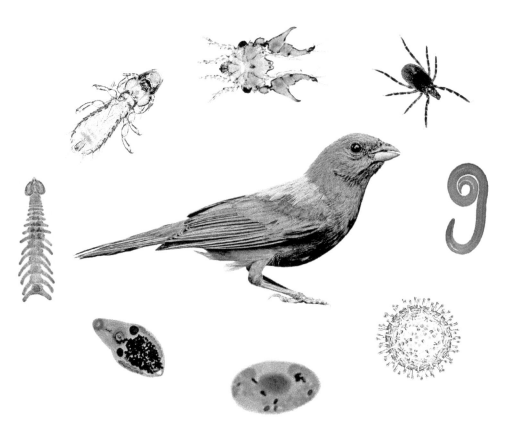

Figure 10.1. Birds can be thought of as ecosystems in-and-of themselves, serving as hosts for a plethora of symbionts from disparate branches in the tree of life.

1996, Combes et al. 1996, Atkinson et al. 2008). Indeed, studies of avian parasites and pathogens have allowed ornithologists to address many important questions, from understanding how avian life history traits are associated with higher or lower prevalence and probability of parasite infection (Clayton et al. 1992, Clayton and Walther 2001, Fecchio et al. 2011, Lutz et al. 2015) to tracking how avian populations have shifted their distributions over time. Because most birds are volant, they have also provided an important system for studying the evolution of virulence in rapidly spreading emerging pathogens (Hochachka and Dhondt 2000) and have brought to light the importance of broadly sampling potential hosts when studying the origin of epidemics (Kilpatrick et al. 2006, Dhondt et al. 2014).

Specimen-based studies of avian symbionts are particularly useful for studying cophylogenetic history and macroevolutionary patterns in avian hosts and parasites (Johnson and Clayton 2003a, Weckstein 2004, Johnson et al. 2011), as well as spatiotemporal relationships between birds and their environments (Parker et al. 2011, Galen and Witt 2014). For example, studies of museum specimens dating back to the early 20th century have allowed researchers to determine when avipoxvirus was first introduced into endemic Galápagos finches and mockingbirds (Parker et al. 2011) and Hawaiian forest birds (Jarvi et al. 2007). In separate studies, the sampling of haemosporidian parasites, which are the causative agents of malaria, across a broad range of avian hosts has led to generalizations about the power of host life history traits to predict rates of parasitism. For example, flocking behavior, nest type, and nest height or foraging stratum have been significantly linked to rates of parasitism for these sorts of malarial parasites in both the Neotropics (e.g., Fecchio et al. 2011, 2013) and Afrotropics (Lutz et al. 2015).

Haemosporidian parasites of birds have been studied for more than two centuries, and the knowledge we have acquired from these model parasites has informed the study of human and other primate parasites that cause malaria. The incredible diversity comprising avian haemosporidians (Valkiūnas 2005), paired with the known selective pressure these parasites impose on their hosts (Samuel et al. 2015), make them an important group to consider when studying certain aspects of avian biology. In extreme cases, avian malaria can have devastating effects when introduced to naïve populations, as occurred on the Hawaiian Islands, where several species of Hawaiian honeycreepers were driven to extinction by introduced malarial parasites (Atkinson and LaPointe 2009). Extinction caused by malaria is an extreme and rare occurrence; in fact, by most outward measures of health, infected individuals appear to suffer little from malaria (Valkiūnas 2005). However, evidence suggests that subtle, long-term fitness effects are at play in wild birds: chronic malaria has been linked to telomere degradation and senescence in Great Reed Warblers (Asghar et al. 2015), as well as reduced quality of offspring and overall lower reproductive success of infected adults (Knowles et al. 2010). Such documented influences of microscopic blood parasites on avian hosts cannot be ignored when considering host ecology or evolutionary biology.

In some cases, parasites may reveal important information about the evolutionary history of their avian hosts. One of the nice attributes of ectoparasites, such as avian chewing lice (Phthiraptera), is that they are permanent ectoparasites, living their entire life cycle on the host (Johnson and Clayton 2003b). Another nice attribute of this system is that the life cycle of a louse from egg to reproduction is about 1 month (Johnson and Clayton 2003b) and thus a single annual cycle of the avian host contains 12 louse annual cycles. As a result, generation times of the parasites are much shorter than the host generation times, and thus the parasites evolve at a faster rate than their hosts (Whiteman and Parker 2005). Both of these characteristics allow ectoparasitic lice to serve as markers of recent host evolutionary history because the parasites are evolving more quickly than their hosts. Indeed, in some cases the DNA of ectoparasitic lice may serve as a better proxy of recent host evolutionary history than the host's own DNA. In the example that follows, we describe a specific instance where ectoparasitic lice infecting sympatric congeneric toucans in

the genus *Ramphastos* can tell us a great deal about recent host evolutionary history.

Weckstein and colleagues have been collecting associated specimens of both *Ramphastos* toucans and their ectoparasites since the 1990s in an effort to understand both their cophylogenetic and cophylogeographic histories (e.g., Weckstein 2004, Price and Weckstein 2005). *Ramphastos* toucans in Amazonian Brazil include two overlapping species complexes, *Ramphastos tucanus* and *Ramphastos vitellinus*, each of which are geographically variable and form hybrid rings around the Amazon basin (Haffer 1974). At any given locality in the basin, both *R. tucanus* and *R. vitellinus* may host the ischnoceran louse species *Austrophilopterus cancellosus* (Weckstein 2004, Price and Weckstein 2005). Within the *R. vitellinus* species complex, geographic variation in coloration clearly shows a break across the mouth of the Amazon River in eastern Amazonia; this break is also indicated by the subspecific taxonomy of this complex, and it is clear that there is no ongoing gene flow between *R. vitellinus* subspecies across this riverine barrier. In contrast, *R. tucanus*, which shows east–west variation in coloration, does not exhibit a plumage coloration break across the mouth of the Amazon River (Haffer 1974). Instead, the eastern Amazonian subspecies *R. tucanus tucanus*, which has a reddish-orange bill, is found on both the north and south banks near the mouth of the Amazon River (Figure 10.2). Thus, one is left to wonder whether the absence of north–south variation in coloration in *R. t. tucanus* across the mouth of the Amazon River is due to ongoing gene flow or, alternatively, recent cessation of gene flow, such that there has not been sufficient time for divergence in plumage coloration to accrue. Genetic data from *R. tucanus* for mitochondrial DNA (mtDNA) indicates a break across the mouth of the Amazon, suggesting cessation of gene flow between north and south bank populations of *R. tucanus*. However, sequences of nuclear introns from north and south bank *R. t. tucanus* have similar or shared haplotypes, which is consistent with a history of either ongoing or recent cessation of gene flow (J. D. Weckstein, unpubl. data). In this case an additional marker might be useful for corroborating the mtDNA results and assessing the alternative hypotheses of ongoing gene flow or recent cessation of gene flow among these populations, because this cannot be tested with the nuclear intron data. Analysis of mtDNA

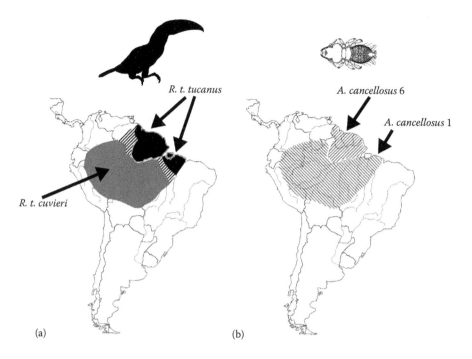

Figure 10.2. (a) Map showing the distribution of *Ramphastos tucanus* subspecies in Amazonia. Gray indicates the range of *R. t. cuvieri*, black indicates the range of *R. t. tucanus*, and hash marks indicate zones of hybridization between these subspecies. (b) Map showing the distributions of the divergent mtDNA lineages (Weckstein 2004) of the toucan louse *Austrophilopterus cancellosus*.

(cytochrome oxidase subunit I) sequences for the ischnoceran chewing louse *A. cancellosus* parasitizing these birds indicates a distinct genetic break across the mouth of the Amazon River, with louse individuals in the Guyanan shield differing by an uncorrected p-distance of 11.2% from those on the south bank of the river mouth (Figure 10.2). Thus, the lice are telling us that there is not ongoing dispersal of *R. t. tucanus* between the north bank and the south bank, corroborating the toucan mtDNA sequence data results and supporting the hypothesis that shared intron haplotypes between host populations on either bank of the Amazon River mouth are therefore the result of retention of ancestral DNA polymorphism on account of recent cessation of gene flow (J. D. Weckstein, unpubl. data). This example is simply one of many that highlight the value of making detailed collections of birds and their associated parasites. One could perform similar studies using myriad parasites with different life history characteristics to reconstruct the evolutionary history and ecology of their avian hosts.

Among avian symbionts, those with parasitic life histories, the major focus of this chapter, are particularly diverse (Windsor 1995), comprising an incredible 30% to 70% of known biodiversity on our planet (Timm and Clauson 1987, de Meeûs et al. 1998, Windsor 1998, de Meeûs and Renaud 2002, Poulin 2005). For a variety of reasons, parasites are important elements in the study of biodiversity (Combes 1996, Combes et al. 1996, Brooks and Hoberg 2000, Brooks et al. 2001, Whiteman and Parker 2005, Parker et al. 2006, Dobson et al. 2008). First, parasites can have important impacts on the health, demography, behavior, and evolution of their avian hosts (Combes 1996, Combes et al. 1996, Parker et al. 2006). Second, parasites are ubiquitous (Combes et al. 1996) with most, if not all, birds carrying many parasite species. For example, an individual bird can harbor lice, mites, ticks, hippoboscid flies, fleas, spiny-headed worms, tapeworms, flukes, roundworms, and protozoans, in addition to a plethora of bacterial, fungal, and viral symbionts. Third, only a fraction of the parasite species on Earth have been identified (Brooks and Hoberg 2001), and historically the effects of parasites on nongame wild avian hosts have been understudied (Parker et al. 2006, Atkinson et al.

2008, Atkinson and LaPointe 2009). Last, many parasites can be successfully used to make inferences about host ecology, population biology, and evolutionary history including historical biogeography (Whiteman and Parker 2005, Nieberding and Morand 2006, Nieberding and Olivieri 2007). Thus, a critical need exists to study the birds and their symbiotic associates (parasitic or otherwise) to understand the interdependencies in the web of life, reconstruct the evolutionary history of life on our planet, and stem the tide of extinction.

Biodiversity inventories of birds and their associated symbionts are a first step toward this end, and proper methods of collection and preservation are essential for correct identification and documentation of both host and symbiont species. A great deal has been written about the importance of avian biodiversity surveys (Lawton et al. 1998, Balmford and Gaston 1999, Norris and Pain 2002, Gregory et al. 2003) and methods for obtaining, preserving, and preparing bird specimens (e.g., Johnson et al. 1984, Proctor and Lynch 1993, Winker 2000). The importance of collecting avian specimens and voucher specimens in general has also long been acknowledged (e.g., Winker 1997, Rocha 2014). However, relatively little has been written regarding how to sample these host specimens for the high diversity of symbionts living on and in them. Although a number of publications have addressed collecting specific groups of parasites and other symbionts (Dubinina 1971, Clayton and Walther 1997, Owen 2011), these publications are scattered across the scientific literature. By far the most comprehensive description of procedures aimed at a complete parasitological investigation of birds was published by Dubinina (1971). This 129-page manual includes an overview of avian anatomy and morphology, step-by-step procedures for examining the entire avian host body for parasites, and directions for proper field fixation and postfixation protocols. However, this manual was published only in Russian and is now difficult to obtain. Furthermore, the introduction of modern research methods and tools since the 1970s has dramatically changed requirements for specimen fixation and preservation, leaving many of the methods presented in this work outdated.

Therefore, we outline here the general workflow, methods, and standards for comprehensive sampling and proper preservation of avian symbionts that we consider to be optimal for a variety of modern and traditional downstream biological research applications. Our goal is to optimize knowledge about each avian host and its symbionts, collected and prepared as traditional museum specimens, by broadly sampling four major symbiont categories: blood parasites (Haematozoa), microbial symbionts (bacteria and viruses), ectoparasites (arthropods), and endoparasites (helminths). The structure of this chapter reflects the order in which samples from these major groups are generally collected in the field workflow. We will not discuss the details of avian specimen preparation methods, as we assume that readers are already familiar with standard avian museum specimen preparation and data collection. If not, then the reader can refer to Winker (2000) or other papers cited earlier, which provide an overview of methods for preparation of bird specimens and the typical data fields that are recorded for each avian specimen. Following the final section on detailed protocols for symbiont sampling and preservation, we summarize a basic sampling workflow that can be applied in most field situations. We hope that this chapter provides a useful resource for avian collectors and field researchers, helping us edge closer to a more complete sampling of each avian "ecosystem."

BEFORE YOU BEGIN: THE FIELD NOTEBOOK

As with host specimens, careful field notes help to capture valuable metadata during avian parasite collection events. We use archival 100% cotton fiber, acid free paper with preprinted data fields and use archival ink (e.g., Pigma) to write notes on these "parasite field notebook" pages, which complement the host catalog notebook pages (Figure 10.3a,b). In these field notes, we record basic data such as host species sampled, locality, date sampled, parasite collector doing the sampling, and whether anything was found; it is also important to note when no parasites are found, so as not to be mistaken for a lack of sampling effort. The notebook pages use a series of checkboxes to denote what sampling was completed and leave room to describe what was collected from a given host specimen. One of the most critical data fields in this parasite field notebook is the host's field number, which is used to link parasite samples (e.g., vials of ectoparasites, blood smears) to a host voucher specimen. However, for this unique identifier to be useful for parasite sampling, it must be assigned to the host specimen *before* the

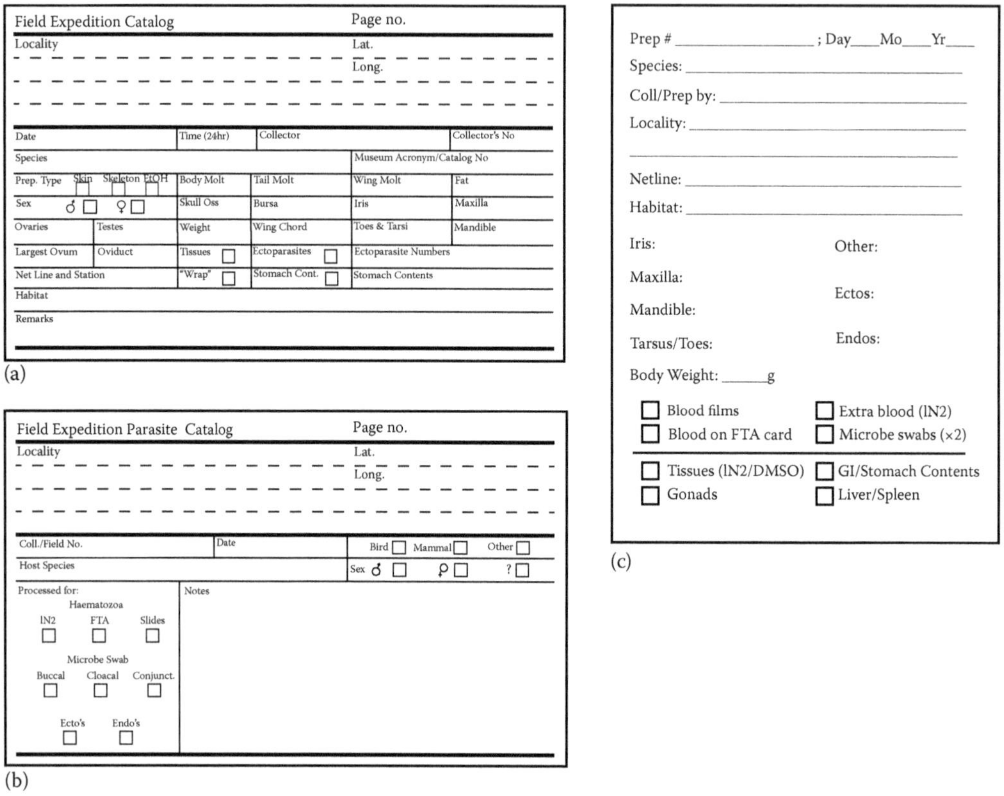

Field Expedition Catalog — Page no. _____

Locality				
Lat. _____ Long. _____				
Date	Time (24hr)	Collector		Collector's No
Species			Museum Acronym/Catalog No	
Prep. Type Skin Skeleton EtOH	Body Molt	Tail Molt	Wing Molt	Fat
Sex ♂□ ♀□	Skull Oss	Bursa	Iris	Maxilla
Ovaries Testes	Weight	Wing Chord	Toes & Tarsi	Mandible
Largest Ovum Oviduct	Tissues □	Ectoparasites □	Ectoparasite Numbers	
Net Line and Station	"Wrap" □	Stomach Cont. □	Stomach Contents	
Habitat				
Remarks				

(a)

Field Expedition Parasite Catalog — Page no. _____

Locality _____ Lat. _____ Long. _____

Coll./Field No.	Date	Bird □ Mammal □ Other □
Host Species		Sex ♂ □ ♀ □ ? □
Processed for:	Notes	

Haematozoa
lN2 □   FTA □   Slides □

Microbe Swab
Buccal □   Cloacal □   Conjunct. □

Ecto's □   Endo's □

(b)

Prep # _____ ; Day____Mo____Yr____
Species: _____
Coll/Prep by: _____
Locality: _____
_____
Netline: _____
Habitat: _____

Iris: Other:
Maxilla:
Mandible: Ectos:
Tarsus/Toes: Endos:
Body Weight: _____ g

□ Blood films   □ Extra blood (lN2)
□ Blood on FTA card   □ Microbe swabs (×2)

□ Tissues (lN2/DMSO)   □ GI/Stomach Contents
□ Gonads   □ Liver/Spleen

(c)

Figure 10.3. The field notebook. (a) Standard host catalog fields. (b) Example of parasite catalog fields. (c) Field sheet to be kept with host specimen as it goes through various stages of sampling and processing.

first parasite sampling begins, and thus before a field preparation number is usually assigned; often a personal catalog number is assigned by the host specimen preparator upon entry into the personal field catalog. In our experience there are multiple ways that this can be handled. One is to immediately assign a tissue or parasite number to each host specimen. Many museums use a separate tissue catalog to track the condition and handling of tissue samples collected in the field, and this number is noted on the voucher data label and host field catalogs. Another option is to assign either a special parasite field number or general host catalog number that follows the host specimen through all steps of sampling, and also note this number on the host field label, host field notebooks, and on a 10 × 15 cm host field sheet that follows the host specimen through parasite sampling and preparation (Figure 10.3c). This host field sheet is a convenient way to maintain notes on which sampling steps have been performed, as well as noting host data (e.g., weight, soft-part

colors) before it is written in the catalog. We typically modify these sheets prior to expeditions to include a country acronym for collection numbers (e.g., "UGA" for Uganda), the year, and fields for specific tissues or samples we may be collecting for various projects.

SAMPLING PROTOCOLS FOR THE STUDY OF BLOOD PARASITES (HAEMATOZOA)

As with other taxonomic groups, the systematic study of avian haematozoans depends on both morphological and molecular data, both of which have their advantages and disadvantages. Phenotypic traits of haematozoan parasites may be convergent (Martinsen et al. 2008) and can be highly plastic depending on the host and the conditions during processing of blood smears (Valkiūnas 2005). Furthermore, haematozoan parasitemia is generally quite low in birds, which can lead to improper diagnosis of infection by microscopic analysis (Richard et al. 2002). The development of molecular protocols has provided more reliable diagnostic methods and has led to the discovery of hundreds of novel haematozoan parasite lineages (Bensch et al. 2009). In addition to improving detection capabilities, molecular methods and the development of phylogenetic markers are proving increasingly useful for studying evolutionary relationships in the haematozoan tree of life (Perkins and Schall 2002, Martinsen et al. 2008, Perkins 2014). However, molecular data are prone to error in cases of multispecies infections, and, alone, are insufficient for the taxonomic description of novel parasites. Therefore, when sampling birds for haematozoan parasites, it is important to collect blood for both morphological and molecular analyses.

Blood Collection and Storage

For live birds, blood can be drawn immediately after recording soft-part colors such as maxilla, mandible, nares, eye-ring, tibiotarsus, and feet. This can be done alone or with the help of a partner, taking personnel experience and the size and vigor of the bird into consideration. The top priorities at this point should be proper handling of the live animal to reduce stress and suffering, and rapid processing of the blood sample once it has been drawn.

Blood from live birds can be obtained from several parts of the body, including the femoral artery, the brachial/ulnar vein, a clipped toenail, or in the case of shot or otherwise dead birds, directly from the wounds, body cavity, or heart. Although blood from a dead bird will still provide useful material for molecular analysis of some parasites, fresh blood is desirable for haemosporidian studies, due to morphological changes elicited in these parasites by a drop in temperature and/or exposure to air (Valkiūnas 2005). We have found brachial and jugular venipuncture to be the most efficient in both small and large birds, as these veins are easily visible, and, in most cases, can be sampled by one person working alone. Sampling blood by clipping the toenail should be avoided, as it frequently leads to the introduction of debris into the sample if not properly cleaned, produces a relatively low volume of blood, and may be quite painful for the animal. The toenail clipping method is not approved for most species by the Ornithological Council, Washington, DC (Fair et al. 2010).

For the majority of bird species, a small gauge needle (22–27 gauge) is best for sampling blood. Smaller gauge needles (larger numbers) reduce the likelihood of hematoma, but may increase the probability of hemolysis, affecting downstream hematocrit measurements and blood smear quality. We typically use 25 to 27 gauge needles. Be sure that your needles are designed specifically for subcutaneous use (frequently denoted "SubQ"), as other needle types (e.g., intradermal use) are blunt-tipped and inappropriate for venipuncture. Before searching for the vein, it is helpful to wet the area with alcohol or water to clear the feathers out of the way, which makes the vein more visible. Some researchers prefer to use petroleum jelly, which holds the feathers out of the way and causes the blood to bead up more effectively, making it easier to draw neatly into a capillary tube. We avoid the use of petroleum jelly due to its matting effect on the feathers of birds that are to be preserved as museum vouchers. The needle should be placed parallel to the vein, bevel side up. With very light pressure, insert the needle ~0.5 to 1 mm into the vein and quickly remove. A small drop of blood will then form and can be collected directly into a heparinized capillary tube. Do not place the capillary tube directly against the vein, as this can inhibit blood flow. Likewise, hyperextension of the wing or leg from which blood is being drawn can restrict blood flow. A typical microhematocrit capillary tube holds about 0.075 ml (75 µl). The volume of blood collected will depend on the size and condition of the bird, but for birds that are to be collected as specimens, 1 to 2 hematocrit tubes (0.075–0.15 ml) is more than sufficient for molecular and morphological analyses of haematozoan parasites (we often rely on <0.05 ml of blood for our studies). If the bird is to be released rather than collected, be sure to take no more than the equivalent of 1% of the bird's body mass in volume of blood (Fair et al. 2010) and check to make sure that the bird is in good condition and that it is alert before releasing.

Once drawn, blood should be stored for both microscopic and molecular analyses. Blood smears for microscopic analysis of parasite morphology should be prepared immediately after drawing blood (see next section). At this point, it is helpful to have a partner to whom you can hand the bird for euthanization. Alternatively, one person can bleed the bird and a second person can make the blood smears. The person taking the blood sample can immediately euthanize the avian specimen after blood has been collected. See the Ornithological Council's guidelines for information on appropriate methods for euthanizing birds for preparation as museum specimens (Fair et al. 2010). Following the preparation of blood smears, multiple methods may be used for preserving whole blood for DNA studies: flash freezing blood in liquid nitrogen (the "gold standard"), storage of blood on Whatman® FTA® Classic Cards, and storage of blood in a DNA preserving buffer (e.g., Queen's lysis buffer, 95% ethanol). We typically place a small amount of blood on an FTA card for quick access in the lab, then store the remainder in liquid nitrogen for long-term storage in a cryogenic facility. Many researchers prefer to use 95% ethanol for the storage of blood for molecular analysis, as it is inexpensive and easily accessible in remote locations. Blood samples on FTA cards (or other filter paper), should be stored in a dry space free of contamination, such as a zip-closing bag with silica beads. FTA cards come with preprinted subsections for applying samples. Because we only require a small amount of blood for molecular analyses, we typically subdivide the cards using a custom-made stamp so that more samples may be stored on an individual card (e.g., we store nine unique blood samples instead of four).

It is important to note that unnecessary handling of birds can lead to a loss of ectoparasites, such as hippoboscid flies, which are volant and may leave the host when they sense a disturbance. Thus, it is best to quickly euthanize the avian host specimen, swab it for microbial symbionts (see following section), and then isolate the carcass in a plastic storage bag containing a fumigant for disabling associated ectoparasites. With this system, blood samples are rapidly prepared and the avian host is quickly relieved of suffering. It is very important to label both the host and blood samples (slides, vials, FTA cards, etc.) with a unique identifier (e.g., host field number or tissue number) before proceeding. This is particularly true if a large number of birds are in queue to be processed, or when multiple researchers are processing the avian host for different parasites and pathogens. Regardless of the circumstances, it is generally good practice to label specimen tubes immediately after sampling, and to tie a leg tag with this unique

identifier directly onto the avian host immediately after it is euthanized.

Preparation and Fixation of Blood Smears

The quick preparation of blood smears is important for two reasons. First, the temperature change of blood can have profound effects on haemosporidian parasite morphology, making subsequent analyses of blood parasites complicated or even impossible. This may be linked to the life history of the parasite, with the temperature change simulating transfer of the parasite from the vertebrate to the invertebrate host, and inducing the progression of the parasite into the next stage of its life cycle (Valkiūnas 2005). Second, even when collected in heparinized microhematocrit tubes, blood can begin to clot, particularly in hot environments. If you are working alone and experience some delay before processing blood, it is helpful to first dab the end of the microhematocrit tube onto the FTA card (or other sterile paper, if planning to store blood in a lysis buffer) before applying a blood drop to the glass slides for preparation of blood smears. This removes any blood that has clotted at the end of the microhematocrit tube, allowing blood to flow more freely from the tube. As has been described in many useful guides (e.g., Gilles 1993, Valkiūnas 2005, Owen 2011), smears should be prepared on clean glass slides. Dust particles, grease, and scratches will significantly decrease the quality of your blood smear, and contaminated slides should be avoided. Unused slides that have been contaminated by dust or debris can be cleaned using ethanol and disposable wipes (e.g., Kimwipes manufactured by Kimberly-Clark) if necessary. A small drop of blood no larger than 3 mm in diameter is all that is needed to produce a good blood smear. A common error in the preparation of blood smears is the use of too much blood. The drop should be placed at one end of the slide, and a second clean "smearing" slide backed up at a 30 to 45 degree angle until it is touching the drop (Figure 10.4a,b), at which point the blood will spread across the back end of the smearing slide via capillary action. The smearing slide should then be pushed forward, briskly and smoothly, with blood trailing behind it (Figure 10.4c). If done properly, the blood smear should be in the shape of a bullet, with densest concentration of blood near the origin of the drop, and the edges of the smear feathering out toward the end (Figure 10.4d).

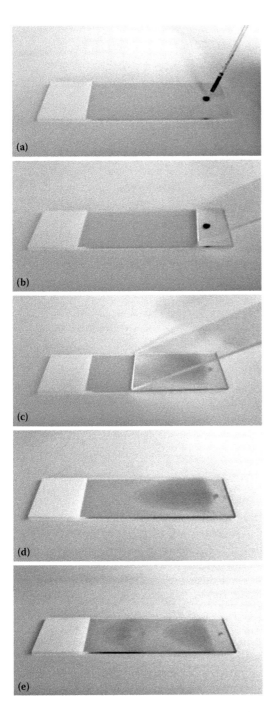

Figure 10.4. How to make a blood smear. (a) Place small drop of blood from microhematocrit tube near end of slide. (b) Back second slide up to drop of blood at a 30- to 45-degree angle. (c) Move the "smearing" slide quickly and smoothly across to spread the blood in a thin film. (d) Single thin blood smear. (e) Example of two blood smears prepared on a single slide.

It is important to produce multiple slides per individual when possible. Not only does this increase the number of fields that can be searched for haemosporidian and other blood parasites, but more important, it will allow for the deposition of slides at different institutions, which is often important when operating with collaborators. In the interest of maximizing the number of searchable fields, while maintaining the ability to share slides with different institutions, it is quite practical to produce multiple blood smears from the same individual on individual glass slides (Figure 10.4e). The ability to do this will vary with level of skill, environmental conditions, and the condition of the bird.

It is best to fix blood smears as soon as possible once they are air-dried, placing them in 100% methanol for 1 minute. If methanol is unavailable, it can be substituted with 96% ethanol and an extended fixation time of 3 minutes. Slides can be placed back-to-back in a Coplin jar containing fresh methanol. Replace the methanol frequently (every two to three batches of slides) to limit the effects of dilution and debris. Allow slides to air-dry face up, and once dry, place the slides back-to-back in a plastic slide box for storage. Alternatively, slides can be individually wrapped with paper, such as a Kimwipe, and bound together using rubber bands. Store fixed blood smears with silica beads, and stain as soon as possible. Although most staining agents containing methyl blue will allow for microscopic detection of Haematozoa, Giemsa remains the gold standard, and is the most commonly used stain for parasitological studies of haematozoan blood parasites. Rapid staining methods used for diagnosis of human malaria (e.g., Field's or Romanowsky stains) are less stable than Giemsa and prone to fading, and therefore are not appropriate for long-term storage and taxonomic studies. It is best to purchase a high-quality Giemsa stain and produce your own staining buffers (see Box 10.1 for formulas and staining protocol).

In many field situations, it is not possible to stain slides on the same day, or even within 1 week, of preparation (which is recommended). Older blood smears, if made and fixed properly in the field, are still useful for taxonomic research. However, it is a good idea to "refix" the slides once back in the laboratory by dipping them again in 100% methanol for 1 minute, and allowing to air-dry (R. Barraclough, pers. comm.). Older slides tend

Box 10.1   Giemsa Staining Protocol for Haematozoan Parasites

1. Prepare alkaline (N_2HPO_4) and acid (KH_2PO_4) stock phosphate buffers as follows:
   a. Buffer A: (9.50 g N_2HPO_4) + (990.50 mL dH_2O) = 1000 mL alkaline stock
   b. Buffer B: (9.07g KH_2PO_4) + (990.93 mL dH_2O) = 1000 mL acid stock
   c. Working buffer: (61 mL Buffer A) + (39 mL Buffer B) = 100 mL working buffer. The working buffer pH should be in the range 7.0 to 7.2.

   Stock buffers can be kept and reused to prepare a working buffer, which should be made fresh every few days. Stock buffers can be stored at room temperature indefinitely. Fresh working buffer should be made every few days and can be stored at room temperature as well.

2. Place a thin layer of high-quality Giemsa stain at the bottom of a Coplin jar, then add working buffer to produce a ~10% buffered Giemsa stain.

3. Slides should have already been fixed in methanol after blood smear preparation (in the field). However, it is a good idea to dip slides in methanol again before staining, particularly if they have been exposed to humidity, dust, and so forth. This additional methanol rinse will produce cleaner and more evenly stained slides.

4. Add slides to Coplin jar and allow to stain for 60 to 90 minutes. The duration of staining time will vary depending on the age of the slides, the quality of the Giemsa stain, and the concentration of the buffered Giemsa stain. Older slides tend to take up stain more readily and are likely to stain too darkly if left for too long. It is a good idea to test one or two slides before processing an entire batch.

5. Once staining is complete, remove slides from the Coplin jar and rinse off residual stain under water.

6. Slides should be labeled archivally, and the label should at minimum include the host voucher number. Once dry, slides should be placed in a secure slide box for long-term storage.

to absorb more stain, so Giemsa staining concentration and staining time should be reduced. It is a good idea to test your staining protocol on one slide before processing an entire batch. If the blood smear is overly dark and blue, simply dilute your stain or reduce the amount of time (add more stain or time if the slide appears too light). Once staining is complete, slides should be placed in a durable slide box and kept in a cool, dry environment for long-term storage. These slides will serve as vouchers and can be referred to at any point for morphological analyses of myriad haematozoan parasites found in avian hosts (Figure 10.5).

SAMPLING PROTOCOLS FOR THE STUDY OF MICROBIAL SYMBIONTS (BACTERIA, FUNGI, AND VIRUSES)

Studies of microbial symbionts in wildlife are in their relative infancy, and methods for sampling bacterial, fungal, and viral symbionts of birds are still being developed and improved. As our understanding of the interplay between microbial symbionts and avian evolution and ecology grows, so too should collections of samples from vouchered birds that are appropriate for studying these microbes (e.g., gastrointestinal tracts, fecal, buccal, and conjuctival swabs, etc.). The collection of such samples will ultimately provide important time series for the study of changes in microbial diversity in birds, which may allow researchers to measure the effects of environmental phenomena such as climate change and anthropogenic habitat disturbance, as well as the impacts of naturally occurring phenomena, such as dispersal and colonization, epidemics, and naturally fluctuating food cycles. Avian gut microbiota have probably received the greatest amount of attention (for a review of the current trends in this area of study, see Waite and Taylor 2015), and recent studies (e.g., van Dongen et al. 2013, Hird et al. 2015) provide useful methodological descriptions for studying the avian gastrointestinal microbiome.

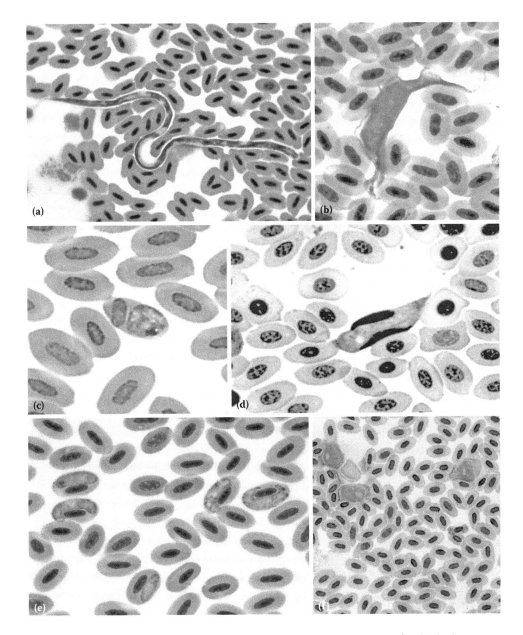

Figure 10.5. Microphotographs from Giemsa-stained blood smears of haematozoan parasites found in birds.
(a) Microfilarial nematode ex *Pycnonotus barbatus*, Vwaza Marsh Wildlife Reserve, Malawi. (b) *Trypanosoma* sp. ex *Pycnonotus barbatus*, Vwaza Marsh Wildlife Reserve, Malawi. (c) *Plasmodium* sp. ex North American passerine. (d) *Leucocytozoon toddi* ex *Meleagris gallopavo*, Ithaca, NY. (e) *Haemoproteus* sp. ex *Ispidina picta*, Vwaza Marsh Wildlife Reserve, Malawi. (f) Coinfection with *Leucocytozoon* sp. and *Haemproteus zosteropis* ex *Zosterops senegalensis*, Nyika National Park, Malawi.

However, microbial symbionts are by no means restricted to the gastrointestinal tract, and other areas to consider when sampling a bird for microbial symbionts include the respiratory tract, conjuctiva, nares, and feathers.

As microbiome studies are still in their relative infancy, and best methods and practices are still being developed, we encourage researchers to check the most recent literature for proper sampling and storage techniques. Methods of preservation for the study of viruses in particular vary substantially, and knowledge of the viral family of interest is important when determining sample preservation methods. Because this area of study is in a state of rapid development, we will provide here only the most basic advice on when and how to incorporate sampling for microbial symbionts into the field workflow, based on our experience.

Sampling for microbial symbionts (bacteria, fungi, viruses) should be conducted immediately after euthanization. Following euthanization, insert separate cotton-tipped sterile applicators into the (a) cloaca, (b) buccal cavity, and (c) conjuctiva, and rotate several times within each region to swab the area as thoroughly as possible. The applicator can then be placed into a sterile collection tube, the handle broken off, and the tube sealed. Proper storage of swabs will depend on the questions being addressed and resources available. Some options include immediate storage in liquid nitrogen, RNA*later*™, or other buffers (see Vo and Jedlicka 2014 for examples on downstream processing methods). We strongly encourage researchers to consider incorporating these simple and relatively inexpensive methods into their sampling regime, as few collections of microbial symbionts from wild birds currently exist, and the benefits and impacts of longitudinal studies are as yet undetermined.

SAMPLING PROTOCOLS FOR THE STUDY OF ECTOPARASITES

Birds are parasitized by a wide variety of ecto-parasitic arthropods including, but not limited to, fleas, hippoboscid flies, lice, mites, and ticks (Figure 10.6; see Clayton and Walther 1997). Here we will focus on methods for sampling these parasites from dead avian host specimens collected during biotic survey field expeditions, although there are a variety of additional methods that can be used to collect ectoparasites from live birds and that can be used in laboratory settings (Clayton and Walther 1997, Walther and Clayton 1997, Clayton and Drown 2001). Clayton and Walther (1997) include a broad review of methods for quantification and collection of avian ectoparasites.

Fumigation and Collection of Ectoparasites

The first rule for collecting ectoparasites from avian host specimens is to not allow dead host specimens to come into physical contact with one another. Each freshly killed or caught bird should be isolated in a separate clean bag. Birds

that are mist-netted can be carried back to the specimen preparation area alive using clean cloth bags. Bags should be washed thoroughly between uses to avoid potential contamination of parasites between individual birds. Birds that are shot can be placed immediately into a plastic storage bag with a note indicating soft-part colors (which may fade very quickly) and a cotton ball with a few drops of ethyl acetate on it. This will begin fumigation, allowing the bird to be immediately ruffled for ectoparasites upon arrival in the field camp. Upon arrival into camp each dead bird specimen in a bag should have a field number assigned to its field sheet (e.g., Figure 10.3c) so that the remaining host and parasite data collected from the bird can be linked to the voucher host specimen. Birds that are caught live will be euthanized after blood samples are collected. These birds should also be placed in a clean plastic storage bag with a cotton ball soaked with a few drops of ethyl acetate and a field sheet indicating field number assigned to that specimen. Ethyl acetate is considered harmless to humans and yet is effective for killing ectoparasites (Fowler 1984).

After fumigating the bird for 15 to 20 minutes, carefully remove the bird from the plastic storage bag over a large sheet of clean white paper. In the field we typically use a lunch tray covered with a large sheet made by taping together two pieces of 8.5″ × 11″ white paper. In windy conditions it is a good idea to use a cardboard box or other windbreak to block the wind. Also, one can tape the paper to the lunch tray to keep the wind from blowing the sheet away, which could result in the loss of ectoparasites. Before ruffling the bird's feathers to dislodge and remove ectoparasites, always check the inside of the plastic storage bag for parasites. If any ectoparasites have fallen off of the host inside the bag, use a paint brush wetted with absolute ethanol to pick up the ectoparasites and place them in a new vial filled with absolute ethanol (do not use denatured ethanol). Hold the bird with one hand and use the other hand and fingers to ruffle all of the bird's feather tracts. Start with the wings, including the primaries and coverts, and then while holding the legs with one hand you can ruffle the feathers of the belly, back, and head. Then hold the head and/ or beak (if the bird has a large bill) and "beat" the bird to loosen attached ectoparasites. For small birds one can also hold them between two cupped hands and shake them up and down like

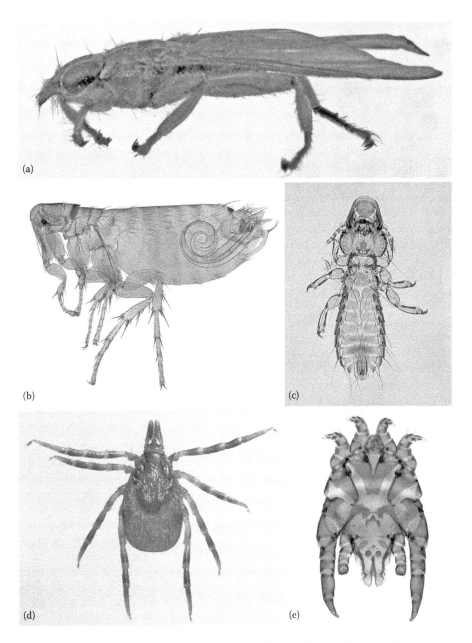

Figure 10.6. Representative images of common ectoparasite groups found on birds. (a) Hippoboscid fly: *Icosta Americana* ex *Accipiter cooperi* (photo by Jason Weintraub). (b) Flea: *Ceratophyllus altus* ex *Campephilus magellanicus* (photo by Michael W. Hastriter). (c) Chewing Louse: *Cotingacola lutzae* ex *Laniocera hyppopyrra* (photo by Michel P. Valim). (d) Tick: *Ixodes brunneus* (photo by Lorenza Beati). (e) Feather mite: *Anomothrix machadoi* ex *Buceros leadbeateri* (photo by Fabio Akashi Hernandes).

dice. This also helps to loosen strongly attached ectoparasites. Furthermore, ectoparasites such as feather lice (Phthiraptera: Ischnocera) can include four ecomorphs that are specialized on different regions of the avian host's body, including the head, wing, body, and generalist ecomorphs (Johnson et al. 2012). Be sure to cover all

of the body carefully to thoroughly sample these different ectoparasites.

Pick up all ectoparasites that fall off of the host onto the paper, using the tip of a fine paintbrush moistened with absolute ethanol. Place these parasites into a vial of absolute ethanol. It is best not to use a forceps to pick up ectoparasites because this may damage morphological features on the specimens. Although many previous papers have suggested using 70% ethanol for preservation of ectoparasites (e.g., Clayton and Walther 1997), we have found that absolute ethanol is best because it preserves both morphology and DNA of the specimen; specimens stored in 70% ethanol will very quickly be useless for DNA extraction. However, if absolute ethanol is unavailable, 95% ethanol can be used in its place for collection and storage of ectoparasites. Place a label made with archival acid-free cotton fiber paper and written using an indelible Pigma Micron pen inside the vial. The label should contain the host taxon name, field collecting number, date of collection, collecting locality, and name of parasite collector. Be sure to note information on the ectoparasite collecting event in the field notes catalog. It is also important to note negative collecting events (when no parasites are found), as these data will allow one to calculate prevalence and intensity of parasitism. After picking up the parasites, continue with several more bouts of ruffling until no parasites fall off of the host. Before moving on to the next host specimen, clean the collecting surface and inspect your hands to be sure there are no contaminant ectoparasites on them.

The ethyl acetate fumigation with postmortem ruffling method outlined earlier will collect most lice, ticks, fleas, hippoboscid flies, and external mites (Figure 10.6). However, this method is not appropriate for quantification for all of these parasites. For permanent ectoparasites, such as lice, which live their entire life cycle on the host, this postmortem ruffling method is quantitative only when conducted to a point of diminishing returns (Walther and Clayton 1997, Clayton and Drown 2001). Moreover, this method is not suitable for quantitatively sampling ectoparasites that live inside the throat pouch, nasal cavities, feather quills, and under the skin. To thoroughly sample avian feather mites, one should visually search through the plumage using a stereomicroscope (although ruffling will allow the collection of some mites). This also allows one to note the locations where each mite taxon is found. For the subset of feather mites that inhabit the wings, one can hold the flight feathers up to the light and look for mites inserted between the feather barbs. One can then use the handle of the paintbrush to disturb and "unzip" the barbs of these feathers so that the mites fall onto the collecting paper. Other mites, such as nasal mites, require flushing the nares

with water into a gallon jar, then pouring through a #200 sieve to filter out the mites. Other quantitative methods are available, such as body washing, which removes an even larger fraction of ectoparasites than postmortem ruffling (Clayton and Drown 2001). However, this method is not practical for field survey situations but is useful for smaller-scale studies when specimens can be processed in the lab (e.g., Koop and Clayton 2013). Sometimes embedded ticks do not fall off the host after fumigation. In this case, use a forceps to grab the tick as close to the skin as possible to dislodge it without damaging its mouthparts.

After ruffling, the ectoparasite collector will pass the host specimen on to a bird skinner who will prepare the bird specimen and gently necropsy the carcass to gather standard internal organ data. The bird skinner then will pass the carcass on to the endoparasite collector for further dissection and collection of endoparasites. The bird skinner can either sample liver and heart tissues at this time or can pass labeled tubes to the endoparasitologist to collect these tissues.

Preparation and Curation of Ectoparasite Specimens

After returning from the field, individual vials of ectoparasite specimens can be examined to determine and quantify contents. We examine specimens in a glass dish filled with absolute ethanol and use a stereomicroscope to observe specimens and manipulate them with a paintbrush and/or bent syringe needle. Sometimes we use a glass bulb pipet to return specimens to the original vial. Always be sure that a pipet is clean before reusing it. Specimen preparation methods for each ectoparasite group are taxon-specific and can be used to produce slides for morphological examination and for slides of voucher specimens, from which DNA has been extracted. For morphological examination and vouchering for molecular projects, lice are mounted in Canada balsam using a clearing and slide mounting technique described by Palma (1978), whereas mites are mounted in Hoyer's medium (Baker and Wharton 1952). For DNA extraction of lice we typically use a sterilized syringe needle to make a cut between the head and the thorax or between the thorax and the abdomen of the louse depending on the taxon (Valim and Weckstein 2011) and then place this specimen into the digestion buffer provided in the QIAamp DNA Micro Kit (Qiagen, Hilden, Germany). We then allow the louse to digest over two nights and then follow the manufacturer's directions. In pipetting the liquid from the digestion to the Qiagen filter, we are careful to leave the louse in the original digestion tube. We then add 70% ethanol to the tube to preserve the louse until we begin the slide mounting process. Depending on the size of the louse we elute the DNA off the filter with 50 to 100 μl of buffer AE. We give each louse a unique identifier that includes abbreviations for the louse taxonomic name, host alpha taxonomic name, the date of extraction, and the tube number in that batch of extractions. Rather than wait a long time after DNA extraction, it is best to start clearing and slide mounting vouchers as soon as possible. Other ectoparasites, such as hippoboscid flies, can be kept in ethanol, pinned, or slide mounted in Balsam depending on the size of the fly.

SAMPLING PROTOCOLS FOR THE STUDY OF ENDOPARASITES

Birds can harbor an astounding diversity of parasitic worms from all major groups of helminths (except monogeneans), namely, the cestodes (Eucestoda), digeneans (Digenea), nematodes (Nematoda), and acanthocephalans (Acanthocephala; Figure 10.7). Birds can be parasitized by both adult and larval stages of various parasitic worms, although birds in general have fewer larval stages of helminths in comparison with other major vertebrate groups because they rarely serve as intermediate or paratenic hosts of helminths. Helminths can be found in virtually every part of the bird's body. Although the majority is parasitic in the gastrointestinal tract (GIT), including somewhat unusual sites such as cloaca, inside of the crop, or under the lining of the gizzard, many parasitize other organs, such as the liver and gall bladder, kidneys, bursa of Fabricius, trachea, eyes, and mouth cavity (Atkinson et al. 2008). Adult filarial nematodes can be found in the body cavity, under the skin, on the brain, on the heart, and inside bones. Finally, blood flukes and larval stages of filarial nematodes (microfilariae) may reside in both the venous and arterial sides of the circulatory system. Thus, a complete helminthological examination of an avian host can be a very time-consuming process, especially when it involves larger birds. It is extremely difficult

to write a unified procedure for all birds due to the great diversity of avian host sizes, anatomical peculiarities, parasite localization, and parasite loads. For example, waterfowl and other aquatic birds usually—although not always—host greater diversity of parasitic worms than terrestrial birds. The dissection protocol in the next section focuses primarily on helminth recovery from the GIT and associated organs, where the majority of parasitic worms are expected to be found. Illustrations and descriptions of bird anatomy can be found in any general ornithology or vertebrate anatomy textbook, and are also readily available through numerous online resources.

Examination of Gastrointestinal Tract and Other Organs for Endoparasites

Birds are most commonly examined for endoparasites as a part of broader ornithological studies that involve the collection of voucher specimens (e.g., study skins and skeletons) for deposition in museum collections. In such cases, a parasitologist typically receives a body or organs of an already euthanized and skinned bird, which they may proceed to dissect.

Dissection techniques may vary. When the host body is received, you should first carefully examine the exterior of the carcass (particularly around the neck) for the presence of filariid nematodes, which should be easily visible. Each avian carcass should be placed in a dissecting tray of appropriate size. In a comprehensive sampling protocol, such as the one recommended in this chapter, an incision already should have been made to access the body cavity for tissue sampling and sexing of the host. However, if the body is not already opened by the host voucher preparator (e.g., in cases where birds have been donated by hunters), then you will need to open the carcass by making an incision on the abdominal side of the body. The incision should be made with scalpel or scissors along the midline of the body on its abdominal side, approximately from the level of sternum to the level close to cloaca, but not reaching the cloaca. When not preserving the skeleton, one can cut through the rib cage to provide easier access to the organs located in the upper part of the body (esophagus, trachea, heart, air sacs). When making any cuts be careful to not cut the GIT, as this will cause gut contents (and possibly helminths) to spill into the body cavity. This may also happen

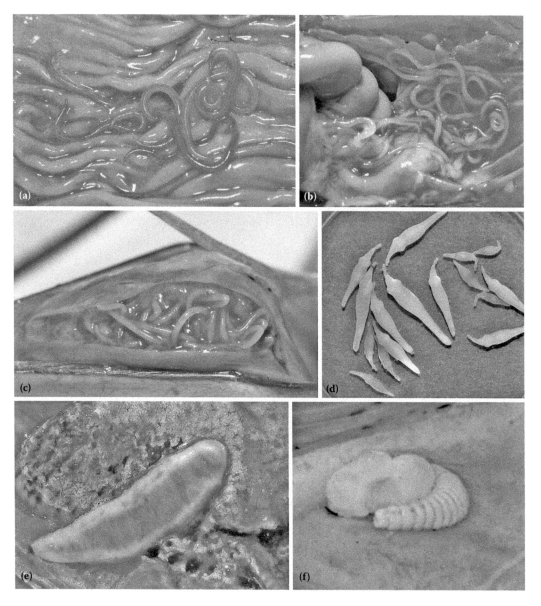

Figure 10.7. Examples of different groups of helminths inhabiting various sites in bird bodies. (a) Spirurud nematodes in stomach of Anhinga *Anhinga anhinga*. (b) Filariid nematodes in body cavity of Barn Swallow *Hirundo rustica*. (c) Dracunculid nematode *Avioserpens* sp. under skin of the chin area of Little Egret *Egretta garzetta*. (d) Acanthocephalans obtained from Northern Shoveler *Anas clypeata*. (e) Hymenolepidid cestode *Cloacotaenia megalops* attached to the wall of cloaca of Northern Shoveler *A. clypeata*. (f) Cyclocoeliid digenean in body cavity (on the lung) of Eurasian Coot *Fulica atra*.

if a bird was shot and the GIT is damaged. As with the exterior of the carcass, you should inspect the interior of the body cavity for visible filariid nematodes once an incision has been made. If blood flukes (Schistosomatidae) are among the targeted parasites, use citrated saline solution (which can be prepared by dissolving 5 g of nimiodized table salt and 3 g of sodium citrate $C_6H_5Na_3O_7$ in 1 L of

water) throughout this procedure. Make sure to pour some citrated saline into the bird body cavity as soon as you open it. If blood flukes are not a target, use regular saline throughout. Any helminths discovered should be kept alive in saline until fixation.

Remove the complete GIT by carefully cutting connective tissues holding it in place. Then cut the mesenteries that hold intestinal coils together. When separating the intestine from the liver do not cut the gall bladder. Cut through skin surrounding the cloaca to keep it intact. In young birds, the bursa of Fabricius may be found on the side of the cloaca. It is best to keep the cloaca and bursa of Fabricius together until examination. Move the entire GIT into a tray of appropriate size (e.g., a large glass petri dish for small passerine birds, or a large glass baking dish for ducks or other large bodied birds). At this point, different parts of GIT (esophagus, stomach, small intestine, ceca, rectum, cloaca, bursa of Fabricius, etc.) can be separated for subsequent examination. Remove the liver and put it in a separate dish with saline. The spleen and pancreas very rarely contain parasites, although digenean infections can be encountered in these organs. Carefully remove the kidneys. This can be done by pulling one end of a kidney upward using a forceps of appropriate size and cutting underneath with scissors. Place kidneys into a separate dish with saline. Remove the trachea and place it in a dish with saline. The trachea very rarely serves as a site for helminths, but large digeneans such as *Orchipedum* in cranes and pathogenic nematodes *Syngamus trachea* in galliforms can be found there.

Disrupt air sacs with a gloved hand or using scissors. Carefully rinse the entire body cavity with citrated or regular saline, and pour it from the body cavity into the pan, and then from the pan into a beaker of appropriate size. Allow the contents to settle. This process, called sedimentation, allows endoparasites and other solids to settle to the bottom and the bloody mixture in the supernatant to be discarded so that parasites may be observed and collected for fixation. Once sedimentation is mostly completed, discard the supernatant into another container by carefully pouring it off. Be sure to pour the supernatant slowly to avoid loss of the materials at the bottom of the beaker, then add fresh saline to the sediment. Shake or stir. Repeat the procedure until the supernatant is reasonably clear. Pour small portions of sediment into a petri dish and examine under a stereomicroscope. Although some digeneans can be large, such as members of the Cyclocoeliidae, others, such as those that fall out of damaged intestine or kidneys, may be much smaller. Blood flukes or their fragments, for example, may be extremely small and transparent. Helminths should be transferred using

pipettes with orifices of different sizes or lifted with curved forceps, curved needles, or similar instruments. It is important to avoid grabbing and holding any helminths using forceps, with the exception of large nematodes and acanthocephalans, which can be taken and transferred using soft forceps. Handling helminths with forceps almost invariably leads to their damage or destruction.

Examination of the intestine usually takes longer than other organs. The order of organ examination depends on the priorities of your study. We usually examine liver and kidneys first. In small birds, the gall bladder may be studied without separation. In larger birds, however, it is best to separate the gall bladder from the liver and cut it open for examination in a separate small petri dish. Liver and kidneys need to be torn into small pieces, which can be done using scissors or tweezers (especially in the case of very small birds). However, we prefer to gently break apart liver and kidneys with gloved fingers, which preserves ducts for examination and careful dissection of parasites, and reduces the probability of parasites being cut or damaged. Some dicrocoeliid digeneans from the liver (e.g., *Brachylecithum*, *Lutztrema*) and members of the family Eucotylidae from the kidneys can be tightly packed in the ducts and may not be easy to recover. The disrupted liver and kidneys of small birds can be examined immediately under a stereomicroscope. In case of larger birds (e.g., aquatic species), process the liver and kidney tissues using the same sedimentation method as described for the body wash: pour the disrupted liver or kidney tissues into a jar or a bottle with a lid, this time shaking the bottle, then pour the liquid into a beaker, allow for sedimentation to clear the supernatant, and retain the solids at the bottom. Repeatedly add new clean saline and carefully pour off until the supernatant is clear. Once clear, the solid contents at the bottom of the beaker can be examined. There is no need to shake tissues more than once. Examine the sediment in small portions under a stereomicroscope. Besides readily visible digeneans (e.g., Dicrocoeliidae and Opisthorchiidae in liver, Eucotylidae and Renicolidae in kidneys), the liver and kidneys are also important target organs to search for blood flukes. They can be extremely small, transparent, and are frequently fragmented, thus requiring particular attention during sediment screening.

The esophagus can be opened with scissors longitudinally. Some helminths, such as larger nematodes, can be readily seen and removed without optics, but the lumen and walls of the esophagus need to be examined under a stereomicroscope. After examination, you can compress the esophagus wall between two pieces of glass (size and thickness vary depending on the size of the bird), because some nematodes can be located in the wall and can be seen only under compression.

The stomach may contain representatives of several nematode families, digeneans, and even cestodes. The proventriculus and gizzard can be separated before examination. Stomach contents need to be removed and examined for the presence of helminths, nematodes in particular. Upon preliminary examination, the proventriculus wall can be scraped with the edge of a microscope slide and the scraped material examined under the stereomicroscope. Nematodes, digeneans, and cestodes may be found under the lining of the gizzard. In small birds, the gizzard wall lining can be easily peeled using forceps. In larger birds, especially large aquatic birds, peeling the gizzard wall can be more difficult, usually resulting in multiple fragments, and thus should be done in saline. After all of the lining is removed and rinsed, the liquid should be examined for helminths. The tapeworm genus *Gastrotaenia*, which are uniquely parasitic under the gizzard wall lining in anseriform birds, are small and easily mistaken for nematodes.

For the following steps, using scissors with rounded/blunt/balled ends, or at least on one end, is strongly recommended (Figure 10.8a,b). The duodenum and small intestine (Figure 10.8c,d) typically contain the highest diversity and numbers of helminths. If blood flukes are among the targeted taxa, then examine the mesenteric veins and veins of the intestinal wall and cloaca for these parasites before opening the intestine or cloaca. Next, one can remove excessive tissue (e.g., fat and connective tissue) around the intestine and open it with a longitudinal incision using scissors while holding the end of the intestine with tweezers/forceps (Figure 10.8e). For easier detection of tapeworm strobilae, it is best to begin the incision at the posterior end (Figure 10.8f). If a tapeworm is detected, the subsequent dissection can give more attention to find the scolex (or scoleces in case of numerous cestodes) buried in the mucous layer of the intestinal wall. After the entire intestine is open, it needs to be inspected for any other helminths readily visible by naked eye or under a stereomicroscope (Figure 10.8g). Acanthocephalans may be deeply embedded in the intestinal wall and need to be carefully removed. Any other visible helminths also should be removed and placed in a petri dish with saline (Figure 10.8h). In some cases, this is not feasible due to a very high number of helminths, especially if they are very small. To dislodge embedded helminths, scrape the intestinal wall with a side of a microscope slide under a shallow layer of saline solution. It is best to secure the end of the intestine with forceps, and it is essential to press the slide with sufficient force to scrape the intestinal wall deep enough to not break helminths that are deeply embedded in the layer of mucus. If the glass only slides on the surface of the mucus, then it is likely that some worms will be broken while others will remain attached to the intestinal wall in the mucus. Following this step, pour the saline with scrapings into a cylinder or bottle with a screw cap, shake vigorously, and empty into a tall beaker (a water bottle with the top cut can be used in the field) and allow to settle for about 2 minutes, depending on the density of the mix, before pouring off the supernatant. Sedimentation should follow the same basic procedure previously outlined. When the supernatant is clear the sediment should be examined under a stereomicroscope or can be fixed with ethanol for subsequent examination in the lab (for a similar procedure, see Justine et al. 2012).

Fixation of Helminths

The choice of fixative for helminth fixation depends on the purpose and future use of the material. In cases when numerous individuals of the same species of helminths are available and time permits, it is always a good idea to use several fixatives, each optimized for a different downstream purpose. For example, morphological light microscopy, transmission electron microscopy (TEM), scanning electron microscopy (SEM), histology, immunology, and molecular methods each require a different fixative. Some of these downstream procedures, such as TEM and immunological/immunohistochemical

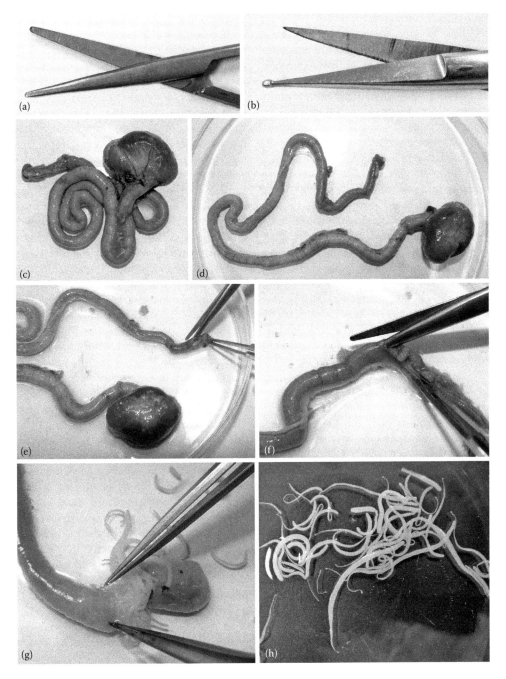

Figure 10.8. Dissection of the intestine of a small bird (American Robin, *Turdus migratorius*). (a, b) Recommended scissors with rounded or balled (so-called artery scissors) ends. (c) GIT removed from bird (esophagus was separated). (d) Straightened intestine with mesentery cut and fat removed. (e, f) Opening of the intestine with scissors starting from posterior end. (g) Gentle tearing of intestine using fine forceps after tapeworms were discovered. (h) Cestodes removed from intestine in saline.

studies, require specialized fixatives, whereas others, such as morphology, SEM, and basic molecular analysis not involving next-generation sequencing (NGS) methods, can use a common fixative. Seventy-percent ethanol is a fixative of choice but with some limitations. Some authors prefer fixation of specimens for future staining with hot (steaming but not boiling) 10% formalin (i.e., 4% solution of formaldehyde) with subsequent storage in formalin. Others fix specimens in formalin with subsequent transfer to 70% ethanol for long-term storage. We have not noticed a significant difference in side-by side comparisons of specimens simultaneously fixed with either fixative (and stained with the same stain). However, the use of heated formalin may have a negative impact on the collector's health due to inhalation of toxic vapors. Furthermore, formalin hinders DNA extraction and subsequent molecular analyses. This is particularly important when considering that morphologically similar ("cryptic") species of helminths are not readily distinguishable in the field and may be present in samples. We therefore advocate the use of 70% ethanol as the single universal fixative equally suitable for "routine" morphological and molecular studies. It is also less hazardous and has fewer transportation restrictions than formalin.

We would like to emphasize that specimens fixed in ethanol (70% or 95%) need to be placed in a freezer or at least a refrigerator as soon as possible and stored long term in a freezer. In the author's lab (V. V. Tkach), 80% ethanol is usually used as the starting concentration in the field because, during pipette specimen transfer to the storage vials, some fluid (water or saline) is inevitably added to the ethanol, which dilutes it to ~70%. Thus by using 80% ethanol we ensure that concentration does not decrease significantly (which would result in poorly preserved specimens). It is recommended that you change ethanol once after initial fixation to ensure sufficient concentration, though this is not always feasible during field collecting trips due to the time and material limitations. As a general rule, for NGS and transcriptomic applications, freezing in liquid nitrogen is the gold standard for field preservation, followed by 95% ethanol for NGS, and RNA*later* (Sigma-Aldrich, St. Louis, Missouri) for studies targeting RNA. Each group of helminths needs to be fixed using a slightly different protocol to properly preserve the morphological features of interest. The following are our recommendations for fixation of live worms using 70% ethanol.

Flatworms (digeneans and small- to medium-sized cestodes) can be heat killed with hot water. Remove most of the saline from a petri dish leaving only a very small amount to cover worms to prevent even momentary desiccation. Pour hot water (steaming, not boiling) onto the worms, and stir the water using propulsion by pipette. Add ambient temperature water immediately to prevent overheating and then transfer the worms into vials with 70% ethanol (again, for practical reasons we use 80%) as soon as possible. In field conditions, a good quality thermos can be used to keep water hot for some period of time rather than reheating it every time one needs to heat kill specimens. Change hot water as needed. Alternatively, flatworms can be pipetted into a petri dish or a small beaker with hot saline (see Cribb and Bray 2010) with subsequent transfer to ethanol. In the case of very thick-bodied digeneans, a subsample can be fixed in ethanol using slight pressure with a cover slip or slide (depending on the worm size). Such specimens can provide a better view of the organization of internal organs. However, specimens fixed in this manner may be distorted and should not be used to make measurements. Tapeworms should never be fixed under pressure. Very large tapeworms usually contract and become less suitable or completely unusable for morphological analysis if the aforementioned heat-killing method is used. Thus, very large tapeworms may be killed and relaxed at the same time by moving them between a petri dish with water (ambient temperature) and a dish with ethanol of weak concentration (10%–15%) using a curved needle or curved tweezers to hold the tapeworm from underneath. Scoleces of large cestodes with armed rostellum can be fixed separately before the strobila is relaxed to avoid loss of rostellar hooks. When the tapeworms die they can be fixed in ethanol. We usually pre-fix tapeworms still in a petri dish before transferring them to a vial of appropriate size. The volume of ethanol in the vial should be at least 4 to 5 times greater than the volume of tissue. Thus, with larger tapeworms, 50 mL falcon tubes may be the preferred storage container rather than vials.

Larger nematodes with a thick cuticle can be heat-killed following the general procedure outlined earlier. However, hot saline has to be used rather than water to prevent nematode bodies from rupturing due to the difference of osmotic pressure. Instead of hot saline being poured onto nematodes, the petri dish containing nematodes already in saline can be heated using an alcohol burner or lighter until the nematodes die. Heat the saline only to the point that it begins steaming and no further. The nematode cuticle may shrink somewhat in ethanol, even if it is only 70% concentration. Higher concentrations of ethanol may distort nematodes irreversibly and are certainly not recommended for specimens to be used for future morphological examination. Neutral buffered 10% formalin can be used for fixation and does not negatively affect morphology, but we tend to not use it for the aforementioned reasons.

Small nematodes with thin cuticle can be fixed with hot saline or hot 70% ethanol. In the latter case they are simultaneously killed and relaxed. One must exercise caution to prevent ethanol from catching on fire during heating.

For an adequate morphological study of acanthocephalans, the proboscis should be fully everted. This is rarely achieved by heat killing. Leaving them to die in water until the prosci are everted usually produces the best results. When possible, a dish containing water and acanthocephalans should be kept at low temperature (e.g., in a refrigerator), but even at ambient temperature (e.g., in a field camp) the desired result is usually achieved. Then acanthocephalans can be transferred into ethanol.

All vials should have internal labels. Writing information with a marker on the outside of the vial is not sufficient and is likely to result in the loss of data, which may render the specimens useless. Even vapors of ethanol inside a vial storage box may dissolve ink on the outside of vials. Therefore, labels to be placed inside of specimen vials need to be made of a paper that resists prolonged soaking in fixative without deterioration (archival acid free cotton fiber paper is best, but there are other alternatives), and should be written by hand or printed using either pencil or alcohol-proof ink or alcohol-resistant printer ink. In the latter case, the ink needs to be tested prior to use, otherwise there is a risk of losing label information. Ethanol should fill almost all of the remaining space inside the vial, leaving just a very small space for potential expansion at higher temperature. Leaving too much air in vials may result in dried specimens during the transportation of the vial boxes.

Preparing Endoparasites for Morphological Study

Morphoanatomy of flatworms (digeneans and cestodes) is usually studied on permanent total microslide mounts. Parasitic nematodes are usually studied on temporary mounts and acanthocephalans can be studied on either permanent total mounts (mostly to study internal organs) or temporary mounts (mostly to study the proboscis armature and egg structure). There are a plethora of recipes for stains that have been used for trematode and cestode total mounts over the last hundred years. They can be found in numerous manuals and special publications (e.g., Dubinina 1971, Ivashkin et al. 1971, Pritchard and Kruse 1982, Georgiev et al. 1986). We refrain here from a discussion of advantages or disadvantages of one or another staining method. Instead, we provide only two stains and corresponding protocols that have been successfully used in Tkach's laboratory for a great diversity of parasitic flatworms (see Boxes 10.2 and 10.3). Iron acetocarmine, Gill's haematoxylin, and Delafield's haematoxylin are useful alternatives that can be found in the aforementioned references. It is important to realize that there is no one-size-fits-all staining procedure for every kind of specimen, and the amount of time and stain concentration may reasonably vary from taxon to taxon of flatworms. Staining and mounting is as much an art as it is science. Tkach's lab usually uses staining protocols with alum carmine (after Dubinina 1971, with minor modifications) and Mayer's haematoxylin (a somewhat modified protocol used in T. Cribb's laboratory; e.g., Miller et al. 2010). Both stains are water based and require specimens to be rinsed in water prior to staining.

Box 10.2   Alum Carmine Staining Protocol for Endoparasites

1. Rinse worms in distilled water. Rinse time depends on the size of the specimens. Water may need to be changed once for larger specimens. When transferred from ethanol to water, specimens will float. Sink them by pipetting water onto them or by using a soft tool such as a paintbrush.

2. Transfer specimens to stain. The stock solution of the stain can be diluted with distilled water immediately prior to staining. The level of dilution is flexible, but a stain that is roughly 2× diluted usually works well. Somewhat longer staining time with more diluted carmine usually produces better results, but it depends on the group of parasites, fixation, and so forth. Test stain a few specimens of lesser value and you will know what works best. Staining time can be from a few minutes for small specimens with concentrated carmine to more than 30 minutes for large worms using more diluted stain.

3. Transfer specimens to water to rinse off the stain.

4. Destain in acid alcohol (0.5%–1% solution of HCl in 70% ethanol) while observing the specimen under a dissecting scope. The body filling tissue (parenchyma) should be generally free of stain, but enough stain should remain to color the internal organs. Large specimens with thick tegument may not be transparent enough for good assessment of coloration. In these cases one has to rely on experience. Replenish the acid alcohol if it becomes too pink. Destaining may take only seconds in some cases, therefore we recommend using 0.5% HCl solution in ethanol to avoid rapid destaining.

5. Transfer specimens to water to rinse off the destaining solution. At this point specimens can be straightened if needed. This can be usually achieved by stretching specimens on a piece of paper while keeping them under a thin layer of water, then add 70% ethanol and keep adding ethanol in small portions (to keep specimen wet at all times) until the specimen is hardened and can be transferred into a beaker for further dehydration.

6. Dehydration. Specimens need to be moved through a series of ethanols of ascending concentration. Ethanols at 50%, 70%, 80%, 90%, and 100% are recommended (95% can be added between 90% and 100%). To ensure complete dehydration, an additional change of 100% ethanol is recommended. Specimens can be moved through a series of beakers or may stay in the same beaker while ethanol is changed. Either way, make sure that specimens are not exposed to air at any point during the procedure to avoid an immediate desiccation and loss of the specimens. Time in each ethanol depends on the size of specimens; 30 minutes in each grade is usually sufficient for small specimens up to 3 or 4 mm in length. Longer times are recommended for larger/thicker specimens. An hour is recommended in 100% ethanol.

7. Clearing. After water has been removed from the specimens by dehydration, they are transferred to a clearing agent (clove oil [eugenol] is recommended). The clearing agent renders the parasite transparent and is miscible with the mounting medium of choice.

8. Mounting. We strongly recommend damar gum as the embedding medium. It is sold by many suppliers and is clear, cheap, relatively fast drying (much faster than Canada balsam), and xylene soluble so the specimen can be remounted if needed. The embedding medium hardens as the solvent evaporates, making a permanent mount of your specimen. To provide a support for specimens we use precut pieces of cover slips placed on both sides of a specimen prior to covering it with cover slip.

Box 10.3   Mayer's Haematoxylin Staining Protocol for Endoparasites (Similar to Alum Carmine Protocol, with a Few Notable Differences)

1. Rinse worms in distilled water as with alum carmine protocol.
2. Transfer specimens to stain. The stock solution of the stain needs to be diluted at least 1:1 with distilled water. Do not use metal instruments when working with haematoxylin. Use only a dedicated pipette to not mix with other chemicals. Somewhat longer staining time with more diluted haematoxylin usually produces better results, but it depends on the group of parasites, fixation, and so forth. Usually staining takes from 15 to 60 minutes.
3. Transfer specimens to water to rinse off the stain.
4. Destain in 1% aqueous solution of HCl while observing the specimen under a dissecting scope. The body filling tissue (parenchyma) should be free of stain, but enough stain should remain to color the internal organs. Large specimens with thick tegument may not be transparent enough for good assessment of coloration. In these cases one has to rely on experience.
5. Transfer specimens to 1% ammonia solution to neutralize destaining process. Coloration will change from red to blue or purple. Use water to rinse off the destaining solution. At this point specimens can be straightened if needed (see Box 10.2).
6. Dehydration. Specimens need to be moved through a series of ethanols of ascending concentration as in Box 10.2.
7. Clearing. After water has been removed from the specimens by dehydration, they are transferred to a clearing agent. In this case either methyl salicylate or clove oil can be used. Methyl salicylate usually produces somewhat more contrasting coloration. Specimens should be first transferred to a methyl salicylate/ethanol mix in 1:1 ratio and then to pure methyl salicylate. See Box 10.2 for directions on using clove oil.
8. Mount specimens as in Box 10.2.

CONCLUSION

The concept of the "extended avian specimen" is strongly embodied by the comprehensive sampling of avian hosts and their symbionts promoted in this chapter. Collection of avian symbiont data has led to many important discoveries in studies of avian disease ecology and conservation (Atkinson and LaPointe 2009, Parker et al. 2011, Samuel et al. 2015), morphology and development (Clayton and Cotgreave 1994, van Dongen et al. 2013), and evolutionary biology (Weckstein 2004, Whiteman and Parker 2005). Such discoveries could not have been made based on the study of host voucher specimens in the absence of symbiont data. Therefore, as technology and sampling methods continue to improve, and as species and their symbionts face ever-increasing threats to their existence, it is crucial for avian biologists to consider not only the birds we are studying, but the plethora of microbes and parasites that are living in and on each bird. By investing in these aspects of the extended specimen, researchers will preserve data that may shed light on many important areas of avian biology, as well as provide data for myriad microbial and parasitic taxa that are relatively poorly studied.

For ornithologists who do not intend to study parasites or pathogens directly, or who do not have the facilities to curate their specimens, many options exist for ensuring that specimens find their way into collections where they will be curated and utilized. We strongly recommend that type and voucher specimens be deposited in museum collections. In the United States, the main collections curating helminth and other endoparasite specimens, and providing loans for studies, are the U.S. National Parasite Collection (now a part of the Smithsonian Institution, Washington, DC), Harold W. Manter Laboratory of Parasitology at the University of Nebraska (Lincoln, NE), and the parasite collection of the Museum of Southwestern Biology (Albuquerque, NM). Most museums with large entomological holdings will also house collections of arthropod ectoparasites and are appropriate places for depositing this material. There are, of course, a large number of other parasitological collections both in the United States and around the world. We recommend submission of specimens to museums that provide specimen loans for examination. Several publications (e.g., Lichtenfels and Prtichard 1982, Lamothe-Argumedo et al. 2010, Zinovieva et al. 2015) provide useful information on location and scope of taxonomic coverage of the most important helminth museum collections.

ACKNOWLEDGMENTS

We would like to thank the organizers of the 143rd American Ornithologists' Union conference and the organizers of the symposium in which our comprehensive sampling workflow was first publicly presented. Specifically, we thank M. Webster and K. Bostwick for their commitment to spreading the idea of the "extended specimen." We would also like to thank our many colleagues who have helped support, implement, and refine many of the protocols in our workflow, including J. Engel, T. Gnoske, J. Bates, S. Hackett, and N. Rice. For photographs of ectoparasites, we thank J. Weintraub (Academy of Natural Sciences of Drexel University), M. W. Hastriter (Monte L. Bean Life Science Museum, Brigham Young University), M. P. Valim (Museu do Zoologia da Universidade de São Paulo), L. Beati (U.S. National Tick Collection, Georgia Southern University), and F. Akashi Hernandes (Universidade Estadual de São Paulo). This work was supported in part by the U.S. National Science Foundation (grants DEB-1503804 (1120054) to JDW, and DEB-1120734 and DEB-1021431 to VVT). This work has also been supported by grants to HLL from the NSF Malaria Research Coordination Network and the Cornell Lab of Ornithology Athena Fund.

LITERATURE CITED

Asghar, M., D. Hasselquist, B. Hansson, P. Zehtindjiev, H. Westerdahl, and S. Bensch. 2015. Hidden costs of infection: chronic malaria accelerates telomere degradation and senescence in wild birds. Science 347:436–438.

Atkinson, C. T., and D. A. LaPointe. 2009. Introduced avian diseases, climate change, and the future of Hawaiian honeycreepers. Journal of Avian Medicine and Surgery 23:53–63.

Atkinson, C. T., N. J. Thomas, and D. B. Hunter. 2008. Parasitic diseases of wild birds. Wiley-Blackwell, Ames, IA.

Baker, E. W., and G. W. Wharton. 1952. An introduction to acarology. MacMillan, New York, NY.

Balmford, A., and K. J. Gaston. 1999. Why biodiversity surveys are good value. Nature 398:204–205.

Bensch, S., O. Hellgren, and J. Perez-Tris. 2009. MalAvi: a public database of malaria parasites and related haemosporidians in avian hosts based on mitochondrial cytochrome b lineages. Molecular Ecology Resources 9:1353–1358.

Brooks, D. R., and E. P. Hoberg. 2000. Triage for the biosphere: the need and rationale for taxonomic inventories and phylogenetic studies of parasites. Comparative Parasitology 67:1–25.

Brooks, D. R., and E. P. Hoberg. 2001. Parasite systematics in the 21st century: opportunities and obstacles. Trends in Parasitology 17:273–275.

Brooks, D. R., M. G. P. van Veller, and D. A. McLennan. 2001. How to do BPA, really. Journal of Biogeography 28:345–358.

Clayton, D. H., and P. Cotgreave. 1994. Relationships of bill morphology to grooming behavior in birds. Animal Behaviour 47:195–201.

Clayton, D. H., and D. M. Drown. 2001. Critical evaluation of five methods for quantifying chewing lice (Insecta: Phthiraptera). Journal of Parasitology 87:1291–1300.

Clayton, D. H., R. D. Gregory, and R. D. Price. 1992. Comparative ecology of neotropical bird lice (Insecta, Phthiraptera). Journal of Animal Ecology 61:781–795.

Clayton, D. H., and B. A. Walther. 2001. Influence of host ecology and morphology on the diversity of Neotropical bird lice. Oikos 94:455–467.

Combes, C. 1996. Parasites, biodiversity and ecosystem stability. Biodiversity and Conservation 5:953–962.

Combes, C., P. Hudson, and M. Hochberg. 1996. Host-parasite co-evolution. Biodiversity and Conservation 5:951.

Cribb, T. H., and R. A. Bray. 2010. Gut wash, body soak, blender and heat-fixation: approaches to the effective collection, fixation and preservation of trematodes of fishes. Systematic Parasitology 76:1–7.

de Meûs, T., Y. Michalakis, and F. Renaud. 1998. Santa Rosalia revisited: or why are there so many kinds of parasites in "the garden of earthly delights"? Parasitology Today 14:10–13.

de Meûs, T., and F. Renaud. 2002. Parasites within the new phylogeny of eukaryotes. Trends in Parasitology 18:247–251.

Dhondt, A. A., J. C. DeCoste, D. H. Ley, and W. M. Hochachka. 2014. Diverse wild bird host range of Mycoplasma gallisepticum in eastern North America. PLoS One 9:e103553.

Dobson, A., K. D. Lafferty, A. M. Kuris, R. F. Hechinger, and W. Jetz. 2008. Homage to Linnaeus: how many parasites? how many hosts? Proceedings of the National Academy of Sciences USA 105:11482–11489.

Dubinina, M. N. 1971. Parasitological investigation of birds. Nauka, Leningrad, USSR (in Russian).

Fair, J. M., E. Paul, and J. J. Jones. 2010. Guidelines to the use of wild birds in research. Special Publication of the Ornithological Council, Washington, DC.

Fecchio, A., M. R. Lima, P. Silveira, É. M. Braga, and M. Â. Marini. 2011. High prevalence of blood parasites in social birds from a neotropical savanna in Brazil. Emu 111:132–138.

Fecchio, A., M. R. Lima, M. Svensson-Coelho, M. A. Marini, and R. E. Ricklefs. 2013. Structure and organization of an avian haemosporidian assemblage in a Neotropical savanna in Brazil. Parasitology 140:181–192.

Fowler, J. A. 1984. A safer anaesthetic for delousing live birds. Ringer's Bulletin 6:69.

Galen, S. C., and C. C. Witt. 2014. Diverse avian malaria and other haemosporidian parasites in Andean house wrens: evidence for regional co-diversification by host-switching. Journal of Avian Biology 45:374–386.

Gaunt, A. S., and L. W. Oring. 1999. Guidlines for the use of wild birds in research. Special Publication of the Ornithological Council, Washington, DC.

Georgiev, B. B., V. Biserkov, and T. Genov. 1986. In toto staining method for cestodes with iron aceto-carmine. Helminthologia 23:279–281.

Gilles, H. M. 1993. Bruce-Chwatt's essential malariology. Hodder Education Publishers, London, UK.

Gregory, R. D., D. Noble, R. Field, J. Marchant, M. Raven, and D. W. Gibbons. 2003. Using birds as indicators of biodiversity. Ornis Hungarica 12:11–24.

Haffer, J. 1974. Avian speciation in tropical South America. Publications of the Nuttall Ornithological Club No. 14.

Hird, S. M., C. Sanchez, B. C. Carstens, and R. T. Brumfield. 2015. Comparative gut microbiota of 59 Neotropical bird species. Frontiers in Microbiology 6:1–16.

Hochachka, W. M., and A. A. Dhondt. 2000. Density-dependent decline of host abundance resulting from a new infectious disease. Proceedings of the National Academy of Sciences USA 97:5303–5306.

Ivashkin, V. M., V. L. Kontrimavicius, and N. S. Nazarov. 1971. Methods of collection and investigation of helminths and terrestrial mammals. Nauka, Moscow, USSR (in Russian).

Jarvi, S. I., D. Triglia, A. Giannoulis, M. Farias, K. Bianchi, and C. T. Atkinson. 2007. Diversity, origins and virulence of Avipoxviruses in Hawaiian Forest Birds. Conservation Genetics 9:339–348.

Johnson, K. P., S. M. Shreve, and V. S. Smith. 2012. Repeated adaptive divergence of microhabitat specialization in avian feather lice. BMC Biology 10:1–11.

Johnson, K. P., J. D. Weckstein, M. J. Meyer, and D. H. Clayton. 2011. There and back again: switching between host orders by avian body lice (Ischnocera: Goniodidae). Biological Journal of the Linnean Society 102:614–625.

Johnson, N. K., R. M. Zink, G. F. Barrowclough, and J. A. Marten. 1984. Suggested techniques for modern avian systematics. Wilson Bulletin 96:543–560.

Justine, J. L., M. J. Briand, and R. A. Bray. 2012. A quick and simple method, usable in the field, for collecting parasites in suitable condition for both morphological and molecular studies. Parasitology Resources 111:341–351.

Kilpatrick, A. M., P. Daszak, M. J. Jones, P. P. Marra, and L. D. Kramer. 2006. Host heterogeneity dominates West Nile virus transmission. Proceedings of the Royal Society B 273:2327–2333.

Knowles, S. C., V. Palinauskas, and B. C. Sheldon. 2010. Chronic malaria infections increase family inequalities and reduce parental fitness: experimental evidence from a wild bird population. Journal of Evolutionary Biology 23:557–569.

Koop, J. A., and D. H. Clayton. 2013. Evaluation of two methods for quantifying passeriform lice. Journal of Field Ornithology 84:210–215.

Lamothe-Argumedo, R., C. Damborenea, L. Garcia-Prieto, L. I. Lunaschi, and D. Osorio-Sarabia. 2010. Guide to helminthological collections of Latin America. Museo De La Plata, La Plata, Argentina.

Lawton, J. H., D. E. Bignell, B. Bolton, G. F. Bloemers, P. Eggleton, P. M. Hammond, M. Hodda, R. D. Holt, T. B. Larsen, N. A. Mawdsley, N. E. Stork, D. S. Srivastava, and A. D. Watt. 1998. Biodiversity inventories, indicator taxa and effects of habitat modification in tropical forest. Nature 391:72–76.

Lichtenfels, J. R., and M. H. Pritchard. 1982. A guide to the parasite collections of the world. Allen Press, Lawrence, KS.

Lutz, H. L., W. M. Hochachka, J. I. Engel, J. A. Bell, V. V. Tkach, J. M. Bates, S. J. Hackett, and J. D. Weckstein. 2015. Parasite prevalence corresponds to host life history in a diverse assemblage of afrotropical birds and haemosporidian parasites. PLoS One 10:e0121254.

Martinsen, E. S., S. L. Perkins, and J. J. Schall. 2008. A three-genome phylogeny of malaria parasites (Plasmodium and closely related genera): evolution of life-history traits and host switches. Molecular Phylogenetics and Evolution 47:261–273.

Miller, T. L., R. D. Adlard, R. A. Bray, J. L. Justine, and T. H. Cribb. 2010. Cryptic species of Euryakaina n. g. (Digenea: Cryptogonimidae) from sympatric lutjanids in the Indo-West Pacific. Systematic Parasitology 77:185–204.

Nieberding, C. M., and I. Olivieri. 2007. Parasites: proxies for host genealogy and ecology? Trends in Ecology and Evolution 22:156–165.

Norris, K., and D. J. Pain. 2002. Conserving bird biodiversity: general principles and their application. Cambridge University Press, Cambridge, UK.

Owen, J. C. 2011. Collecting, processing, and storing avian blood: a review. Journal of Field Ornithology 82:339–354.

Palma, R. L. 1978. Slide mounting of lice: a description of the Canada balsam technique. New Zealand Entomologist 6:432–436.

Parker, P. G., E. L. Buckles, H. Farrington, K. Petren, N. K. Whiteman, R. E. Ricklefs, J. L. Bollmer, and G. Jimenez-Uzcategui. 2011. 110 years of Avipoxvirus in the Galápagos Islands. PLoS One 6:e15989.

Parker, P. G., N. K. Whiteman, and R. E. Miller. 2006. Conservation medicine on the Galápagos islands: partnerships among behavioral, population, and veterinary scientists. Auk 123:625–638.

Perkins, S. L. 2014. Malaria's many mates: past, present, and future of the systematics of the order Haemosporida. Journal of Parasitology 100:11–25.

Perkins, S. L., and J. J. Schall. 2002. A molecular phylogeny of malarial parasites recovered from cytochrome b gene sequences. Journal of Parasitology 88:972–978.

Poulin, R. 2005. Evolutionary trends in body size of parasitic flatworms. Biological Journal of the Linnean Society 85:181–189.

Price, R. D., and J. D. Weckstein. 2005. The genus Austrophilopterus Ewing (Phthiraptera: Philopteridae) from toucans, toucanets, and araçaris (Piciformes: Ramphastidae). Zootaxa 918:1–18.

Pritchard, M. H., and G. O. Kruse. 1982. The collection and preservation of animal parasites. Technical Bulletin No. 1. The Harold W. Manter Laboratory. University of Nebraska Press, Lincoln, NE.

Proctor, N. S., and P. J. Lynch. 1993. Manual of ornithology. New Haven, CT.

Richard, F. A., R. N. Sehgal, H. I. Jones, and T. B. Smith. 2002. A comparative analysis of PCR-based detection methods for avian malaria. Journal of Parasitology 88:819–822.

Rocha, L. A., A. Aleixo, G. Allen, F. Almeda, C. C. Baldwin, M. V. Barclay, J. M. Bates, A. M. Bauer, F. Benzoni, C. M. Berns, M. L. Berumen, D. C. Blackburn, S. Blum, F. Bolaños, R. C. Bowie, R. Britz, R. M. Brown, C. D. Cadena, K. Carpenter, L. M. Ceríaco, P. Chakrabarty, G. Chaves, J. H. Choat, K. D. Clements, B. B. Collette, A. Collins, J. Coyne, J. Cracraft, T. Daniel, M. R. de Carvalho, K. de Queiroz, R. Di Dario, R. Drewes, J. P. Dumbacher, A. Engilis Jr., M. V. Erdmann, W. Eschmeyer, C. R. Feldman, B. L. Fisher, J. Fjeldså, P. W. Fritsch, J. Fuchs, A. Getahun, A. Gill, M. Gomon, T. Gosliner, G. R. Graves, C. E. Griswold, R. Guralnick, K. Hartel, K. M. Helgen, H. Ho, D. T. Iskandar, T. Iwamoto, Z. Jaafar, H. F. James, D. Johnson, D. Kavanaugh, N. Knowlton, E. Lacey, H. K. Larson, P. Last, J. M. Leis, H. Lessios, J. Liebherr, M. Lowman, D. L. Mahler, V. Mamonekene, K. Matsuura, G. C. Mayer, H. Mays, Jr., J. McCosker, R. W. McDiarmid, J. McGuire, M. J. Miller, R. Mooi, R. D. Mooi, C. Moritz, P. Myers, M. W. Nachman, R. A. Nussbaum, D. Ó. Foighil, L. R. Parenti, J. F. Parham, E. Paul, G. Paulay, J. Pérez-Emán, A. Pérez-Matus, S. Poe, J. Pogonoski, D. L. Rabosky, J. E. Randall, J. D. Reimer, D. R. Robertson, M. O. Rödel, M. T. Rodrigues, P. Roopnarine, L. Rüber, M. J. Ryan, F. Sheldon, G. Shinohara, A. Short, W. B. Simison, W. F. Smith-Vaniz, V. G. Springer, M. Stiassny, J. G. Tello, C. W. Thompson, T. Trnski, P. Tucker, T. Valqui, M. Vecchione, E. Verheyen, P. C. Wainwright, T. A. Wheeler, W. T. White, K. Will, J. T. Williams, G. Williams, E. O. Wilson, K. Winker, R. Winterbottom, and C. C. Witt. 2014. Specimen collection: an essential tool. Science 344:814–815.

Samuel, M. D., B. L. Woodworth, C. T. Atkinson, P. J. Hart, and D. A. LaPointe. 2015. Avian malaria in Hawaiian forest birds: infection and population impacts across species and elevations. Ecosphere 6:1–21.

Valim, M. P., and J. D. Weckstein. 2011. Two new species of Brueelia Kéler, 1936 (Ischnocera, Philopteridae) parasitic on Neotropical trogons (Aves, Trogoniformes). ZooKeys 128:1–13.

Valkiūnas, G. 2005. Avian malarial parasites and other Haemosporidia. CRC Press, Boca Raton, FL.

van Dongen, W. F. D., J. White, H. B. Brandl, Y. Moodley, T. Merkling, S. Leclaire, P. Blanchard, É. Danchin, S. A. Hatch, and R. H. Wagner. 2013. Age-related differences in the cloacal microbiota of a wild bird species. BMC Ecology 13:1–12.

Vo, A. T., and J. A. Jedlicka. 2014. Protocols for metagenomic DNA extraction and Illumina amplicon library preparation for faecal and swab samples. Molecular Ecology Resources 14:1183–1197.

Waite, D. W., and M. W. Taylor. 2015. Exploring the avian gut microbiota: current trends and future directions. Frontiers in Microbiology 6:673.

Walther, B. A., and D. H. Clayton. 1997. Dust-ruffling: a simple method for quantifying ectoparasite loads of live birds. Journal of Field Ornithology 68:509–518.

Weckstein, J. D. 2004. Biogeography explains cophylo-genetic patterns in toucan chewing lice. Systematic Biology 53:154–164.

Whiteman, N. K., and P. G. Parker. 2005. Using parasites to infer host population history: a new rationale for parasite conservation. Animal Conservation 8:175–181.

Windsor, D. A. 1995. Equal rights for parasites. Conservation Biology 9:1–2.

Windsor, D. A. 1998. Most of the species on earth are parasites. International Journal for Parasitology 28:1939–1941.

Winker, K. 1997. The role of taxonomy and systematics. Conservation Biology 11:595–596.

Winker, K. 2000. Obtaining, preserving, and preparing bird specimens. Journal of Field Ornithology 71:150–297.

Zinovieva, S. V., N. N. Butorina, V. Z. Udalova, O. S. Khasanova, L. V. Filimonova, V. G. Petrosyan, and A. N. Pel'gunov. 2015. World collections of parasitic worms. Biology Bulletin 42:540–545.

CHAPTER ELEVEN

Student-Led Expeditions as an Educational and Collections-Building Enterprise*

*David W. Winkler, Teresa M. Pegan, Eric R. Gulson-Castillo,
Joseph I. Byington, Jack P. Hruska, Sophia C. Orzechowski,
Benjamin M. Van Doren, Emma I. Greig, and Eric M. Wood*

Abstract. As natural history collections are enjoying a renaissance in their uses as referents for modern genomic studies, they are also expanding their scope by adding or linking to digital media on behavior, vocalizations, and anatomy. At the same time, a new generation of students, uncommonly well-versed in natural history and biodiversity, is hungering for demanding fieldwork in out-of-the-way localities. This combination of developments creates an historic opportunity for student-led expeditions to study taxa of phylogenetic interest by collecting "complete specimen packages," traditional museum specimens coupled with digital documentation of their behaviors and ecologies. This chapter summarizes our experiences with such expeditions, from the preparatory steps, through activities during the expedition, to the challenges of post-expedition work. We hope that our successes and failures can inform similar programs at other institutions.

Key Words: collections, complete specimen package, digital media, expeditions, exploratory science, natural history, ornithology, specimens, undergraduate students.

EXPANDING THE SCOPE OF NATURAL HISTORY MUSEUMS

Since their inception, natural history collections and museums have been the foundation of comparative biology. These libraries of the natural world house physical information on organismal phenotypes from all over the world and also provide a window into the past. Natural history museums continue to play a central role in modern comparative biology as the repository for type specimens, vouchers, and comparative material, and as the referents for the functional genomics of organismal form (Bradley et al. 2014). Recently, some natural history collections have begun to broaden their mandate to include collections of

* Winkler, D. W., T. M. Pegan, E. R. Gulson-Castillo, J. I. Byington, J. P. Hruska, S. C. Orzechowski, B. M. Van Doren, E. I. Greig, and E. M. Wood. 2017. Student-led expeditions as an educational and collections-building enterprise. Pp. 185–200 in M. S. Webster (editor), The Extended Specimen: Emerging Frontiers in Collections-based Ornithological Research. Studies in Avian Biology (no. 50), CRC Press, Boca Raton, FL.

video and sound. The power and sophistication of photographic and audio recording equipment have increased as their costs have decreased, and the availability of digital media is increasing in parallel with our capacity for storage and curation. Digital collections such as the Macaulay Library of Natural Sounds (http://macaulaylibrary.org), the Borror Lab of Bioacoustics (https://blb.osu.edu), the Internet Bird Collection (http://ibc.lynxeds.com), and xeno-canto (http://www.xeno-canto.org) are growing and have amassed an impressive sample of sounds and behaviors from an increasingly comprehensive representation of world avian diversity. The increasing availability of digital scientific media raises the opportunity for traditional collections to reference their physical specimens with media curated in a distinct digital archive, seamlessly available through the Internet in the digital catalog of the museum holding the physical specimen. The same can now be done routinely with specimens being referenced to their gene sequences stored in GenBank. And the value of physical specimens, particularly full-body fluid specimens, is increasing with the refinement of computed tomography (CT) scanning, which has made many aspects of organismal structure available for study as never before (see http://www .biotech.cornell.edu/brc/imaging-facility/services /high-resolution-x-ray-ct-services). Museums are thus serving as a storehouse and portal for an ever-increasing wealth of data about organisms; as the volume of digital "extended specimens" and our ability to curate them increase, we have the opportunity to discover an ever-richer tale about variation and evolutionary change in organisms across time and space. Specimens in collections have always been used in exciting new ways that past curators never envisioned and, as the diversity of phenotypic information that can be effectively stored increases, so does the variety of future potential uses.

In many ornithological collections, a key focus of the past 30 years has been the construction of phylogenies for the birds of the world. Remarkable progress has been made, making a broad understanding of the relationships of all birds possible on a common molecular ground (e.g., Hackett et al. 2008, Jarvis et al. 2014, Prum et al. 2015). These studies have extended even to studies of the phylogenetic position of enigmatic birds like the Malaysian Rail-babbler (*Eupetes macrocerus*) and Sapayoa (*Sapayoa aenigma*), which are now in their own distinct families, or the philentomas (*Philentoma* spp.), which end up allied to a larger group (see summary of relationships in Winkler et al. 2015). The spectacular success of molecular approaches to phylogeny construction has, in some circles, led to the notion that the need for specimen collection has disappeared, and that all that is necessary for study is a sample of blood, skin, or feather. However, it is important to stop and take stock. Phylogenies are not an end for all biologists: their enormous value is realized when they teach us something about organisms themselves. Phylogenies can help us understand the nature of differences and similarities among taxa, and they can help us explore how behavioral or morphological phenotypes vary over space and time. For studies of morphology we still require the physical specimen, which allows us to quantify its dimensions and other aspects of its physical being. Increasingly, with new digital techniques, we can preserve behavioral traits as well. As the richness of media and other data contained and referenced in natural history collections grows, we can make use of molecular phylogenies to re-explore the structure, behavior, and evolution of all animals.

Here, we present our experiences putting this broader view into practice, including ideas about how this can lead to exciting new training opportunities, through a series of student-led collecting expeditions. The goal of these expeditions was to collect museum specimens and digital media on as many individuals as possible of a target taxon, collecting whenever possible both digital media and physical specimens of the same individuals to form "Complete Specimen Packages" (see "The Complete Specimen Package"). At the same time, we wanted to cultivate a new generation of leaders in museum collections, undergraduates who could enter graduate school with a wealth of independent expedition fieldwork behind them. Our experiences with these undergraduate-led expeditions have been varied and educational for everyone involved, and they ultimately allowed us to gather large volumes of information using a diversity of techniques. We hope that this account will inspire and inform other student expeditions at other educational institutions.

We first explain our concept of the "Complete Specimen Package," and, after summarizing the diverse material collected in our expeditions, we present a set of guidelines to encourage and facilitate student expeditions like ours in the future.

Most of these guidelines deal with aspects of the student expeditions that are pertinent at all times, but we separate some aspects that apply only to the periods before, during, or after an expedition. We finish with some general observations on these expeditions and their future usefulness.

The Complete Specimen Package

Museums and digital media repositories are both superb resources, but their value is greatest when they can be linked together. To do so, we can associate a museum specimen with videos, sound recordings, and photographs of the same individual while it was still alive, thus capturing multiple aspects of the individual phenotype. We call such a set of physical and digital data from the same bird a "Complete Specimen Package" (CSP). Ultimately, the CSP facilitates making the most of every bird collected, and thereby increases the benefits to be gained from the sacrifice of an individual. These digitally extended specimens ideally include as many behaviors as possible, photographs of habitat and nests that could not be physically collected, and copies of any original field notes relevant to the specimen taken by expedition members. Biologists hoping to gather a CSP will often wish to color band or radio tag individual birds to allow detailed media collection, territory mapping, and nest finding, and the documentation of parental care patterns recorded during regular nest watches. Collecting natural history data is a challenging and time-consuming endeavor, requiring a much longer and larger investment in time and energy than traditional museum collecting. Increasing the complexity of each specimen package inherently means that fewer specimens can be collected on a given expedition. A bird captured in a mist-net and fitted with color bands or a radio tag, rather than being euthanized and processed immediately, is clearly reducing the specimen-collection efficiency of an expedition. Many expeditions may not be able to afford this extra investment in time and resources that a focus on the CSP requires; indeed, in practice we found this framework challenging to follow perfectly. But it provided a useful model to guide our work, and we believe that many in the museum community recognize and appreciate the increased depth of phenotypic and accession data that can be stored once digital media and other ancillary materials are embraced as part of a museum's holdings. Exciting research opportunities, such as comparing variation in signals produced by different individuals to variation in their morphology, are just beginning to be explored.

The CSP is a tremendous supplement to the value of museums because it multiplies the number of possible future uses for specimens. In our own expeditions, we have structured our activities to increase the likelihood that future studies can benefit from our efforts by taking as many samples as feasible from each individual. Ideally we begin by taking video and audio recordings of the bird's natural behavior in the field, and then attempt to capture the same individual we have recorded. We preferred preparing fluid specimens instead of round skins, as they allow for morphological studies using CT scans (Figure 11.1). Fluid specimens provide us with stunning access to internal morphology that is impossible with round skins. Before preserving the fluid specimen, we typically removed a wing and prepared it as a spread wing (see Chapter 8, this volume), took feather samples from various body parts, a blood sample, a wing muscle sample, body measurements, and photographs. Taking as many samples as possible provides material for many different types of studies, including genetic, spectrometric, isotopic, and morphometric. By storing so much of the bird digitally as audio and video recordings, and later CT scans, the CSP also makes it possible to give colleagues in host countries a rich source of data for their own use in education and research.

This broadened scope of ornithological expeditions—collecting both specimens and digital media—makes them especially attractive to students. In our experience, not many biology undergraduates come to university wanting to collect specimens, but a great many come already captivated by birds, interested to learn more about them and wanting to observe them in the field. Most of these students are also passionate about discovery, enthralled by the idea of traveling to a poorly known place and describing a new species or nest. They are keen to learn how to record the sounds and sights of birds, and they are often quick to pick up the detailed skills required to be excellent recordists, photographers, or videographers. Once students are introduced to the wonders of modern CT investigations of morphology, or once they grow interested in details of molt or sex and age differences, the value of

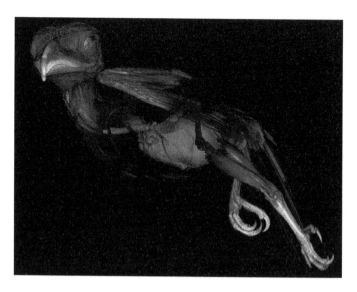

Figure 11.1. A computed tomography (CT) image of what appears to be the first fluid specimen of Dusky Broadbill, *Corydon sumatranus*, which we collected at Tawau Hills Park in Sabah, Malaysian Borneo, in 2012. We have highlighted the bird's digestive tract and heart, which are placed in context with the remainder of the body to demonstrate the potential of this technology for morphological studies. Image created by Eric R. Gulson-Castillo and Claire H. Fox using *OsiriX* 6.0.2, 64-bit software (Rosset et al. 2004).

museum collections becomes much more tangible to them. Collecting, which they may have thought was something they would never do, can become a valued activity, conducted in the most humane way possible, so as to gather as much information as possible from each bird they encounter.

Cornell Expeditions in Field Ornithology and Ivy Expeditions

The expanded view of the nature of museum collections, brought together with the need to educate the next generation of organismal biologists, gave us a dual set of goals for museum expeditions: to collect CSPs, and, at the same time, to encourage students to lead museum expeditions of discovery. We gathered a group of enthusiastic Cornell undergraduates and their advisers, which we called Cornell Expeditions in Field Ornithology (CEFO). CEFO organized the logistics for three separate expeditions to Sabah, Malaysian Borneo, in 2012, 2013, and 2014, and one to Panama in 2014. These expeditions were sponsored by a friend of the Cornell Lab of Ornithology, who was inspired by a previous undergraduate expedition that first explored the remote Cerros del Sira of eastern Peru in 2008. This friend of the Cornell Lab wished to foster and fund similar expeditions of discovery for other undergraduates

and established the Ivy Expeditions Fund (Ivy is a family name) to support such expeditions and other student field experiences. The expeditions reported here were supported predominantly by the Ivy Expeditions Fund, and we will refer to them simply as "Ivy Expeditions."

The members of the Cerros del Sira expedition—Benjamin Winger, Michael Harvey, and Glenn Seeholzer—displayed exceptional independence and ingenuity in scraping together funds for their trip and seeing it all the way through to many new distributional records and the description of a new species of barbet (*Capito fitzpatricki*) from their expedition (Seeholzer et al. 2012). Moreover, that expedition propelled each of the participants into successful graduate and academic careers. Realizing how much the independent expedition experience enhanced their careers, Winkler, the primary adviser, embraced a hands-off approach to advising the Ivy Expeditions from the start, recruiting and training a team that was expected to be self-motivated throughout the process. Trusting that they would make mistakes and learn from them, Winkler and the advisers he recruited endeavored to steer the group lightly and keep its mistakes from being dangerous or threatening to further expedition work. The presence of advisers was important to make sure that the group stayed functional and safe, but the details of each expedition and how it spent its time were ultimately up to the students.

These Ivy Expeditions yielded good numbers of both traditional museum specimens and digital media on the behavior of species in its target clade, the Old World Suboscines, plus a good number of other species (Tables 11.1 and 11.2). The specimens have been made broadly available for research, with physical specimens stored at the Cornell Museum of Vertebrates (Ithaca, NY), the Sabah Parks Vertebrates Collection (Sabah, Malaysia), and the Museo de Vertebrados of the Universidad de Panamá (Panama City, Panama). Digital specimens are archived and available online at the Macaulay Library. At the same time, the expeditions provided unique field experiences for 14 students from Cornell. At the time of writing, we have published one report on a previously undescribed sonation from Black-crowned Pittas (*Erythropitta ussheri*; Pegan et al. 2013) and one description of Sapayoa (*Sapayoa aenigma*) nesting behavior (Dzielski et al. 2016); one expedition report has been published (Hruska et al.

2016) and two more are in review (Gulson-Castillo et al., in review; Pegan et al., in review). Six more papers are in preparation from these expeditions: a paper analyzing parental care in *Erythropitta ussheri*, a paper on song variation in *Erythropitta ussheri*, a comparative paper on Pitta movements and wing morphology, an expedition report from Serinsim, and two papers on broadbill thermoregulation.

Based on our experience in these four expeditions we here offer observations and suggestions that we hope will help make future expeditions

TABLE 11.2
Summary of all specimens and media collected.

Physical specimens, Borneo 2012	32
Physical specimens, Borneo 2013	56
Physical specimens, Panama 2014	8
Non-suboscine species with media	256
Non-suboscine species collected	54

TABLE 11.1
Summary of specimens and media collected on Old World Suboscines.

Species	Fluids	Nest-stages captured	Audio cuts	Video cuts	Localities
Black-and-yellow Broadbill (*Eurylaimus ochromalus*)	7	Building, incubation	15	118	Tawau Hills, Serinsim, Sepilok Reserve (video only)
Banded Broadbill (*Eurylaimus javanicus*)	4	Building	4	44	Tawau Hills, Serinsim, Sepilok Reserve (video only)
Dusky Broadbill (*Corydon sumatranus*)	1	Building	10	16	Tawau Hills
Black-and-red Broadbill (*Cymbirhynchus macrorhynchos*)	1	Inactive nests	1	3	Serinsim, Sepilok Reserve (video only)
Green Broadbill (*Calyptomena viridis*)	3	Inactive nests	7	0	Tawau Hills, Serinsim
Whitehead's Broadbill (*Calyptomena whiteheadi*)	0	Feeding	0	1	Kinabalu National Park
Black-crowned Pitta (*Erythropitta ussheri*)	5	Feeding, fledging	49	110	Tawau Hills, Serinsim
Blue-banded Pitta (*Erythropitta arquata*)	0	None	2	7	Tawau Hills
Blue-headed Pitta (*Hydrornis baudii*)	0	None	4	18	Tawau Hills
Bornean Banded Pitta (*Hydrornis schwaneri*)	0	None	7	2	Tawau Hills
Hooded Pitta (*Pitta sordida*)	2	None	2	1	Tawau Hills, Serinsim, Sepilok Reserve (salvage)
Sapayoa (*Sapayoa aenigma*)	6	Incubation, feeding	13	180	Darién NP

of this sort more efficient and effective. After presenting the themes and ongoing activities that ran through all phases of our expeditions, we present those that were distinctive to the periods before, during, or after our expeditions. Finally, we conclude with valuable lessons learned and thoughts for the future of student-led CSP expeditions.

GENERAL THEMES AND ONGOING ACTIVITIES OF THE EXPEDITIONS PROGRAM

Setting the Framework

Within the broader context of university education, CSP student expeditions work best if they have an intellectual foundation and focus. One that we found useful was to choose a target clade and explore patterns of trait distribution across that clade, mindful of the comparative challenge of teasing apart phylogenetic and ecological influences on trait variation. Our four expeditions were focused on a clade of particular phylogenetic interest: the Old World Suboscines (pittas, broadbills, and allies). This focus permitted the students to prepare ahead of time, familiarizing themselves with everything that was known about the natural history of each of the species of Old World Suboscine likely to be encountered at the target sites and to study what was known about the phylogenetic relationships among members of this group. Indeed, these relationships (Figure 11.2) provided much of the initial fascination for the students, especially the fact that the broadbills are not a single clade but two (Eurylaimidae and Calyptomenidae), and the two broadbill families may be no more closely related to one another than they are to the morphologically different Pittidae (Irestedt et al. 2006, Moyle et al. 2006).

Another fascination was biogeographic, as the same phylogenetic work revealed that the monotypic genus *Sapayoa* of Central and South America is embedded well within the Old World Suboscines, the only Neotropical bird bearing this distinction.

With the detailed study made possible by this targeted focus, students could enter the field as naturalist-scholars. Acquainted with all that was known about their target taxa, they were able to appreciate every new discovery by knowing whether an observation was novel and how it compared with observations of other members of the clade. When young naturalists find themselves in a tropical forest for the first time, it can be difficult to focus on any one thing. We did not require expedition members to ignore other species and record data and collect specimens of only the target taxa; however, we did expect them to pursue and prioritize opportunities to engage with the target taxa.

With growing familiarity of the target clade, comparative questions began to emerge associated with topics such as pitta wing patches, the large bills of the broadbills, and patterns in breeding biology. In these comparisons and many others, the students encountered the richness and frustration of making comparisons across groups. For example, African members of the Eurylaimidae and Calyptomenidae seem to have switched phenotypic patterns in the many millions of years that separate them from their Asian relatives: in contrast to the boldly patterned eurylaimids and green calyptomenids of Asia, the only eurylaimid in Africa is green with a small bill and lacks pattern, and the only African calyptomenid genus is patterned in black and white and brown with a largish bare bill. Clearly, though groups delineated by modern molecular studies generally

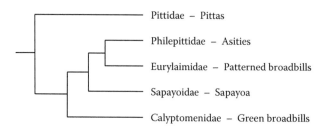

Figure 11.2. A tree representing the phylogeny of the Old World Suboscine families. Members of Calyptomenidae were once considered to be part of the Eurylaimidae, and the single species in Sapayoidae, which lives in the Neotropical Chocó, is actually more closely related to the Old World broadbills than to any of the New World Suboscines. (Tree adapted from Irestedt et al. 2006 and Moyle et al. 2006.)

retain a cohesiveness in phenotypes, the role of selection over many millions of years and many thousands of kilometers can overcome the influence of common ancestry.

The most striking intellectual difference between the Ivy Expeditions and other opportunities for undergraduate research is that the expeditions were exploratory and documentary in nature, rather than targeted toward testing a specific hypothesis or answering a preexisting question. This focus on a clade, rather than a question, provided a distinct contrast to the training that many undergraduates typically receive in science. The vast majority of curricula from high school onward are focused on portraying the core activity of science as hypothesis testing, despite the importance of discovery and pattern recognition in zoology (Greene 2005) and other realms of science, including physics (Xu et al. 2015), computer science (Rudin et al. 2014), and genomics (Sporns 2015). The "science as only hypothesis testing" view leaves out the critical early phases of discovery science, the gathering of observations that must be strung together into a pattern. Only once a pattern has been described can hypotheses to explain it be adduced. It may be tempting to believe that in the modern day there remain few undiscovered patterns to reveal, leaving only hypothesis-driven research as valuable; but this is certainly untrue. In the case of birds, one of the best-documented groups in the world, we know far more about how most of them are related to one another evolutionarily than about any other aspect of their biology. For most species, hypotheses are difficult to formulate because the patterns of variation across and between groups have usually not been described. The expeditions to Borneo and Panama exposed the students to this early pattern-generating phase of scientific discovery. Coming from university labs, they sometimes questioned the lack of a hypothesis to be tested, but daily discussions of field observations left them with a growing appreciation of the generalizations and patterns that they were beginning to construct during their time in the forest. Indeed, early expeditions spawned nascent patterns that led to hypothesis testing in later field expeditions.

Responsibilities of the Students

Preparations began 6 to 8 months before each expedition. One or two student leaders were chosen early on to help organize the group, and they also helped choose expedition members. For our expeditions, potential crew members were asked to submit two essays, one explaining their interest in participating and the other on the skills and expertise they could contribute. Once the other members were chosen, duties were assigned by the leaders based on the skills and preferences of the students. Within the crews, each member had a defined role that put her or him in charge of a number of related tasks: one person dealt with permitting and legal issues; another organized travel and lodging; another handled finances; another prepared media curation templates and routines; and yet another organized first-aid kits and safety plans. In our smaller expeditions, multiple roles were assigned to each student. In most of our expeditions, the student in charge of health and safety was certified for Wilderness First Aid or as a Wilderness First Responder. There was plenty to do on all these fronts, and the students learned a lot about details they may never have imagined would be important.

As in all other aspects of the expeditions, students were the leaders of the post-expedition process as well. The tasks assigned before the expeditions carried over until after the field work: for example, the finance student was responsible for submitting receipts, the permit student ensured that specimens were exported and imported properly, and the media curation student helped others make sure that their media and notes were fully curated and archived.

Responsibilities of the Advisers

The student-led nature of these expeditions, with a light adviser hand, distinguished our expeditions from other undergraduate research opportunities at Cornell. Allowing the students to lead their expeditions, and insisting that they do so, was one of the most difficult and important roles of the advisers to these expeditions. The adviser necessarily needed to stay on top of permits, risk management, and budget, as well as monitor group dynamics, but we recognized that greater student involvement means greater learning. We intentionally left as much as possible to the students. Generally, we thought this model worked well, but that occasionally too little guidance could be detrimental to students' experiences. We therefore stress that communication between students and advisers is critical throughout the process, so that both parties can find an appropriate and effective balance.

Winkler served as the primary adviser, and one to two other advisers accompanied students into the field. Other advisers to our expeditions included two postdocs, a graduate student, and experienced fieldwork leaders. On campus, Winkler served as a conduit to necessary resources for training in health and safety, risk management, climbing methods, and animal ethics; sources of funding; and interactions with other faculty, staff, or potential advisers. He also oversaw student decision-making, logistics, and overall organization of duties. He worked with students to craft and enforce a reasonable code of conduct for expedition members to develop explicit lists of expectations and responsibilities. In the field, he and the other advisers helped with fieldwork and taught undergraduates techniques with which they were not yet sufficiently familiar. They helped the expedition's student leader keep track of ongoing activities, noticing when some had been neglected, encouraging curation of data, and providing perspective to help keep morale high. After expeditions returned, Winkler oversaw the curation of specimens and digital data, and, as the group's experience with Sabah's birds increased, he encouraged students to start making connections and finding patterns in their observations. This naturally led to greater emphasis on writing the results of the expeditions. Retaining a hands-off style, advisers helped the students to think about which of the possible projects would be the most valuable additions to the literature, and helped them with statistics that they needed but had not yet studied in classes.

Specimens, Their Curation, and Important Logistics

Collecting specimens internationally always carries with it logistical challenges, such as filling out paperwork, interacting with government officials, and ensuring the safety of the specimens while traveling. Of all of our experiences, this is where we depended most upon the guidance and help of our in-country collaborators and key government officials.

Permit work began early on, well before the expedition, because unexpected forms or letters sometimes became necessary during the process and had to be requested before arriving at the field site. Early on, we placed a student in charge of this paperwork to prioritize the necessary tasks, often the student leader. During the process, this student communicated frequently with the faculty adviser and other leadership members of the team, as well as with local collaborators once a site had been chosen, to determine exactly what and when information was required. It proved valuable to have one person take charge of permits because it minimized the number of people that foreign institutions had to interact with. One of the most important decisions that needed to be made was the number of specimens and species that would be requested for permission to collect from the host country. Although conservative numbers increase the likelihood that the permits will be granted, we found it important to think this through thoroughly lest unforeseen opportunities for collecting specimens of great scientific interest be precluded by the limited scope of a collecting permit.

We made efforts early in the expedition to visit the museums and collaborators in the host country that assisted with the permit application process. In this way, the team could arrange logistics that would become important later after birds had been collected and it was time to leave the country. These visits also served as a friendly gesture that solidified the team's relationship with collaborators, and as an opportunity to interact with the people under whose jurisdiction the team would be working. In our experiences in Sabah, these relationships were absolutely critical to navigating a permitting system that was being revised and changed even as we were in the field.

Export permits required that we knew the number of specimens to be transported, which could often not be known until the last few days of the expedition. Therefore, we obtained export permits at the end of or after the expeditions. However, when the permit is only approved after the expedition has returned home, the local museum needs to ship the specimens once all permits are finalized. Thus, we learned that it was essential to leave everything well-organized, segregated, and labeled with an associated spreadsheet to ensure that the correct specimens were shipped and that those intended to be deposited in the local museum were left behind with their associated data.

Upon returning to the home institution, specimen importation and curation had to be carefully scheduled and coordinated with the museum curators in charge of their final disposition. Importation generally required careful communication with in-country collaborators to ensure for the last time that all the paperwork was attached to the specimens as they traveled over international borders. Once they arrived, specimen curation was straightforward, and it helped a great deal if some members of the expedition had already been working as student employees in the museum before the expedition so that they could personally guide the specimens into their accession in the home museum.

Data Curation

We found that setting up a good template for curating and keeping data both during and after the expeditions made the difference between an efficient process and a painfully protracted ordeal. This was especially true for media and specimen collecting, and everything ran much more smoothly when crew members knew from the curators exactly which metadata should be collected and associated with each specimen. The curators could also often suggest formats or software for keeping the necessary data organized in the field. We have found that students benefit from coaching on taking effective notes in the field. It is easy for them to believe that they can remember notable details of observations until the evening or even until the end of the expedition, but it is essential to write notes when they are freshly in mind and include as much detail as possible, enriching the papers that eventually make use of field observations. Similarly, media should be curated as soon as possible after they are collected.

After the expeditions, it was critical to have the students digitize and centralize their data: field notebooks were most useful if they were scanned immediately upon return (and many found it useful to take photos of journal pages every night as an in-field digital backup); scribbled numeric values, like measurements, nest activity times, and GPS points, were really only useable by others if they were organized into spreadsheets; and hastily described observations that were fleshed out in the field while memories were fresh could be added to compilations of notes to add to an expedition report or short notes paper. Once everyone digitized his or her notes, individual spreadsheets

and documents could be compiled into centralized versions to be duplicated and shared with all expedition members. Some types of data were digitized and "published" at the same time: For example, eBird checklists for daily lists or point counts were fully curated online as soon as they were submitted. For long-term storage of data and large audio and video files that were not digitized, we used cloud-based storage in addition to hard drives. After a few years, hard drives may fail unexpectedly, and if data analysis and curation have not been completed by then, cloud-based data storage in addition to hard drives can prevent catastrophe.

EXPERIENCES IN PRE-EXPEDITION WORK

Seminar

A weekly seminar for connecting students with expedition interests was an integral part of our efforts. Any interested students were welcomed, including those who did not go on expeditions—these students could still participate in learning about the process and, later, analyzing the data. Early on, we were focused on building what we termed "Target Taxa Sheets," information sheets summarizing all that was known for each of the species in our target clade, ranging from careful reading of the previous compilations of natural history data (e.g., Lambert and Woodcock 1996, Erritzoe and Erritzoe 1998, del Hoyo et al. 2003) to unpublished data on media archives, blog posts, and birding tour reports. As the first expedition approached, more of our time at seminar was devoted to progress reports from each of the students responsible for various logistical preparatory tasks. Not only did this involve all the students in planning an expedition (and thus learning what goes into such planning), it provided each of the students with regular deadlines for reports to their peers, which generally helped to keep all preparations running smoothly and on schedule. At the end of each semester, we shared the results of practice digital recording and photography excursions that had been proceeding locally all term (see "Proficiency in Field Skills"), and discussed how to improve both the aesthetic and scientific quality of our media. Once the early expeditions were concluded, more of the weekly seminar time was devoted to discussions of data analyses and manuscript preparation. We found that even students who had no firm prospect of participating in an expedition could help a great deal with data entry, proofing, and analyses, and they were welcomed as coauthors of manuscripts as they developed.

Picking Field Sites and Seasons

Picking a proper field site and visiting it at a proper time is critical not only for ensuring access to birds of interest, but also for keeping an expedition crew comfortable and safe, maintaining contact with the outside world, keeping equipment usable, and basically facilitating the success of the entire expedition. We used the framework provided by the Old World Suboscine phylogeny and our pursuit of CSPs to guide our choice of field sites. Borneo is a hotspot of Old World Suboscine diversity, with multiple representatives from three families (Pittidae, Eurylaimidae, and Calyptomenidae) generally common in the rain forests. The Darién of Panama is part of the small range of the Sapayoa. The demands of the technology necessary for gathering digital media led us to prefer sites in Borneo that provided access to excellent habitat with electricity and adequate shelter from the rain, rather than choosing the most remote sites available. In Panama, we were more constrained by access to our study species, so we had to make additional arrangements, including obtaining a portable electric generator, to meet our media collection goals. In general, however, we recommend choosing a field site with electricity for technology-focused expeditions such as ours.

One challenge of working in the tropics with undergraduates was that the peak breeding season often overlaps with the academic semester. Our two summer expeditions to Borneo occurred after the bulk of breeding for most rainforest species at our sites, and the 2013 expedition in northern winter was by far the most productive in terms of nests found and data collected. It was worthwhile to go to extra trouble to be in the field during the breeding season, as the CSP potential then, with the prospect of breeding information on target taxa, is so much higher. Accordingly, students needed to take the semester off to go into the field, and some of our members were recent graduates who could attend without interfering with their studies. Further work with the university could help to make a semester away from campus, devoted solely to intensive field work, a much smaller impediment to continued academic progress.

Proficiency in Field Skills

As students became experts on the written knowledge about the target species, they also trained in field skills that would be used during the expedition. Using an expedition as training for totally inexperienced students is detrimental to the productivity of the expedition, and so some pre-expedition training was extremely useful. The field skills necessary for an exploratory CSP expedition depend on the specific focus of the expedition, but they can include sound and video recording, still photography, mist-netting, banding and radio tagging, euthanasia, and specimen preparation techniques. We had a local training season in forests near Ithaca for students not selected for the first expedition, and this was excellent preparation (one of the participants went on to become the leader of a later expedition), but it required having experienced field biologists available to teach the skills. The Macaulay Library loaned video and audio equipment to our expeditions, and its staff also held training sessions for our students bound for an expedition. If equipment was being purchased for an expedition, it was important to purchase it far enough in advance that crew members could practice using it extensively before entering the challenging environment of the rainforest. Universities sometimes also offer classes that can be useful as expedition preparation. Cornell offered classes in bird banding, specimen preparation, and tree climbing, all of which helped prepare students for our Ivy Expeditions. However, there really is no substitute for tropical fieldwork, which is why we found it important when setting expedition goals to recognize that no matter the preparation, expeditions would still be a learning experience for students.

Health and Risk Management

Managing health and safety is important for any expedition, but especially so when undergraduates are involved. Institutional risk management and health and safety groups were involved with our expeditions from the early planning stages. Risk management concerns were addressed as early as possible before each expedition, as the staff and local situations were different for each. The Environmental Health and Safety office at

Cornell ran safety trainings for each crew, tailoring the presentations to the risks most likely to be faced by each group. We encouraged students to obtain Wilderness First Aid or Wilderness First Responder certifications to help everyone on expedition be safer and ready if a terrible accident occurred. Mosquito-borne diseases were our most challenging threat, and we had to ramp up standard practices for vector control, with religious use of DEET or "bug-bafflers" (thin, hooded shirts of mosquito netting to wear over clothing).

Managing Travel and the Budget

Expedition planning began well in advance of the fieldwork, and the purchase of plane tickets usually served as the breaking point for an expedition: they were the single largest expense of all the expeditions, the cost of which would only rise as the departure date approached. They could not be purchased until the budget had been firmed up and approved and the student participants identified. Although students were in charge of these details, the input of advisers was particularly important during budget development, especially for those budget items that are obvious to seasoned field biologists but often unconsidered by undergraduates without such experience. Once the plane tickets were purchased, the expeditions took on increasing momentum. The student in charge of travel and lodging had to calculate potential charges for excess luggage and incorporate those into the budget. He or she also learned how difficult it can be to confirm the availability of lodging and food at a target site without actually being present. The student in charge of finances needed to curate each receipt according to the rules prescribed by the accounting staff at the home institution, grappled with obtaining cash and monitoring its flow while on expedition, ensured that everything was fully itemized for reimbursement, determined which expenses were the responsibility of individuals versus the expedition's funding, and submitted everything to accounting staff soon after the return of the group.

EXPERIENCES DURING THE EXPEDITIONS

Time Management and Good Habits in the Field

Exploratory research is challenging. Although at first it may seem simple enough and intellectually unsophisticated, it actually requires ample skill, patience, careful interpretation, and intellectual resourcefulness in both on-the-spot situations and in the day-to-day planning that occurs between forays into the field. Target taxa were difficult to see and track in dense tropical forest. In the jungle, the myriad things happening all around were significantly distracting, and the enormity of a relatively little known fauna was at times overwhelming. Nest finding, an important aspect of natural history research, took up many hours on most days, and it was easy to work hard and simply not find anything, wondering the whole time if the targeted birds even had an active nest. We thus used several strategies to track our activities, stay organized, and make the most of discoveries we made during our expeditions. We detail some of our experiences and strategies later.

We limited our group meals to a supper every evening, and in our larger expeditions we had help from locals with cooking, while the smaller crews took turns cooking. Mealtime proved a great time to get bird lists compiled, leaving time for more thoughtful work after supper. We generally compiled a group daily bird list representing estimates of the total number of individuals recorded of each species. We submitted these lists to eBird (eBird.org), and they thus became immediately available and public as a record of the presence of the birds we encountered.

Most of our planning took place during the evenings, when we had group meetings to summarize daily achievements and discoveries as well as "failures" or goals that we were unable to complete during the day. We would then plan the next day's activities based on these achievements and the most important of our long-term goals. Our meetings allowed us to maintain a balance among media collection of target taxa, finding new territories or spot mapping, finding nests and nest watching, and netting birds. Nest finding, nest watching, and netting were the most time-intensive activities, so planning was centered on these activities, with other activities such as media collection and territory mapping fit in between.

To keep things running smoothly on a day-to-day basis, we emphasized the importance of good field habits. Many of these routines were applicable to doing work at any field site; one of the first orders of business was to get to know the terrain and landmarks of the study area well enough to pursue birds off trail without getting lost. Keeping

GPS units (with fresh batteries) with crew members at all times was imperative for safety and also useful for georeferencing field observations. Crew members linked by a buddy system, with regular check-ins with each other via text messages or two-ray radios, allowed independent fieldwork to be done safely. Keeping batteries fresh and camera cards backed up prevented disastrous data losses, though the occasional losses we did experience were sometimes a necessary wake-up call to solidify our routines. Other habits were specific to particular circumstances. For example, in places prone to mosquito-borne diseases, making sure that team members are using insect repellent and bug-bafflers became a group responsibility as interrupting fieldwork for visits to a hospital is a serious burden on the entire group. At our rainforest sites, maintaining and keeping rain gear on everybody, and their equipment, was essential given the near-daily downpours. Our reliance on a small gasoline-powered generator in Panama added to the challenge of sticking to these routines.

Collecting the Complete Specimen Package and Other Natural History Data

Our activities in the field tended to be more varied than usual for a museum expedition because of the diversity of the data we sought. One of our foremost goals was nest finding, because detailed observations at a nest over time provide valuable natural history knowledge for comparisons with related taxa. Still, nest searching was often a low-yield investment of time, and it was interspersed with other activities. We also devoted some days entirely to catching birds, which we would either measure and band, or collect. Whenever our birds had distinct territories, we spent hours collecting audio and video samples from each individual throughout the study site. We also mapped territories, which required discipline throughout the field season as we tried to make a habit of stopping to record the coordinates of every target bird we heard every time we heard one. Color banding allowed us to confirm the ranges of individuals.

Each of these activities can lead to valuable data on its own, although it is most valuable when targeted at the same individual to build a CSP. The breeding behavior of very few tropical species has been studied in detail, which means that nest finding would be valuable in many places and

for many potential target taxa. Banding and collecting are both fruitful endeavors in the tropics; banding is especially valuable if expeditions are likely to return to the same area, given the paucity of banding data in the Old World Tropics compared to other regions. Collecting species other than the target taxa is also useful if permits can be obtained, particularly when taking fluid specimens, which are generally poorly represented in collections.

Outreach

When possible, there are great benefits to integrating outreach into an expedition. One of our most rewarding activities in Borneo was visiting nearby schools to talk to the children about birds and to play games with them patterned after those in Cornell's BirdSleuth program (www.birdsleuth.org). Local schools were happy to let us devise activities for the children as a way for them to practice English, and we used this opportunity to let them try out our binoculars and recording equipment, and to talk to them about what we were learning from the birds in their area. We felt we were giving back to the local communities by telling them about what we were doing, and it was great to see the amazement on the kids' faces when they got their first glimpse of a colorful bird like a Collared Kingfisher (*Todiramphus chloris*) in a spotting scope. At the same time as giving the kids a new perspective on the forests in their backyard, we also learned informally from them about how their communities viewed the forests and the birds.

EXPERIENCES IN POST-EXPEDITION WORK

Outcomes: Specimens and Papers

The expeditions led to personal growth and a greater awareness of the natural history of birds to all people involved, but to guarantee the success of the expedition, crew members must actively make the outcomes of the expedition available to the scientific community as a whole. We see basically two scientific outcomes of an exploratory collecting expedition: the specimens (physical and digital) and the papers. Papers are a critical part of the CSP if they describe the natural history of the birds studied and collected. However, we found that the post-expedition period was also when student commitment began to decline. We learned how important it is to keep students mindful of, and enthusiastic about, the goals of the expedition, and steer their commitment to ensuring the scientific outcomes are fulfilled.

Writing papers after the expedition offered a special challenge for undergraduates. Unlike graduate students, undergraduates did not have the prospect of a dissertation motivating them to work on papers, and they also had the added pressures of a full-time course load. They had the natural human tendency to give most attention to the tasks with intrinsic deadlines: crew members had to clean and store the gear to get it off the floor of the lab; they had to deal with the financial reimbursements or someone would have lost money; and external grants required reports by certain dates. Curators were good at reminding students of the need to curate physical or media specimens, and, at the very least, to submit metadata in an organized fashion to accompany each accession. But papers do not speak up for themselves, and students had to learn to cultivate the inner voice that reminds them that papers will not happen unless they make them happen. The adviser provided that voice at first, but the goal was to develop the ability to generate those reminders internally in the students' own minds.

Our expedition group found itself grappling with the challenge of developing papers from earlier expeditions at the same time that it felt compelled to plan the next expedition. Divided interest and the inherent urgency of expedition preparation tasks compared to data analysis inevitably slowed the progression of our work toward submitting publications. Although an obvious solution to this problem is to extend the time between expeditions, the rate of turnover at universities means that to do so greatly dilutes the pool of experienced older students available to help train new members and lead expeditions. All but the first expedition leaders were crew members on earlier expeditions, and running each expedition with an entire crew of novice members would be very inefficient. This challenge made us realize that, if possible, the program could be improved in two ways. First, a project of this size should probably not be attempted by a single faculty member as the only constant member of the group, and the availability of more advisers would do much to improve the sustainability and ongoing productivity of a student expedition program.

Having more advisers would take the burden of training new members off the experienced group members who may leave campus before subsequent expeditions. Second, student commitment to post-expedition work needs to be secured by enrollment in a post-expedition course for credit (even if it is for directed independent research) to complete all curation and bring manuscripts to submission. It may sometimes be advisable to have a separate post-expedition seminar for working on papers if there is another expedition being planned, full of pre-expedition needs, at the same time. A seminar, or regular weekly meetings of paper-based research groups with the adviser(s), is a direct way to integrate deadlines into the process, and enrolled students who must fulfill expectations to get credit for the work can profit from this additional nudge of motivation.

CONCLUSIONS

Four expeditions have yielded us enough experience to assess our CSP goals, the student-led character of our expeditions, and the value of exploratory museum expeditions more generally.

The Complete Specimen Package proved an elusive goal. Of all the target taxa specimens listed in Table 11.1, only for the Sapayoa and the Black-crowned Pittas did we collect digital media from the same individuals that were collected as physical specimens. The two main reasons for this are the extreme difficulty of taking media of and then capturing individuals, and the challenges associated with relocating captured birds to take media. In our particular situation, the reticence of pittas, the unpredictability of green broadbills, the low population densities of Sapayoas, and the high-canopy habits of the patterned broadbills in the tall trees at Tawau Hills were all hurdles to achieving individual-oriented CSPs. We could have been much more successful with more easily captured species. However, in cases such as these it is still usually possible for an expedition focused on CSPs to yield reasonable numbers of both physical specimens and digital media for the expedition's target taxa (for our results, see Table 11.1). Even though the media may not be of the same individuals that are later collected, both physical specimens and digital media were collected from the same locality at the same time. It clearly depends on the natural history of the bird whether it is practical to collect CSPs, and for birds and localities where this is possible, the CSP remains a valuable goal. But even when the CSP is not attainable, "population CSPs," with rich media and specimens collected of the same species from the same locality and time are still extremely valuable (see Chapter 13, this volume).

Focus fieldwork and choose sites carefully. Our targeted attention on the Old World Suboscines generally worked well to focus our fieldwork, although we also learned that it is important to carefully consider the accessibility of the target species in the research sites available. We found in Borneo that broadbills and pittas were sufficiently abundant and accessible for study; in the Darién, however, Sapayoas were relatively difficult to find. With only the uncommon Sapayoa to study in Panama, the success of that expedition hung by a much narrower thread and was more dependent on luck than that of the expeditions in Borneo, which could afford to be flexible if one of their target taxa proved scarce (indeed, Green Broadbills were difficult to find in Tawau Hills Park, but Black-and-yellow Broadbills and Black-crowned Pittas were abundant). Choosing focal taxa that are both scientifically interesting and amenable to exploratory fieldwork by students—who, no matter how thorough their training, are still sure to be relatively inexperienced field biologists—is an important consideration; thus, having a *too* narrow focus may be just as detrimental as having no focus at all.

Our student-led expeditions were a great success. Thirteen of the fourteen students on our expeditions are continuing to pursue a career in ecology or evolutionary biology, and the Ivy Expeditions gave our students an uncommonly rich experiential learning opportunity. Experiencing all the challenges of planning and executing an expedition, in-depth exposure to the wonders and roadblocks of working in tropical forests, familiarity with modern digital ornithology techniques, negotiating both day-to-day arrangements and permits in another culture: all of these experiences are ones that will stay with the students for a lifetime, and they will be finding applications of these experiences to challenges in their futures no matter where those future careers take them. Still, we cannot honestly report that our program seamlessly turned students into mature scientists. Relatively inexperienced undergraduates were in charge of massive logistical challenges and trying to produce results both in the field and upon

returning to the university. Hiccups in the process were bound to happen. Some of the challenges we faced were an essential part of learning in our hands-off advising mode; they were an integral part of why it was so valuable to allow undergraduates to lead these projects. Others were detrimental to the success of the expeditions and the experience of the students and should be avoided. In the first 4 years of developing this program we have learned precious lessons about the potential kinks of working in situations like ours, and we hope, with this report of our experiences and recommendations, that we can save others time in the future.

Exploratory expeditions are important. In the face of today's world of high-impact journals and complex, multi-institutional experiments, some may fail to see the significance of a paper that describes the daily habits of a small bird in a far-off place. But students can be reminded that, when they are making discoveries about the natural history of a bird, they are contributing to the most fundamental knowledge base of the world's best-known group of vertebrates. Papers they publish may not appear in what are judged as high-impact journals today, but natural history endures. While researching target taxa sheets, students can be encouraged to note the publication dates of the papers they are reading or the dates on the museum tags of specimens they are measuring: many of these may be decades to centuries old, and we are still learning from them. In contrast, even high-profile journal papers that are important today can quickly become outdated. Exploratory expeditions are adding to the venerable and durable storehouse of data on biodiversity in museum collections. If expeditions are pursuing CSPs, they are solidifying the addition of richer digital media to that tradition. Students from these expeditions were doing what museum scientists have always done: carefully gathering information about the natural world to preserve it for posterity, feeding the hypotheses of future scientists, and offering an irreplaceable window into a particular time and place (Bradley et al. 2014). Far from being irrelevant, natural history and collections work is all the more important today, when entire ecosystems are on the brink of being lost forever. Indeed, in a hundred years, the record left by natural history expeditions may be all that is left of what was once an ancient, complex, and exhilarating place.

ACKNOWLEDGMENTS

All authors of this manuscript participated in the expeditions, and we thank our fellow expedition members: J. M. Hite, M. Stager, J. A. Kapoor, D. Y. Gu, B. R. Magnier, E. M. Kelly, K. Chalkowski, K. S. Lauck, and S. A. Dzielski. For technical support in the United States, we thank M. Medler, B. Clock, K. Bostwick, C. Dardia, M. Holton, and I. J. Lovette. For support in Malaysia, we thank A. Biun, F. Tuh, M. bin Lakim, R. Repin, J. Rumpadon, Mr. Simon, Mr. Jool, J. Mas, J. Guntabid, C. Y. Chung, and A. F. bin Amir. For support in Panama, we thank M. Miller, L. Camacho, D. Buitrago, P. Castillo, I. Pissaro, S. Gladstone, A. Hruska, C. Hruska, Autoridad Nacional del Ambiente de Panamá, and the Smithsonian Tropical Research Institute. A. R. McCune and M. S. Webster provided helpful comments on the manuscript. Funding for the expeditions was provided by the Ivy Expeditions Fund, Cornell University, National Geographic Young Explorer's Grant number 9505-14, The Explorer's Club, the Rawlings Cornell Presidential Research Scholars Program, the Morely Student Research Grant, and a gift from T. Wickenden. All the expeditions took place under Cornell IACUC Protocol No. 2001-0051 to D. W. Winkler.

LITERATURE CITED

Bradley, R. D., L. C. Bradley, H. J. Garner, and R. J. Baker. 2014. Assessing the value of natural history collections and addressing issues regarding long-term growth and care. BioScience 64:1150–1158.

del Hoyo, J., A. Elliott, and D. A. Christie. 2003. Handbook of birds of the world. Volume 8: Broadbills to Tapaculos. Lynx Edicions, Barcelona, Spain.

Dzielski, S. A., B. M. Van Doren, J. P. Hruska, and J. M. Hite. 2016. Reproductive biology of the Sapayoa (*Sapayoa aenigma*), the "Old World suboscine" of the New World. Auk 133:347–363.

Erritzoe, J., and H. B. Eritzoe. 1998. Pittas of the world: a monograph on the pitta family. Lutterworth Press, Cambridge, UK.

Greene, H. W. 2005. Organisms in nature as a central focus for biology. Trends in Ecology and Evolution 20:23–27.

Gulson-Castillo, E. R., T. M. Pegan, J. R. Shipley, J. M. Hite, S. C. Orzechowski, J. P. Hruska, and D. W. Winkler. (in review). Observations on the broadbills (Eurylaimidae, Calyptomenidae) and pittas (Pittidae) of Tawau Hills Park, Sabah, Malaysian Borneo. Forktail.

Hackett, S. J., R. T. Kimball, S. Reddy, R. C. K. Bowie, E. L. Braun, M. J. Braun, J. L. Chojnowski, W. A. Cox, K. Han, J. Harshman, C. J. Huddleston, B. D. Marks, K. J. Miglia, W. S. Moore, F. H. Sheldon, D. W. Steadman, C. C. Witt, and T. Yuri. 2006. A phylogenomic study of birds reveals their evolutionary history. Science 27:1763–1768.

Hruska, J. P., S. A. Dzielski, B. M. Van Doren, and J. M. Hite. 2016. Notes on the avifauna of the northern Serranía de Pirre, Panamá. Bulletin of the British Ornithologists' Club 136:224–242.

Irestedt, M., J. I. Ohlson, D. Zuccon, M. Källersjö, and P. G. P. Ericson. 2006. Nuclear DNA from old collections of avian study skins reveals the evolutionary history of the Old World suboscines (Aves, Passeriformes). Zoologica Scripta 35:567–580.

Jarvis, E. D., S. Mirarab, A. J. Aberer, B. Li, P. Houde, C. Li, S. Y. W. Ho, B. C. Faircloth, B. Nabholz, J. T. Howard, A. Suh, C. C. Weber, R. R. da Fonseca, J. Li, F. Zhang, H. Li, L. Zhou, N. Narula, L. Liu, G. Ganapathy, B. Boussau, M. S. Bayzid, V. Zavidovych, S. Subramanian, T. Gabaldón, S. Capella-Gutiérrez, J. Huerta-Cepas, B. Rekepalli, K. Munch, M. Schierup, B. Lindow, W. C. Warren, D. Ray, R. E. Green, M. W. Bruford, W. Zhan, A. Dixon, S. Li, N. Li, Y. Huang, E. P. Derryberry, M. F. Bertelsen, F. H. Sheldon, R. T. Brumfield, C. V. Mello, P. V. Lovell, M. Wirthlin, M. P. Cruz Schneider, F. Prosdocimi, J. A. Samaniego, A. M. Vargas Velazquez, A. Alfaro-Núñez, P. F. Campos, B. Petersen, T. Sicheritz-Ponten, A. Pas, T. Bailey, P. Scofield, M. Bunce, D. M. Lambert, Q. Zhou, P. Perelman, A. C. Driskell, B. Shapiro, Z. Xiong, Y. Zeng, S. Liu, Z. Li, B. Liu, K. Wu, J. Xiao, X. Yinqi, Q. Zheng, Y. Zhang, H. Yang, J. Wang, L. Smeds, F. E. Rheindt, M. Braun, J. Fjeldså, L. Orlando, F. K. Barker, K. A. Jønsson, W. Johnson, K.-P. Koepfli, S. O'Brien, D. Haussler, O. A. Ryder, C. Rahbek, E. Willerslev, G. R. Graves, T. C. Glenn, J. McCormack, D. Burt, H. Ellegren, P. Alström, S. V. Edwards, A. Stamatakis. D. P. Mindell, J. Cracraft, E. L. Braun, T. Warnow, W. Jun, M. T. P. Gilbert, and G. Zhang. 2014. Whole-genome analyses resolve early branches in the tree of life of modern birds. Science 346:1320–1331.

Lambert, F., and M. Woodcock. 1996. Pittas, Broadbills, and Asities. Pica Press, Mountfield, Sussex, UK.

Moyle, R. G., R. T. Chesser, R. O. Prum, P. Schikler, and J. Cracraft. 2006. Phylogeny and evolutionary history of Old World suboscine birds (Aves: Eurylaimides). American Museum Novitates 3544:1–22.

Pegan, T. M., E. R. Gulson-Castillo, J. M. Hite, J. R. Shipley, J. A. Kapoor, E. I. Greig, and D. W. Winkler. (in review). An annotated checklist of the birds of Tawau Hills Park, Sabah, Malaysian Borneo, with observations on notable species. Forktail.

Pegan, T. M., J. P. Hruska, and J. M. Hite. 2013. A newly described call and mechanical noise produced by the Black-and-Crimson Pitta *Pitta ussheri*. Forktail 29:160–162.

Prum, R. O., J. S. Berv, A. Dornburg, D. J. Field, J. P. Townsend, E. M. Lemmon, and A. R. Lemmon. 2015. A comprehensive phylogeny of birds (Aves) using targeted next-generation DNA sequencing. Nature 526:569–573.

Rosset, A., L. Spadola, and O. Ratib. 2004. OsiriX: an open-source software for navigating in multidimensional DICOM images. Journal of Digital Imaging 17:205–216.

Rudin, C., D. Dunson, R. Irizarry, H. Ji, E. Laber, J. Leek, and T. McCormick. 2014. Discovery with data: leveraging statistics with computer science to transform science and society. A Working Group of the American Statistical Association.

Seeholzer, G. F., B. M. Winger, M. G. Harvey, A. D. Cáceres, and J. Weckstein. 2012. A new species of barbet (Capitonidae: *Capito*) from the Cerros Del Sira, Ucayali, Peru. Auk 12:551–559.

Sporns, O. 2015. Enabling discovery science in human connectomics. Science Bulletin 60:139–140.

Winkler, D. W., S. M. Billerman, and I. J. Lovette. 2015. Bird families of the world: an invitation to the spectacular diversity of birds. Lynx Edicions, Barcelona, Spain.

Xu, S.-Y., I. Belopolski, N. Alidoust, M. Neupane, G. Bian, C. Zhang, R. Sankar, G. Chang, Z. Yuan, C. Lee, S. Huang, H. Zheng, J. Ma, D. S. Sanchez, B. Wang, A. Bansil, F. Chou, P. P. Shibayev, H. Lin, S. Jia, and M. Z. Hasan. 2015. Discovery of a Weyl fermion semimetal and topological Fermi arcs. Science 349:613–617.

CHAPTER TWELVE

Biodiversity Informatics and Data Quality on a Global Scale*

Carla Cicero, Carol L. Spencer, David A. Bloom, Robert P. Guralnick,
Michelle S. Koo, Javier Otegui, Laura A. Russell, and John R. Wieczorek

Abstract. Exciting developments in biodiversity informatics, and particularly the growth of aggregated data networks such as ORNIS and VertNet, have stimulated new areas of research in ornithology and other disciplines that would not have been possible otherwise. In addition to aggregating data from across natural history collections, VertNet provides valuable tools and services for assessing and improving data quality through pre- and postdata publication workflows as well as through feedback mechanisms, and for providing usage statistics back to participating collections. Furthermore, VertNet's outreach to the collections community has resulted in national and international training in georeferencing and biodiversity informatics that has had a broad impact. Rapid access to high-quality data has made possible studies that explore large and complex datasets reflecting the global and temporal distributions of species. Specific examples include studies of niche evolution and speciation, risk assessment of invasive species, phenotypic evolution and the colonization of novel habitats, latitudinal gradients and evolutionary divergence, and biodiversity and conservation planning. These studies showcase the value of natural history collections, and highlight the need for ongoing support to protect and grow those collections and to enable efficient access to their data. The role of natural history collections in addressing challenges faced by society is more important than ever before. In the modern era of sharing data, collections hold exciting promise for future and emerging avenues of research that extend the value of specimens beyond what we have ever dreamed before.

Key Words: biodiversity, conservation planning, Darwin Core, data aggregation, data networks, data standards, evolution, natural history museum, niche modeling, VertNet.

Museum collections provide extensive and irreplaceable information about avian diversity, evolution, biogeography, and natural and life histories. Specimens and related materials (e.g., media and archives) in these collections provide the foundation for many types of ornithological and cross-disciplinary studies, and also strongly inform science and conservation issues related to changes in species distributions, phenologies, phenotypes, and genotypes (Graham et al. 2004, Moritz et al. 2008, Tingley et al. 2009, Nufio et al. 2010, Erb et al. 2012). When data

* Cicero, C., C. L. Spencer, D. A. Bloom, R. P. Guralnick, M. S. Koo, J. Otegui, L. A. Russell, and J. R. Wieczorek. 2017. Biodiversity informatics and data quality on a global scale. Pp. 201–218 in M. S. Webster (editor), The Extended Specimen: Emerging Frontiers in Collections-based Ornithological Research. Studies in Avian Biology (no. 50), CRC Press, Boca Raton, FL.

from museum collections are compiled and integrated, they provide irreplaceable information for directing and prioritizing conservation efforts from local to national and global scales (e.g., IUCN Standards and Petitions Working Group 2008, Fajardo et al. 2014). In addition, improved discovery of data from specimen preparations such as tissues is essential for modern studies of genetic diversity (Wandeler et al. 2007), especially where collecting new samples is prohibitive.

Despite the well-documented value of collections for science and society (Krishtalka and Humphrey 2000, Suarez and Tsutsui 2004), sequestered data in analog formats have hampered the ability of researchers and policy makers to discover and utilize these resources. Digitization and mobilization of such data, including label data, audio clips and images, field notes, illustrations, and gene sequences, largely removes impediments to discovery, interoperability, and enhancement (Edwards et al. 2000, Canhos et al. 2004, Soberón and Peterson 2004, Guralnick et al. 2009). However, the mass mobilization of museum data presents both technical and social challenges (Pennisi 2005, Berendsohn et al. 2010, Vollmar et al. 2010).

Data mobilization is a cooperative task that requires community data standards (e.g., the Darwin Core; Wieczorek et al. 2012), distributed knowledge, and scalable tools to aggregate, integrate, improve, and redistribute those records (e.g., Robertson et al. 2014). Vertebrates represent the most extensively digitized and data-rich group, and have laid the groundwork for mobilizing specimen-related data from other taxa as they become digitized. VertNet and its four predecessor vertebrate networks (ORNIS, HerpNET, MaNIS, and FishNet; Constable et al. 2010, Guralnick and Constable 2010) have been at the forefront of developing the infrastructure, standards, community training, and capacity needed for data publishing, sharing, and improvement. In this chapter, we briefly review data sources and standards, discuss past and current efforts to mobilize data through VertNet, and stress the importance of quality for discovering data in collections-based research. Although we focus on bird specimen data, we recognize the great value of parallel citizen science informatics efforts (e.g., eBird, http://ebird.org). We also address how high-quality data can enhance and extend the value of specimens in novel ways, and provide

examples of emerging ornithological research and conservation uses that would not have been possible without efficient access to large amounts of data from museum collections.

DATA SOURCES, STANDARDS, AND SHARING

Efficient access to data through a distributed network system depends largely on three factors: (1) the community's willingness to provide open access to data, (2) community standards for data sharing, and (3) quality of the data. The willingness to share publicly is particularly challenging, and is complicated by concerns over data quality and a perception that local collection staff will lose control over the data they steward (Erike et al. 2012, Huang et al. 2012). The reluctance to share data was identified as a potential impediment early in the development of the first biodiversity data networks. As a result, aggregated data networks have been built on the foundation that curators or collection managers must maintain local control of data for their individual collections. Through such control, collection personnel can better check data against original specimen documentation (thereby improving data quality), choose what restrictions to place on their data, and update their data as new information becomes available (Graham et al. 2004).

Collections use a variety of applications for data capture at the local level, ranging from spreadsheets (e.g., Excel) to simple database packages (e.g., Filemaker Pro, Access) and more sophisticated collection management systems (e.g., Arctos, Specify, KE EMu). These local sources vary tremendously in their implementation and use of authorities and controlled vocabularies, with data fields subject to free text input or more strict controls. Furthermore, collections are staffed by people with a wide range of expertise and knowledge. Data quality is positively associated with the expertise of collection staff and the implementation of databases that are well-structured and have stricter controls over data capture and input. Because of this heterogeneity, the community has faced a hard challenge in sharing data across collections and making those data broadly discoverable.

The Darwin Core was established to address this problem by developing a standard for aggregating heterogeneous data sources to improve sharing and interoperability (Wieczorek et al. 2012). Although continually evolving, this standard has been accepted and implemented widely by the global informatics and museum communities. It is important, however, to distinguish between standardized data *terms* and data *values*. While the Darwin Core applies standardization to data terms (see the quick reference guide at http://rs.tdwg .org/dwc/terms), it does not address standardization of values captured for those terms (e.g., how values for sex, preparations, country, etc. are entered). Community adherence to standardization of values, as much as possible, is needed to make data truly discoverable. For example, researchers interested in finding specimens from a particular country or state are more likely to succeed if the data follow a standardized representation of geographic names, and likewise if data are entered by selecting rather than typing, and potentially misspelling, names. The development of global standards for sharing data, along with technological advances and community buy-in, has made access to data from museum collections easier than ever and has also helped to create a more knowledgeable global community.

BIRDS IN THE CLOUD: INCREASING DATA ACCESS THROUGH VERTNET

Twenty years ago, museum databases were just beginning to come online, but researchers and others searching for museum data were often hindered by the need to contact collections individually to request even the simplest information. Records were received in various formats from paper copies to computer printouts or data files. The difficulty in gaining access to data, curatorial effort required, slow response times, and heterogeneous data formats created severe bottlenecks in the ability to discover specimens and data, let alone conduct specimen-based research.

In response to the growing need for such data, along with a push by vertebrate communities, the U.S. National Science Foundation (NSF) funded the Distributed Information Network for Avian Biodiversity Data in 1998. This project resulted in the Species Analyst prototype, which was designed to access and visualize biodiversity data from systematic collections and observational datasets. Subsequent NSF funding led to the development of four taxon-based networks that were based initially on this prototypical work: Ornithological Information System (ORNIS, 2004–2010, http://ornisnet.org); HerpNET (2002–2008,

http://herpnet.org); Mammal Networked Information System (MaNIS, 2001–2006, http://manisnet .org); and FishNet (1999–2000) and FishNet2 (2004–2010, http://www.fishnet2.net). These early networks were created with four core objectives: (1) to facilitate open access to specimen data on the web (e.g., Peterson et al. 2005), (2) to enhance the value of specimen collections, (3) to conserve curatorial resources, and (4) to use a design that could be adopted easily by other disciplines with similar needs. As each new vertebrate network was created, it adopted these objectives and contributed to the networks collectively with the development of new tools and services. For example, MaNIS added georeferencing tools and guidelines, HerpNET added georeferencing workshop training and taxonomic synonym resolution, and ORNIS added more data-quality services. Resources were shared across the networks to create economies of scale. The networks grew rapidly (Figure 12.1), from 17 institutions in 2001 to 212 institutions (696 collections) in 2016, and they had a profound impact on biodiversity informatics by mobilizing data, developing technologies to improve access to those data, and building active communities of users. Ultimately, these open and collaborative relationships allowed each network to survive despite a growing technological strain.

The architecture of the four networks involved a three-tiered system of servers (Figure 12.2). Users accessed the first layer (the portals) via web browsers, and portals were connected to a second layer of servers using the Distributed Generic Information Retrieval (DiGIR) protocol. These DiGIR servers, in turn, were connected to a third layer of collection database servers. Both the DiGIR and database servers were managed by individual data contributors (publishers) at each institution or collection. Because this design required hundreds of distinct database and web connections, each of which was a potential point of failure, performance and responsiveness were often unpredictable and unstable. Increasing numbers of simultaneous queries and requests for large datasets overloaded the servers, and successful queries were slow to return results. In addition, the high cost of this architecture—especially the technical expertise needed for server installation, administration, and continued maintenance—was unaffordable for many collections and proved to be unsustainable overall. Other challenges included poor scalability with increasing growth of the networks

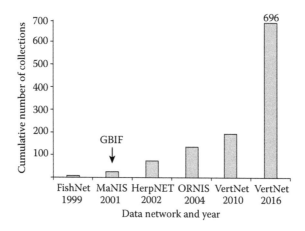

Figure 12.1. Chronology and growth of the vertebrate data networks, and establishment of the Global Biodiversity Information Facility (GBIF). Except for 2016, the year indicates when each network was launched. The cumulative number of collections in VertNet (696) reflects the total number that have requested participation in the network as of June 2016.

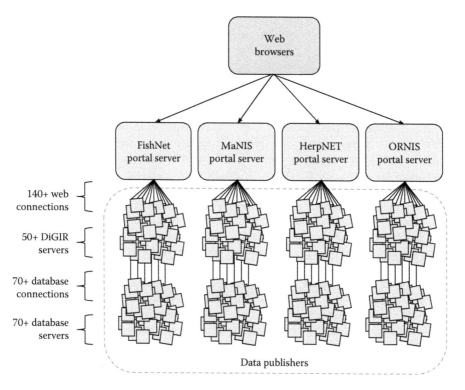

Figure 12.2. Simplified data architecture of the four taxon-based networks, illustrating the layers of servers and web or database connections required to return records from data publishers to data users.

(Figure 12.1), an unbridgeable gap between what users sought and what contributors provided due to nonstandard values across primary data sources, and constraints on the integration of new applications and services that require fast access to all data simultaneously.

VertNet (http://vertnet.org) was created in 2008 through a collaborative effort to find solutions to the challenges faced by a growing demand for standardized online biocollections and an infrastructure that was not capable of meeting that demand. Using feedback from a voluntary community-wide survey of vertebrate network users, a Steering Committee with representatives from the biodiversity informatics and natural history museum communities established three major goals for VertNet: (1) evaluate ways to address issues of performance, scalability, usability, and long-term sustainability of the networks; (2) develop new data applications and services to meet the needs of the community, including ways to enhance data quality and provide usage information and user feedback; and (3) continue to provide broad outreach to the communities served. The VertNet collaborative development project (Table 12.1) has addressed all of these goals

TABLE 12.1
VertNet at a glance (as of June 2016).

Institutions	212
Collections: total/vertebrates only[a]	696/636
Records published[b]	232,233,952
Records viewed on portal[c]	108,058,577
Records downloaded from portal[c]	517,159,866
Citations referencing portals or data[d]	1002
Training workshops and webinars (see Table 12.3)	65

[a] Includes all collections that have requested participation in the network, of which 326 were published as of June 2016 in the VertNet portal.
[b] Includes records published in the VertNet portal (19,087,118; Aves: 5,726,458), eBird records published by VertNet to GBIF (211,883,652), and other records published to GBIF and iDigBio (1,263,182; includes plants, invertebrates, and parasites). Excludes records published on non-VertNet instances of the Integrated Publishing Toolkit, for which VertNet may provide technical support.
[c] Includes records viewed and downloaded from the VertNet portal from April 2014 through June 2016 (no data available for April 2015). Does not include views or downloads from other portals that aggregate from sources mobilized by VertNet. For example, eBird is accessed through GBIF, not through the VertNet portal.
[d] Includes citations for VertNet and its predecessor networks, based on a Google Scholar search for citations containing the network URLs dated 30 June 2016.

and provided a new model for both data sharing and data quality (Constable et al. 2010).

To address the first goal, VertNet implemented a data mobilization workflow (Figure 12.3) that allows distributed publishing in an integrated, low-cost way via a cloud-based platform for harvesting and aggregating data. Records are published via Darwin Core Archive files using the Integrated Publishing Toolkit (IPT; http://www.gbif.org/ipt; Robertson et al. 2014), a free and open-source web application developed by the Global Biodiversity Information Facility (GBIF) with significant support from the broad informatics community, including VertNet. Darwin Core Archives from the IPT are harvested, stored, and indexed using cloud services for searching by VertNet or other web portals that are global (GBIF, http://www.gbif.org), national (iDigBio, https://www.idigbio.org), or project based (e.g., Dimensions of Amazonian Biodiversity, http://www.amazoniabiodiversity.org). Through its own portal, VertNet also implemented an array of filters to improve the efficiency of searches for data that otherwise would be difficult to find due to the heterogeneity of data values. For example, users can filter searches by limiting records to data that contain tissues, media, fossils, type specimens, and geographic coordinates for mapping. In addition, VertNet produces snapshots of all data for a particular taxonomic class and makes those accessible for research through large-scale data repositories (CyVerse, http://www.cyverse.org; DataONE, https://www.dataone.org) and taxon-specific portals (e.g., AmphibiaWeb, http://amphibiaweb.org). VertNet's platform for publishing data has achieved four key advances over its predecessor networks: it is fast, easy, scalable, and inexpensive.

VertNet has continued to work closely with natural history collections to make their data accessible to a broad range of users and disciplines (Table 12.2). Its success with respect to community support, participation, and utility to science is reflected in its user community and growth of the participant list of publishers (Figure 12.1). In 2008–2009, VertNet had 70 collections online from 65 institutions, with another 34 collections waiting to join. As of June 2016, VertNet is working with 696 collections housed at 212 institutions from 38 countries, with 326 collections served through its portal. Over 50% of all animal records and 63%

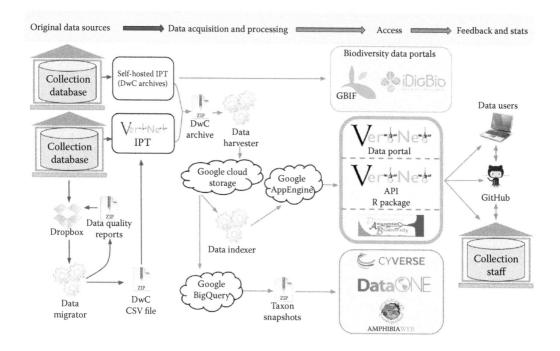

Figure 12.3. Schematic diagram of the VertNet workflow. Red arrows show migration of data from original sources (i.e., collection databases) to the Integrated Publishing Toolkit (IPT) via either a self-hosted or VertNet-hosted path. Many collections pass data through VertNet migrators that are customized for each resource and include data quality reports and improvement; some collections (e.g., Arctos) are published directly to the VertNet IPT through database connections. Green arrows show the VertNet workflow for data acquisition and processing, including harvesting, storing, indexing, and publishing. All published data conform to Darwin Core (DwC) standards through a DwC Archive. Data can be accessed through the VertNet portal, API, and R package, as well as through other biodiversity portals (GBIF, iDigBio, Dimensions of Amazonian Biodiversity) and taxon-based snapshots (hosted at CyVerse, DataONE, AmphibiaWeb). Blue arrows indicate VertNet connections with the user community, including user feedback and usage statistics to collections via GitHub. Some portals (e.g., Dimensions of Amazonian Biodiversity) make use of VertNet architecture and services for data acquisition, processing, and feedback.

of all bird records in GBIF come from VertNet mobilization efforts. This unsolicited demand to participate clearly demonstrates that natural history collections and the biodiversity community have a strongly positive perception about publishing data to VertNet.

The current push to digitize collections and make data accessible online (e.g., Bartlett et al. 2013, Hanken 2013), combined with the increasing use of big data for research, have contributed to this rapid network growth. Moreover, this growth demonstrates a sociological shift from skepticism to enthusiasm for data sharing within the ornithological and other biodiversity communities. Accessibility to data increases the reach and impact, and therefore extends the value of specimen collections for science and conservation (Peterson et al. 2005, Giorgi et al. 2014).

VertNet is now a key part of mainstream biodiversity research.

VertNet's second goal of providing new services to meet the needs of the community has focused on both data-quality issues and new tools to enhance usage information and collaborative work, especially through the integration of community data standards and common applications. Data quality issues are at the heart of accessibility and discovery for publishers and users, yet data can be messy in single resources and overwhelmingly so in aggregate. Publishers work with VertNet in part because of the value and promise of improving their data. Likewise, users rely on standardized records in order to find the data and specimens that they need. For example, researchers who are interested in delayed plumage maturation in birds (e.g., Hawkins et al. 2012)

TABLE 12.2

VertNet user base, disciplines served, and examples of data usage.

Category of use	Examples
	User base
Academic disciplines	Animal Behavior, Agriculture, Anthropology, Behavioral Ecology, Biogeography, Biology, Botany, Conservation, Ecology, Entomology, Environmental Science, Evolutionary Biology, Forestry, Genomics, Herpetology, History, Image and Media, Ichthyology, Libraries and Archives, Mammalogy, Ornithology, Paleontology, Parasitology, Plant Ecology, Sociology
Agencies and nongovernmental organizations (NGOs)	City/Regional Planners, International Union for Conservation of Nature, Nature Conservancy, NatureServe, U.S. Army Corps of Engineers, U.S. Bureau of Land Management, U.S. Fish and Wildlife, U.S. Forest Service, U.S. Geological Survey, U. S. National Park Service, U.S. Navy
Professions	Educational or Consumer Software Developers (open source, commercial), Interpreters, K–12/University Educators, Museum Educators, Naturalists, Statisticians, Web Developers
	Data usage
Collections	Collections Management, Database Management, Data Publication, Data Quality Issues, Georeferencing, Image and Media Sharing, Loan Verification
Science	Climate and Environmental Change, Collection Gaps (time, geography), Comparative Analysis of Range and Distribution vs. Human Activity, History of Science, Niche Models, Species Absence, Species Distributions (past, present), Temporal Change
Management/Policy	Conservation, Economic Analysis, Industrial Development, Land Use, Military Training, Trade Routes, Wildlife and Ecosystem Management

and need only male specimens would be challenged to find what they want. In an early subset of ~2.7 million data records in the VertNet data store, the simple Darwin Core field "sex" contained 189 distinct values that could be interpreted unambiguously as "male," 184 distinct values that mean "female," and 331 variations that were ambiguous or undetermined. More complicated data such as geography, taxonomy, people's names, and specimen parts present even greater challenges.

VertNet addresses the issue of data quality through three mechanisms: (1) data migrators that are customized for collections that publish data through the VertNet IPT (Figure 12.3)—these migrators transform the source data into Darwin Core fields, analyze and improve the data, and provide data quality reports to the publisher for improvement at the source; (2) a mechanism for feedback and annotation (Figure 12.3) that allows users to submit data issues through the portal to a GitHub (https://github.com) repository, where issues are archived and collections' staff can interact with users to address these issues; and (3) a real-time spatial quality report for each data record in the portal that provides information on data consistency, completeness, and accuracy.

Together, these mechanisms have led to significant improvements in data quality that benefit both the data publishers and data users.

VertNet also responded to a clear need for publishers to track usage of their data by implementing a mechanism through GitHub that provides monthly usage statistics to contacts for each data resource. Detailed statistics are provided on the number of records queried and downloaded, monthly and cumulatively, as well as the number of queries by country and date, and the specific terms used in the query. These statistics are critical for institutional reporting and applications to funding agencies for collection support, and are important for institutional and public perceptions of the value of natural history collections generally (e.g., Abrahamson 2015).

The third major goal has focused on outreach, training, and capacity building. VertNet personnel have provided these key services with 60 workshops and five webinars that have brought together 1,488 participants from 73 distinct countries since 2003 (Table 12.3). Thus, we have trained people from over 37% of the world's nations. Activities have focused on a broad range of topics in georeferencing and informatics, from coding hackathons and the Integrated Publishing Toolkit to data quality, data

TABLE 12.3
Summary of workshops and webinars presented by VertNet and its predecessor networks, including ORNIS (2003–June 2016).

	Total	Georeferencing	Informatics
Workshops	60	36	24
Webinars	5	–	5
Participants trained	1488	716	772
Distinct countries of participants	73	51	56
Distinct host countries	12	8	8

cleaning and publishing, Darwin Core Standard and ontologies, tools and resources for biogeography and global change biology, and training for students and early career scientists in biodiversity informatics and the use of VertNet data in research. The development of standardized protocols and training materials in both English and Spanish for georeferencing has had one of the broadest impacts, with 36 workshops on georeferencing and 8 workshops in Spanish-speaking countries including Spain, Argentina, and Costa Rica. Georeferencing is a fundamental data-quality task that provides latitude and longitude as geospatial coordinates with measures of uncertainty based on locality descriptions. Georeferencing workshops have had a tremendous impact on community building by following the Train-the-Trainers philosophy, whereby participants are trained to lead subsequent workshops in their home countries or disciplines using workshop materials. This training philosophy has been promoted by GBIF and adopted by iDigBio as part of its U.S. effort to digitize biological collections. Outside of formal training activities, VertNet has advised other communities (e.g., VectorMap, http://www.vectormap.org) on how to establish their own distributed data networks and collaborative georeferencing projects. Interdisciplinary studies such as those involving birds and vector-borne disease pathogens (e.g., West Nile Virus; Kilpatrick 2011) benefit greatly from synergism between VertNet and these other communities.

The work to accomplish these three goals and its inclusive approach has generated unanticipated national and international collaborations. In addition to creating and implementing tools, best practices, and workflows that benefit biodiversity science, it has provided opportunities for technical innovation and advancement, improved the discoverability and quality of shared data, and solidified the social connections between institutions and disciplines. The integration of feedback mechanisms (Figure 12.3) and the synergy with institutions and other community projects such as Arctos (http://arctos.database.museum) are just two examples where VertNet has enabled connectivity among data publishers and users. Likewise, its openness to provide services to all collections, and not just vertebrates, from any participating institution has opened interdisciplinary lines of communication and rallied diverse interest groups around the core theme of data mobilization. VertNet's services and open source data platform also have served as springboards for creativity and innovation, driving new approaches to science. The "rvertnet" package (Chamberlain et al. 2015), for example, enables users to query VertNet through the R-environment (https://ropensci.org) on multiple platforms, and thus further increases access to biodiversity data held in museum collections.

DATA QUALITY: FALLACIES AND TRUTH

Data from natural history collections are being used for increasingly diverse applications, yet the data are known to have errors and biases that require assessment prior to use (Graham et al. 2004). In the previous section, we provided an overview of data-quality issues and the approaches that VertNet has taken to improve the quality of published records. This is a significant topic given both the complexity of issues involved and its importance as one of the strongest incentives for participation in VertNet (Constable et al. 2010). From an informatics perspective, the global acceptance of the Darwin Core Standard (Wieczorek et al. 2012), the development of data-quality improvement services through VertNet and other projects, and the availability of look-up authorities for certain types of data such as geography or taxonomy, have facilitated the technical process of data assessment and cleaning prior to online publication. Nonetheless, sociological challenges remain. Next we discuss three fallacies that have been especially hard to overcome.

Data Perfection

The *fallacy* is that data must be perfect before they can be made public. Although the collections community has embraced participation in online biodiversity data networks, there remains uneasiness for some about publishing data before they have been checked and errors have been corrected. The *truth*, however, is that data will never be perfect in every aspect. More important, increasing access to the community leads to improved data quality through feedback and annotation mechanisms (Peterson et al. 2005). By publishing their data, collections can take advantage of collective community expertise to identify errors, improve data, and even enhance records by adding information such as georeferences associated with occurrence records. Although the challenge of high-quality data can be sobering, especially for complex data such as taxonomy where less than 47% of the name strings in VertNet are valid (Zermoglio et al. 2016), the community benefits from new tools for improvement. VertNet has implemented one such mechanism through GitHub (https://github.com) that allows interactive communication between data publishers and data users (Figure 12.3). For example, VertNet annotations submitted to the Museum of Vertebrate Zoology (Berkeley, California) bird and egg collections have led to corrections in collectors, collecting dates, georeferenced coordinates, and species identifications. These are in addition to data issues submitted to the MVZ through Arctos, its own collection management information system, and through e-mail communications. Such improvements benefit not only the collections, but also anyone accessing the data for research, conservation, or other uses.

Data Petrification

The *fallacy* is that data do not change once they have been captured into a catalog ledger or collection database. The *truth* is that data are dynamic and require curation. The University of Illinois Graduate School of Library and Information Science (http://

www.lis.illinois.edu) defines data curation as "the active and ongoing management of data through its life cycle of interest and usefulness to scholarship, science, and education. Data curation activities enable data discovery and retrieval, maintain its quality, add value, and provide for reuse over time." The curation of data is an emerging field within the informatics and information science disciplines. It is important to recognize that databases are dynamic, living entities that evolve, and that curators and collection managers are *de facto* data curators. Although data may be relatively clean when captured, they can always be improved or enhanced. By providing online access to specimen data, data curators allow feedback that ultimately leads to higher data quality, thereby increasing the value of their collections (Peterson et al. 2005).

Data Fitness for Use

The *fallacy* is that fitness for use of data depends on how and why those data were collected. The *truth* is that the value of data is in the eye of the beholder, and fitness for use depends on the questions being asked: data that are not useful to some, such as localities with coordinates of high uncertainty, may be very useful to others depending on scale and purpose. According to the International Organization for Standardization (ISO 1994), quality is defined as "the totality of characteristics of a product that bear on its ability to satisfy stated and implied needs." Quality can be broken down into intrinsic and extrinsic components (Veregin 1999). Intrinsic quality refers to characteristics of the data themselves, whereas extrinsic quality (i.e., fitness for use) is a measure of how closely the data meet a user's needs in a specific context. Fitness for use of data has received a fair amount of attention in the context of geospatial analyses (e.g., De Bruin et al. 2001, Devillers et al. 2007, Hill et al. 2010, Pôças et al. 2014), but this concept applies broadly to a wide range of studies. Furthermore, fitness for use has an important temporal component. The value of data collected 50 to 100 years ago for studies that use museum skins to examine isotopes (Chapter 6, this volume) or genomics (Chapter 9, this volume) was unknown then because the technology did not exist to extract such data out of the specimens. Such studies provide important data on historic changes in populations, such as in seabird numbers and demography (e.g., Beissinger and

Peery 2007, Norris et al. 2007, Peery et al. 2010). Although data of higher intrinsic quality always have a higher fitness for use, even low-quality data can be useful to some. Thus, curators of museum data should not restrict publication of data unless there are other valid reasons for doing so, such as data that are being actively studied or because the data relate to endangered species protections.

To address and improve data quality, VertNet has identified three categories of data quality issues: (1) *data incompleteness*, which refers to the absence of certain fields within a record (e.g., gaps in taxonomic classifications or higher geography); (2) *data inconsistencies*, which identify mismatches between different fields or sources of data such as variable entries for sex or preparations; and (3) *data errors* within specific fields of a record, including inaccurate georeferences, incorrect higher geography, misidentifications, or spelling errors. Some fields clearly lend themselves to data improvement more than others. In the simple sex example previously presented, textual values interpreted as "male" but that are recorded differently (e.g., "m" or "macho") can be standardized to improve discoverability of male records. In the case of more heterogeneous values such as specimen parts, variable textual descriptions interpreted as representing tissue samples (e.g., "+t" or "liver" or "higado") can be combined into a look-up table against which new data values are compared and then used to filter for records containing "tissue" as a part. In a subset of 2.7 million data records, nearly 3,000 distinct preparation terms were narrowed to 26 values that could be used to filter records with "tissue" through the VertNet portal; however, many values contained in tissue fields from source data (e.g., "vodka!00" or "stst01") could not be reliably interpreted as a tissue term. Incompleteness and errors in other types of fields such as geography and taxonomy also can be improved through comparison with available look-up authorities; examples include adding country for states or taxon order for family.

In an effort to enhance the quality of aggregated data, VertNet developed a set of prepublication tools for data cleaning that address inconsistencies, incompleteness, and errors for 75 of the 169 terms (44%; Table 12.4) in the Simple Darwin Core (http://rs.tdwg.org/dwc/terms/simple/index .htm). The Simple Darwin Core is limited to

TABLE 12.4

Simple Darwin Core terms improved by VertNet through the process of data migration and publication.

Darwin Core term category	Darwin Core term
Record-level (14/19)	type, modified, language, license, accessRights, references, institutionID, collectionID, institutionCode, collectionCode, ownerInstitutionCode, basisOfRecord, informationWithheld, dynamicProperties
Occurrence (13/21)	occurrenceID, catalogNumber, recordedBy, sex, lifeStage, reproductiveCondition, establishmentMeans, occurrenceStatus, preparations, associatedMedia, associatedSequences, otherCatalogNumbers, occurrenceRemarks
Organism (2/7)	associatedOccurrences, previousIdentifications
Event (6/18)	eventDate, startDayOfYear, endDayOfYear, year, month, day
Location (24/44)	higherGeography, continent, waterbody, islandGroup, island, country, countryCode, stateProvince, county, municipality, locality, verbatimLocality, minimumElevationInMeters, maximumElevationInMeters, minimumDepthInMeters, maximumDepthInMeters, locationRemarks, decimalLatitude, decimalLongitude, geodeticDatum, coordinateUncertaintyInMeters, verbatimCoordinateSystem, georeferenceProtocol, georeferenceVerificationStatus
Identification (3/8)	identificationQualifier, typeStatus, dateIdentified
Taxon (13/33)	scientificName, higherClassification, kingdom, phylum, class, order, family, genus, subgenus, specificEpithet, infraspecificEpithet, taxonRank, nomenclaturalCode

NOTE: See http://rs.tdwg.org/dwc/index.htm for the definition of Darwin Core terms. VertNet improves 75 terms out of 169 terms total (44%) in the Simple Darwin Core. Numbers in parentheses indicate number of terms improved by VertNet versus the total number of terms in that category. Additional Simple Darwin Core Term categories that are not improved by VertNet include Material Sample (1 term) and Geological Context (18 terms).

terms that fit in a flat, two-dimensional data structure of rows (records) and columns (fields). The remaining terms do not easily fit in a flat data structure. Furthermore, VertNet addresses post-publication data quality in four ways. First, each specimen record retrieved through the VertNet portal is associated with a spatial quality report that is generated on the fly and presented visually to the data user. This report uses validation tools developed by the Map of Life Project (http://mol.org) to evaluate inconsistencies, completeness, and errors in spatial data. Second, VertNet ranks the records retrieved through searches according to specific criteria, with the most complete (i.e., highest quality = highest ranking) data rising to the top of search results (Table 12.5). Third, VertNet applies filters to certain data types (e.g., basis of record, preparations, type status, geo-references, media) to make the data more easily searchable by users. Finally, users can submit feedback about data errors directly to publishers using GitHub via the "Submit Data Issue" button associated with each record on the VertNet portal. Prepublication and postpublication processing of data improves both the quality and discoverability of data for research, and allows users to better assess fitness of use for specific questions.

BIODIVERSITY DATA AND EMERGING RESEARCH

Rapid access to large amounts of biodiversity data has enabled exciting new and more interdisciplinary directions in ornithological research. Furthermore, access to *high-quality* data is at least as important as simply large quantities of data. Examples from the following five topics illustrate emerging research applications in ornithology that have been made possible by VertNet and other data aggregators, thus extending the extrinsic value of specimens.

Niche Evolution and Speciation

The avian family Thraupidae is highly speciose, containing 371 species of Neotropical birds that have radiated across Central and South America. In addition to being one of the most prominent groups of birds in the region, they occupy a broad diversity of terrestrial habitats and elevational zones (Parker et al. 1996). Using a robust phylogeny of the family (Burns et al. 2014), along with over 340,000 occurrence records downloaded from ORNIS and eBird, Title and Burns (2015) tested whether species richness in the major clades is associated with differences in rate of climatic niche evolution. To deal with the data, they developed a methodological pipeline for

TABLE 12.5

Ranks assigned to VertNet records on the basis of presence or absence of certain data types.

Rank	Binomial	Valid coordinates	Coordinate uncertainty (m)	Year	Month	Day
12	X	X	X	X	X	X
11	X	X	X	X	X	
10	X	X	X	X		
9	X	X	X			
8	X	X		X	X	X
7	X	X		X	X	
6	X	X		X		
5	X	X				
4	X			X	X	X
3	X			X	X	
2	X			X		
1	X					
0						

NOTE: An "X" indicates presence of data for that field. More complete (i.e., better quality) records are ranked highest, and thus rise to the top of the results in VertNet searches.

processing and filtering large numbers of occurrence records. Their major findings were threefold: (1) rates of climatic niche evolution are positively correlated with clade species richness, (2) larger clades occupy environmental space that is disproportionately larger than expected given the geographic area occupied by that clade, and (3) species in more diverse clades have narrower niche breadths. These results suggest that clades that are able to diversify successfully across climatic gradients have a greater potential for speciation and/or are more buffered from the risk of extinction. Acquisition of the data for this project took less than a day; without aggregated access through VertNet or other such portals, it would have taken weeks to accumulate the records and standardize them to the same fields, and some data would not have been accessible at all (P. Title, pers. comm.). Data access has important implications for avian conservation, yet a study of this depth and breadth would be unrealistic without efficient methods of getting data from distributed sources.

Risk Assessment of Invasive Species

Invasive species pose a threat to ecosystems and biodiversity worldwide (e.g., Sala et al. 2000, Simberloff 2013), and thus have garnered the attention of scientists, conservation biologists, wildlife managers, policy makers, economists, and others. Biodiversity databases can play a crucial role in providing both historical and modern occurrence data for studying invasive species. In one recent example, Strubbe et al. (2015) examined the invasion risk of Ring-necked Parakeets (*Psittacula krameri*) using museum data from ORNIS and GBIF, other sources of occurrence records, and mitochondrial DNA assessment of phylogeographic structure. This species is native to Africa and Asia, and is most abundant near human settlements and agricultural areas (Khan et al. 2004). As a globally widespread invasive species, it competes with native birds and bats and causes damage to crops (Hernández-Brito et al. 2014; Peck et al. 2014). Findings from this study showed that (1) phylogenetically distinct groups occupy different portions of the climate space in their native range, (2) native (i.e., Africa, Asia) versus invasive (i.e., Europe) populations have low niche overlap, and (3) Ring-necked Parakeets have expanded into climates colder than their native range. Furthermore,

incorporating human modification of habitats into climate-based models significantly improved the prediction of invasion risk. Association with humans in the native range of Ring-necked Parakeets apparently allows this species to persist where it has invaded outside of its native climatic niche. Studies such as this one depend on access to aggregated georeferenced data of high quality for niche modeling (see Chapter 7, this volume).

Phenotypic Evolution and the Colonization of Novel Habitats

Phenotypes are fundamental for describing biological diversity, and represent one of several dimensions driving divergence in populations (e.g., Winker 2009). Furthermore, studies of phenotypic evolution are important for understanding how individuals are able to adapt to changing environments or to colonize new habitats. Two recent examples nicely illustrate the value of aggregated museum data for exploring the context for phenotypic evolution in native and invasive birds, respectively. The Lesser Woodcreeper (*Xiphorhynchus fuscus*) occupies well-preserved habitat in the Atlantic Forest of Brazil, where it has diverged into four genetically distinct lineages that do not agree with phenotypic subspecies designations (Cabanne et al. 2008). This forest has experienced dynamic range fluctuations associated with glacial cycles (Behling 1998). Although the effects of such fluctuations on genetic diversity and gene flow have been studied in a number of taxa (e.g., Fitzpatrick et al. 2009, Maldonado-Coelho 2012), the impacts on phenotype are less well-understood. To address this issue, Cabanne et al. (2014) captured trait data from museum specimens and used records from ORNIS to map areas of historical instability. They found that forest dynamics during the Pleistocene played an important role in shaping phenotypic variation, and that vicariant events were important but only when isolation was complete. In addition, not all historic events influencing genetic divergence seemed to affect the evolution of phenotype traits in the same direction and intensity. Using these data, they concluded that natural selection must have been a major evolutionary driver in the Atlantic Forest.

On a more recent timescale, Bitton and Graham (2015) examined phenotypic evolution in European Starlings (*Sturnus vulgaris*) in the context of their rapid westward expansion in North America following their introduction to New York in 1890–1891. Specifically, they hypothesized that individuals with longer and more pointed wings would lead the colonization front because of their ability to disperse longer distances, and that successive generations of colonizing birds would evolve increasing wing length and pointedness. They also predicted that wings would become shorter and more rounded following colonization because of potential benefits for foraging and predator avoidance. They used records downloaded from ORNIS to identify starling specimens in museum collections, and used both georeferenced localities and collection dates to categorize specimens as colonizers or noncolonizers. Colonizers were collected within the first 3 to 5 years of a newly colonized area, whereas noncolonizers were those collected more than 5 years after the colonization wave. They also included sex as a variable in their analyses. Results showed that wing length and shape did not change during colonization, although wing pointedness decreased 3.8%, that is, wings became more rounded over the past century. Bitton and Graham (2015) concluded that afforestation of open habitat, and the need for navigating through dense vegetation along edges, could be one factor driving this observed phenotypic change. Furthermore, because European Starlings have expanded in association with urbanization, they suggested that higher concentrations of predators and the need for increased take-off ability to escape predators in urban areas would favor the evolution of rounder wings.

Latitudinal Gradients and Evolutionary Divergence

Divergent ecological selection is an important driver of trait evolution and speciation (Schluter 2000), but few studies have taken a broad continental approach to address whether divergent selection is strongest in the tropics, where speciation rates and species richness are higher, or at high latitudes where rates of paleoclimatic change are greater. To address this question, Lawson and Weir (2014) used ~275,000 georeferenced occurrence records from ORNIS and GBIF to estimate the climatic niche of 111 pairs of New World passerine sister species across a latitudinal gradient, and tested whether rates of climatic-niche divergence, a proxy for divergent selection, are correlated with evolutionary rates in behavioral (song) and ecological (body mass) traits. They predicted the fastest rates of evolution in sister pairs with the fastest divergence in realized climatic niche. In addition, they expected that rates of change in these traits would be elevated at high latitudes, where climate-induced divergent selection would be strongest. Overall, climatic-niche divergence was faster at high latitudes and was correlated with rates of change in song and body mass traits. Thus, climate-mediated selection is important in driving evolution and reproductive isolation at high latitudes.

Biodiversity and Conservation Planning

Museum collections play an important role in conservation biology (Kitchener 1997), and aggregated data from these biodiversity repositories are crucial for identifying and prioritizing areas for habitat protection. Yet, their extended value in this regard is often understated. Two examples from megadiverse countries with limited resources showcase the need for efficient access to biodiversity data for conservation planning. In Peru, Fajardo et al. (2014) used ORNIS and other biodiversity occurrence data to develop species distribution models for 1,163 species of birds along with other vertebrate, invertebrate, and plant groups. They applied these data, in conjunction with specific conservation goals and gap analyses, to identify unprotected areas that, if protected, would make the network of protected areas more connected and representative of Peruvian biodiversity. Their data highlighted conservation gaps and important areas in need of protection.

In Brazil, Georgi et al. (2014) developed a framework for conservation planning in Brazil's Atlantic Forest using ecological niche models developed for 23 endemic bird species. This region is a global biodiversity hotspot, and is highly threatened due to habitat change and exploitation of natural resources. They found that existing protected areas exclude 10% of the habitat of each endemic species, and they proposed conservation areas that would expand protections of the avian biodiversity. Although their occurrence data were not obtained from aggregated sources, they noted the value of efficient access to such data for expanding their effort to more taxa, updating location data, and undertaking conservation planning assessments at regular intervals.

CONCLUSIONS AND FUTURE NEEDS

Aggregated data obtained from VertNet and other sources have broad value for both data publishers and users. On the publisher side, museum collections benefit from making their data available in two key ways: (1) open access leads to higher impact and appreciation of the collections, which is important for public perception of collections and collecting (Winker et al. 2010), institutional support, and funding for collection preservation and improvement; and (2) increased access results in improved data quality through feedback mechanisms, which in turn adds fitness for use of the data. On the other side, users benefit from rapid and efficient access to data for a broad range of purposes, from evolutionary biology to invasive species and conservation planning (see earlier).

VertNet and the informatics and museum communities have come a long way since the early development of biodiversity data networks. Technological advances, combined with a growing eagerness and reduced reluctance for sharing data, have fueled the growth and efficiencies of data access that are needed for biodiversity research and conservation on a large scale. Additional work is still needed to optimize the value of these data for the future, however, and three possibilities are especially relevant for future enhancements.

First is the need for better standardization of data capture in the field and of workflows to bring those data into museum databases using standardized terms, vocabularies, and identifiers. Although VertNet has made substantial progress in developing tools for assessing and improving data quality, not all records are subject to these improvements (e.g., data that do not pass through VertNet migrators; see Figure 12.3). By standardizing the ways in which data are captured initially, such as using globally unique identifiers (Guralnick et al. 2015) and limiting values for sex, preparations, higher geography, and other controlled fields (versus free text input), records will be set to a higher standard *prior* to publication, thus facilitating processing, improving discovery, and enhancing fitness for use. All collections should be encouraged to process their data through quality checks (even if they are self-published) before migration into large data aggregators like VertNet and GBIF.

Second, the large-scale aggregation of global biodiversity data provides a unique opportunity to take advantage of geographic coordinates for retrospective georeferencing. Although ORNIS and other vertebrate database projects focused on georeferencing data for collections that participated in those projects, the current state is that many collections have not been georeferenced. Furthermore, for those that have been georeferenced, curators or collectors have not validated many of the coordinate and uncertainty data. Another issue is that collections that were georeferenced have not always repatriated the values back to their databases. Thus, users must either georeference localities without coordinates anew or exclude them from analyses, and georeferenced records need to be updated in both the original and aggregated data sources so that future users can benefit from those efforts. Development of a locality service that integrates georeferencing services and facilitates repatriation of coordinate data would greatly benefit the museum and research communities.

Third, the value of specimens can be greatly enhanced by adding *context that extends beyond the specimen itself*. For example, audio or video recordings provide useful data on the behavioral phenotype of an individual at a particular time (see Chapters 4 and 5, this volume), whereas field notes add critical ecological information and are essential for validating and improving data. Although aggregated occurrence records include information on associated media, and the Darwin Core can be augmented with an Audubon Core media extension (Morris et al. 2013), current infrastructures do not provide a means for searching by specific media type across collections, for example, all audio recordings of *Pipilo maculatus* held in different institutions. Likewise, no single portal includes digitized field notes associated with collectors and specimens distributed among museums. Although some collection databases have begun to associate their digitized field notes with individual specimens (e.g., Museum of Comparative Zoology, Harvard University, Cambridge, Massachusetts; California Academy of Sciences, San Francisco, California), and partnerships have emerged to increase access to field notebook content (e.g., the U.S. National Museum of Natural History Field Book Project, http://www.mnh.si.edu/rc/fieldbooks), no efforts exist to link specimens in VertNet to collectors' field notes archived in different collections. Rapid and efficient access to field notes and other types of media through the VertNet network would reap enormous benefits for research and the history of science.

The development of aggregated data sources is only half of the solution for data access. Equally important is the need to sustain these sources through ongoing funding, training, and community engagement. Professional rewards for sharing data are also important for recognizing the value of these resources to the research community (Erike et al. 2012). VertNet has reduced the infrastructural cost of participation and maintenance by an order of magnitude, from approximately $196,000 to less than $20,000 annually. Collections realize low costs because they no longer need servers, specialized software, and technical administrators for distributing their data. However, the services provided to the collections community by VertNet require some level of support for installing new resources, updating resources, maintaining infrastructure, and keeping pace with technological changes, not to mention new innovations. Long-term sustainability cannot depend on government funding cycles or other tenuous support. More community discussion and action is needed on how to support these efforts financially, technologically, and intellectually in the future.

In a rapidly changing world, the role of natural history collections in addressing challenges faced by society is more important than ever before (Thomson 2005; Chapter 13, this volume). Exciting developments in biodiversity informatics led by such collections over the past 15 years, and particularly the growth of aggregated data networks such as VertNet, have stimulated new areas of research in ornithology and other disciplines that would not have been possible otherwise. These studies are broadly applicable to our understanding of evolutionary biology, invasive species, emerging diseases, conservation biology, changing species distributions, and other important topics in the 21st century. Furthermore, they showcase the extended value of natural history collections for science and society (Krishtalka and Humphrey 2000, Suarez and Tsutsui 2004), and highlight the need for ongoing support to protect and grow those collections and to enable efficient access to their data. In the modern era of sharing data, collections hold exciting promise for future avenues of research that are just beginning to emerge.

ACKNOWLEDGMENTS

We thank the National Science Foundation for its long-term support of VertNet (DBI-1062193, DBI-1062148, DBI-1062200, DBI-1062271) and its five predecessor vertebrate networks (Species Analyst, DBI-9808739; FishNet, DBI-9985737, and FishNet2, DBI-0415600; MaNIS, DBI-0108161; HerpNET, DBI-0132303; and ORNIS, DBI-0345448). The National Science Foundation also funded the Map of Life project with which we collaborated to develop spatial validation tools for VertNet (DBI-0960549, DBI-1262610, DBI-1535793, DBI-1262396, DBI-1262600, DBI-0960550). We also thank the National Biological Information Infrastructure (NBII), Global Biodiversity Information Facility (GBIF), and Integrated Digitized Biocollections (iDigBio) for their collaboration, support, and funding. C. Moritz, J. Hanken, L. Trueb, A. T. Peterson, and the rest of the VertNet Steering Committee have been instrumental in their guidance and encouragement through VertNet activities and personnel demands, especially when funding was uncertain. We thank A. Steele and R. Kraft for their important technical contributions to VertNet, and A. Steele and P. Zermoglio for their assistance with earlier versions of Figures 12.2 and 12.3, respectively. Scores of collection staff, students, and volunteers worked to georeference the data in VertNet. Finally, we are most grateful to all of the data publishers and users with whom we have worked over the years to develop, expand, and improve VertNet and share their data.

LITERATURE CITED

Abrahamson, B. L. 2015. Tracking changes in natural history collections utilization: a case study at the Museum of Southwestern Biology at the University of New Mexico. Collection Forum 29:1–21.

Bartlett, C., J. Beach, N. Cobb, J. Cook, C. Dietrich, L. S. Ford, J. Fortes, S. Graves, C. Gries, R. Gropp, G. Guala, J. Hanken, K. Joyce, M. A. Mares, R. McCourt, L. McDade, A. Neill, C. Norris, L. Page, C. Parr, G. Riccardi, N. Rios, K. Seltmann, D. Smith, B. Thiers, and Q. Wheeler. 2013. Implementation plan for the Network Integrated Biocollections Alliance. American Institute of Biological Sciences, Reston, VA.

Behling, H. 1998. Late Quaternary vegetational and climatic changes in Brazil. Review of Paleobotany and Palynology 99:143–156.

Beissinger, S. R., and M. Z. Peery. 2007. Reconstructing the historical demography of an endangered seabird. Ecology 88:296–305.

Berendsohn, W. G., V. Chavan, and J. A. Macklin. 2010. Recommendations of the GBIF Task Group on the global strategy and action plan for the mobilisation of natural history collections data. Biodiversity Informatics 7:1–5.

Bitton, P.-P., and B. A. Graham. 2015. Change in wing morphology of the European Starling during and after colonization of North America. Journal of Zoology 295:254–260.

Burns, K. J., A. J. Shultz, P. O. Title, N. A. Mason, F. K. Barker, J. Klicka, S. M. Lanyon, and I. J. Lovette. 2014. Phylogenetics and diversification of tanagers (Passeriformes: Thraupidae), the largest radiation of Neotropical songbirds. Molecular Phylogenetics and Evolution 75:41–77.

Cabanne, G. S., F. M. d'Horta, E. H. Sari, F. R. Santo, and C. Y. Miyaki. 2008. Nuclear and mitochondrial phylogeography of the Atlantic Forest endemic Xiphorhynchus fuscus (Aves: Dendrocolaptidae): biogeography and systematics implications. Molecular Phylogenetics and Evolution 49:760–773.

Cabanne, G. S., N. Trujillo-Arias, L. Calderón, F. M. d'Horta, and C. Y. Miyaki. 2014. Phenotype evolution of an Atlantic Forest passerine (Xiphorhynchus fuscus): biogeographic and systematic implications. Biological Journal of the Linnean Society 113:1047–1066.

Canhos, V. P., S. Souza, R. Giovanni, and D. A. L. Canhos. 2004. Global biodiversity informatics: setting the scene for a "New World" of ecological modeling. Biodiversity Informatics 1:1–13.

Chamberlain, S., C. Ray, and V. Barve. [online]. 2015. rvertnet: An R interface for the VertNet database. <https://github.com/ropensci/rvertnet> (December 9, 2016).

Constable, H., R. Guralnick, J. Wieczorek, C. Spencer, A. T. Peterson, and the VertNet Steering Committee. 2010. VertNet: a new model for biodiversity data sharing. PLoS Biology 8:e1000309.

de Bruin, S., A. Bregt, and M. Van de Ven. 2001. Assessing fitness for use: the expected value of spatial data sets. International Journal of Geographical Information Science 15:457–471.

Devillers, R., Y. Bédard, R. Jeansoulin, and B. Moulin. 2007. Towards spatial data quality information analysis tools for experts assessing the fitness for use of spatial data. International Journal of Geographical Information Science 21:261–282.

Edwards, J. L., M. A. Lane, and E. S. Nielsen. 2000. Interoperability of biodiversity databases: biodiversity information on every desktop. Science 289:2312–2314.

Erb, P., W. McShea, and R. P. Guralnick. 2012. Anthropogenic influences on macro-level mammal occupancy in the Appalachian Trail Corridor. PLoS One 7:e42574.

Erike, N., A. Thessen, K. Bach, J. Bendix, B. Seeger, and B. Gemeinholzer. 2012. The user's view of biodiversity data sharing—investigating facts of acceptance and requirements to realize a sustainable use of research data. Ecological Informatics 11:25–33.

Fajardo, J., J. Lessmann, E. Bonaccorso, C. Devenish, and J. Muñoz. 2014. Combined use of systematic conservation planning, species distribution modelling, and connectivity analysis reveals severe conservation gaps in a megadiverse country (Peru). PLoS One 9:e114367.

Fitzpatrick, S. W., C. A. Brasileiro, C. F. Haddad, and K. R. Zamudio. 2009. Geographical variation in genetic structure of an Atlantic Coastal Forest frog reveals regional differences in habitat stability. Molecular Ecology 18:2877–2896.

Giorgi, A. P., C. Rovzar, K. S. Davis, T. Fuller, W. Buermann, S. Saatchi, T. B. Smith, L. F. Silveira, and T. W. Gillespie. 2014. Spatial conservation planning framework for assessing conservation opportunities in the Atlantic Forest of Brazil. Applied Geography 53:369–376.

Graham, C. H., S. Ferrier, F. Huettman, C. Moritz, and A. T. Peterson. 2004. New developments in museum-based informatics and applications in biodiversity analysis. Trends in Ecology and Evolution 19:497–503.

Guralnick, R. P., N. Cellinese, J. Deck, R. Pyle, J. Kunze, L. Penev, R. Walls, G. Hagedorn, D. Agosti, J. Wieczorek, T. Capatano, and R. D. M. Page. 2015. Community next steps for making globally unique identifiers work for biocollections data. Zookeys 494:133–154.

Guralnick, R. P., and H. Constable. 2010. VertNet: creating a data-sharing community. Bioscience 60:258–259.

Guralnick, R. P., H. Constable, J. Wieczorek, C. Moritz, and A. T. Peterson. 2009. Sharing: lessons from natural history's success story. Nature 462:34.

Hanken, J. 2013. Biodiversity online: toward a Network Integrated Biocollections Alliance. Bioscience 63:789–790.

Hawkins, G. L., G. E. Hill, and A. Mercadante. 2012. Delayed plumage maturation and delayed reproductive investment in birds. Biological Reviews 87:257–274.

Hernández-Brito, D., M. Carrete, A. G. Popa-Lisseanu, C. Ibáñez, and J. L. Tella. 2014. Crowding in the city: losing and winning competitors of an invasive bird. PLoS One 9:e100593.

Hill, A. W., J. Otegui, A. H. Ariño, and R. P. Guralnick. [online]. 2010. GBIF position paper on future directions and recommendations for enhancing fitness-for-use across the GBIF network, Global Biodiversity Information Facility, Copenhagen. <http://www.gbif.org/resource/80623> (December 9, 2016).

Huang, X., B. A. Hawkins, F. Lei, G. L. Miller, C. Favret, R. Zhang, and G. Qiao. 2012. Willing or unwilling to share primary biodiversity data: results and implications of an international survey. Conservation Letters 5:399–406.

International Organization for Standardization. 1994. ISO 8402. Quality management and quality assurance—vocabulary. International Organization for Standardization, Geneva, Switzerland.

IUCN Standards and Petitions Working Group. 2008. Guidelines for using the IUCN red list categories and criteria. Version 7.0. Prepared by the Standards and Petitions Working Group of the IUCN SSC Biodiversity Assessments Sub-Committee in August 2008.

Khan, H. A., M. A. Beg, and A. A. Khan. 2004. Breeding habits of the rose-ringed parakeet *Psittacula rameri* in the cultivations of central Punjab. Pakistan Journal of Zoology 36:133–138.

Kilpatrick, A. M. 2011. Globalization, land use, and the invasion of West Nile virus. Science 334:323–327.

Kitchener, A. C. 1997. The role of museums and zoos in conservation biology. International Zoo Yearbook 35:325–336.

Krishtalka, L., and P. S. Humphrey. 2000. Can natural history museums capture the future? Bioscience 50:611–617.

Lawson, A. M., and J. T. Weir. 2014. Latitudinal gradients in climatic-niche evolution accelerate trait evolution at high latitudes. Ecology Letters 17:1427–1436.

Maldonado-Coelho, M. 2012. Climatic oscillations shape the phylogeographical structure of Atlantic Forest fire-eye antbirds (Aves: Thamnophilidae). Biological Journal of the Linnean Society 105:900–924.

Moritz, C., J. L. Patton, C. J. Conroy, J. L. Parra, G. C. White, and S. R. Beissinger. 2008. Impact of a century of climate change on small-mammal communities in Yosemite National Park, USA. Science 322:261–264.

Morris, R. A., V. Barve, M. Carausu, V. Chavan, J. Cuadra, C. Freeland, G. Hagedorn, P. Leary, D. Mozzherin, A. Olson, G. Riccardi, I. Teage, and G. Whitbread. 2013. Discovery and publishing of primary biodiversity data associated with multimedia resoures: the Audubon Core strategies and approaches. Biodiversity Informatics 8:185–197.

Norris, D. R., P. Arcese, D. Preikshot, D. F. Bertram, and T. K. Kyser. 2007. Diet reconstruction and historic population dynamics in a threatened seabird. Journal of Applied Ecology 44:875–884.

Nufio, C. R., C. R. McGuire, M. D. Bowers, and R. P. Guralnick. 2010. Grasshopper community response to climatic change along an elevational gradient. PLoS One 5:e12977.

Parker, T. A., D. F. Stotz, and J. W. Fitzpatrick. 1996. Ecological and distributional databases for Neotropical birds. Pp. 113–436 in D. F. Stotz, J. W. Fitzpatrick, T. A. Parker, and D. K. Moskovits (editors), Neotropical birds: ecology and conservation. University of Chicago Press, Chicago.

Peck, H. L., H. E. Pringle, H. H. Marshall, I. P. F. Owens, and A. M. Lord. 2014. Experimental evidence of impacts of an invasive parakeet on foraging behavior of native birds. Behavioral Ecology 25:582–590.

Peery, M. Z., L. A. Hall, A. Sellas, S. R. Beissinger, C. Moritz, M. Bérubé, M. G. Raphael, S. K. Nelson, R. T. Golightly, L. McFarlane-Tranquilla, S. Newman, and P. J. Palsbøll. 2010. Genetic analyses of historic and modern marbled murrelets suggest decoupling of migration and gene flow after habitat fragmentation. Proceedings of the Royal Society of London B 277:679–706.

Pennisi, E. 2005. Boom in digital collections makes a muddle of management. Science 308:l87–189.

Peterson, A. T., C. Cicero, and J. R. Wieczorek, 2005. Free and open access to bird specimen data: why? Auk 122:987–990.

Pôças, I., J. Gonçalves, B. Marcos, J. Alonso, P. Castro, and J. P. Honrado. 2014. Evaluating the fitness for use of spatial data sets to promote quality in ecological assessment and monitoring. International Journal of Geographical Information Science 28:2356–2371.

Robertson, T., M. Döring, R. Guralnick, D. Bloom, J. Wieczorek, K. Braak, J. Otegui, L. Russell, and P. Desmet. 2014. The GBIF Integrated Publishing Toolkit: facilitating the efficient publishing of biodiversity data on the Internet. PLoS One 9:e102623.

Sala, O. E., F. S. Chapin, J. J. Armesto, E. Berlow, J. Bloomfield, R. Dirzo, E. Huber-Sanwald, L. F. Huenneke, R. B. Jackson, A. Kinzig, R. Leemans, D. M. Lodge, H. A. Mooney, M. Oesterheld, N. L. Poff, M. T. Sykes, B. H. Walker, M. Walker, and D. H. Wall. 2000. Biodiversity—global biodiversity scenarios for the year 2100. Science 287:1770–1774.

Schluter, D. 2000. The ecology of adaptive radiation. Oxford University Press, New York.

Simberloff, D. 2013. Invasive species: what everyone needs to know. Oxford University Press, New York.

Soberón, J., and A. T. Peterson. 2004. Biodiversity informatics: managing and applying primary biodiversity data. Philosophical Transactions of the Royal Society of London B 359:689–698.

Strubbe, D., H. Jackson, J. Groombridge, and E. Matthysen. 2015. Invasion success of a global avian invader is explained by within-taxon niche structure and association with humans in the native range. Diversity and Distributions 21:675–685.

Suarez, A. V., and N. D. Tsutsui. 2004. The value of museum collections for research and society. Bioscience 54:66–74.

Thomson, K. S. [online]. 2005. Natural history museum collections in the 21st century. American Institute of Biological Sciences. <http://www.actionbioscience.org/evolution/thomson.html> (December 9, 2016).

Tingley, M. W., W. B. Monahan, S. R. Beissinger, and C. Moritz. 2009. Birds track their Grinnellian niche through a century of climate change. Proceedings of the National Academy of Sciences USA 106:19637–19643.

Title, P. O., and K. J. Burns. 2015. Rates of climatic niche evolution are correlated with species richness in a large and ecologically diverse radiation of songbirds. Ecology Letters 18:433–440.

Veregin, H., 1999. Data quality parameters. Pp. 177–189 in P. A. Longley, M. F. Goodchild, D. J. McGuire, and D. W. Rhind (editors), Geographical information systems (2nd ed.). John Wiley & Sons, New York.

Vollmar, A., J. A. Macklin, and L. S. Ford. 2010. Natural history specimen digitization: challenges and concerns. Biodiversity Informatics 7:93–112.

Wandeler, P., P. E. A. Hoeck, and L. F. Keller. 2007. Back to the future: museum specimens in population genetics. Trends in Ecology and Evolution 22:634–642.

Wiezcorek, J., D. Bloom, R. Guralnick, S. Blum, M. Döring, R. Giovanni, T. Robertson, and D. Vieglais. 2012. Darwin Core: an evolving community-developed biodiversity data standard. PLoS One 7:e29715.

Winker, K. 2009. Reuniting phenotype and genotype in biodiversity research. Bioscience 59:657–665.

Winker, K., J. M. Reed, P. Escalante, R. A. Askins, C. Cicero, G. E. Hough, and J. Bates. 2010. The importance, effects, and ethics of bird collecting. Auk 127:690–695.

Zermoglio, P. F., R. P. Guralnick, and J. R. Wieczorek. 2016. A standardized reference data set for taxonomic name resolution. PLoS One 11:e0146894.

CHAPTER THIRTEEN

Ornithological Collections in the 21st Century*

Michael S. Webster, Carla Cicero, John Bates, Shannon Hackett, and Leo Joseph

Abstract. Recent advances in the technologies used to access and extract data from specimens have created a renaissance for ornithological collections. In this chapter we provide a broad overview of the new methods being used to extend the use of traditional specimens, as well as new specimen types, ancillary materials, and data that are now being collected. These various specimen and data types are highly complementary, in that each captures a unique aspect of the individual phenotype/genotype at a particular place and time. We also briefly review recent developments in informatics that are enhancing the ability of researchers to access specimens and data that they need. These various advances generate exciting new research opportunities, creating an "extended specimen" that further solidifies and strengthens the central role of collections-based research in ornithology. At the same time, these exciting developments present substantial challenges for both the field collector and collection curator. The future of collections-based ornithological research is bright, but considerable public, institutional, and governmental support will be needed to ensure that ornithological collections are preserved and nurtured, so that they can rise to meet their growing research potential.

Key Words: biodiversity, data curation, informatics, museums, natural history collections, ornithology, specimens.

Ornithological collections have long been the backbone of basic and applied research in ornithology. They provide the raw, foundational material for studies uncovering and synthesizing data relevant to the patterns and processes underlying the diversity, distribution, abundance, ecology, evolution, and life histories of birds (Remsen 1995, Collar et al. 2003, Joseph 2011, Holmes et al. 2016), as well as strategies for monitoring and conserving them (e.g., Hickey and Anderson 1968, Becker and Beissinger 2006, Green 2008, Murphy et al. 2011, Kearns et al. 2016). This is because the core mission of ornithological, and other natural history, collections is to preserve, organize, and make accessible specimens and biodiversity data that document patterns—spatial and temporal, large and small—in the diversity of life on earth. Collections-based ornithological research continues to build on this long and rich history, and is entering a renaissance due to recent developments that make collections even more versatile and broadly useful to a diverse set

* Webster, M. S., C. Cicero, J. Bates, S. Hackett, and L. Joseph. 2017. Ornithological collections in the 21st century. Pp. 219–232 in M. S. Webster (editor), The Extended Specimen: Emerging Frontiers in Collections-based Ornithological Research. Studies in Avian Biology (no. 50), CRC Press, Boca Raton, FL.

of research questions than they have ever been in the past.

The chapters in this volume highlight the growing potential for collections-based research in ornithology. First, emerging new technologies in genetics, chemistry, and computer imaging now make it possible for researchers to use "traditional" museum specimens to address new questions in considerable detail and with a temporal dimension. Second, it is now possible to collect new types of specimens and ancillary materials that open doors to whole new areas of inquiry—such as animal behavior, coevolution, disease, ecotoxicology, gene expression, and epigenetics—that traditionally have been beyond the scope of collections-based research (e.g., Graham et al. 2004, Winker 2004, Wandeler et al. 2007, DiEuliisa et al. 2016). These specimen types and materials extend the specimen beyond the physical specimen brought back from the field. Finally, advances in digitization and information technologies now allow researchers to access and analyze huge datasets from biological specimens in museum collections, making it possible to address questions at global scales (e.g., Jetz et al. 2012). These same trends also are responsible for a shift in how we grow and view our research collections. Consequently, museums and their research collections are being used in ways unimaginable just a decade ago.

Ironically, the current explosion in the utility and value of ornithological collections comes at a time when opposition to specimen collecting, a somewhat cyclical phenomenon, is again on the rise. Some within the scientific community have called into question the value or need for specimens (e.g., Minteer et al. 2014; but see Winker et al. 2010, Clemann et al. 2014, Rocha et al. 2014), and recent high-profile media coverage (e.g., Kaplan and Moyer 2015) has fueled public and political debate about the need for continued collecting. For some, there seems to be a sense that the need for collecting has been met, and that now we can turn our attention to other matters, such as digitizing specimens already in collections (Olsen 2015). Moreover, public opposition to collecting specimens is only one challenge facing museums of the 21st century. The exciting new technological advances summarized in this volume also present their own technical, logistical, and social challenges. In particular, different specimen and data types, along with the

push to digitize data for increased online access, often require specialized, advanced, and ever-changing technology and expertise, thus creating challenges for the collector, the curator, and the institution alike.

The various chapters in this volume discuss exciting new ways that ornithological collections can and are being used for modern areas of inquiry, and also highlight some of the challenges. In this chapter we extract and synthesize the key take-home messages from those other chapters. First, we briefly summarize the many exciting new developments in the analysis of traditional museum specimens, new specimen types, and ancillary data associated with those specimens (Chapter 1, this volume). These new developments are covered in more detail in the other chapters. Accordingly, we focus on the big picture that stretches across a spectrum of specimen types and how they can be used to tackle ever bigger and more pressing questions, including climate change (see also Bates et al. 2004, Suarez and Tsutsui 2004, Gardner et al. 2014, Rocha et al. 2014, Lujan and Page 2015). The key message is that the value of ornithological specimens is higher now than it has ever been, as is the continued need to collect more specimens to fill geographic, temporal, and taxonomic gaps. We reiterate that collection of modern specimens also provides data from today for tomorrow's researchers. A common theme is that the value of ornithological specimens will continue to grow as newer technologies and research questions are developed, exactly as has happened in the past. Thus, our ornithological collections must constantly evolve to meet these ever-expanding research needs. Second, we discuss the relationships between traditional, and not-so-traditional, ornithological specimens and other types of data not usually associated with museum collections, at least in popular perception. We argue that, in the era of big data and data-driven "informatics" science (Hochachka et al. 2012, Michener and Jones 2012, Parr et al. 2012; see also Chapter 12, this volume), the distinctions between specimens and data are being blurred, and ornithological collections should embrace that fact to enhance even further the value of their specimens and other holdings. Finally, we provide a broad overview of the many exciting opportunities, and also the challenges, that face ornithological collections as they prepare themselves to address new research questions in the 21st century.

THE ONCE AND FUTURE SPECIMEN

Physical specimens such as study skins, skeletal preparations, fluid specimens, and eggs have always been the core of ornithological research collections. These specimens will continue to be critical for studies in taxonomy, systematics, anatomy, and biogeography (Collar et al. 2003, Bates et al. 2004, Suarez and Tsutsui 2004, Kiff 2005), and also for delineating key features of avian life histories (e.g., Pyle 1997, 2008). It is precisely because birds are relatively so well-known taxonomically that future work is set to resolve fine details of taxonomic and genetic structure and their relationship to landscapes. Indeed, studies in these and other areas have grown in size and scope thanks to recent developments in digitization and data publication technologies. In particular, data aggregators like VertNet (see Chapter 12, this volume) and the Atlas of Living Australia (http://www.ala.org.au) allow researchers to locate relevant specimens and core data quickly from across a global network of natural history collections. This growing ability to locate and use specimens and data from across institutional boundaries is arguably the most obvious recent achievement facilitating collections-based research. The museum ornithological community has been at the forefront of building the resources that make this possible.

A second major recent achievement in collections-based research is the development of technologies for collecting new kinds of data from "traditional" ornithological specimens, such as study skins, skeletons, and fluid preparations. The diversity of ways that these specimens can be used for meaningful biological research has expanded dramatically in recent years (Chapter 1, this volume). Today, researchers can use physical specimens to examine feather coloration using reflectance spectrophotometry and pigmentation analyses of skins as well as eggs (Starling et al. 2006, Brulez et al. 2016; Chapter 3, this volume); feather and pigment microstructure using electron microscopy and other imaging technologies (e.g., Prum 2006, Vinther et al. 2008, Stavenga et al. 2011, Lee et al. 2016); detailed morphology of both skeletal and internal soft parts, as well as eggshell ultrastructure, using an array of computer-based imaging technologies (Grellet-Tinner et al. 2016; Chapter 2, this volume); diet and large-scale movement patterns using chemical isotopic analyses of the feathers and other body parts (Chapter 6, this volume); and population genetics/genomics using DNA extracted from specimens of extant or extinct and highly endangered populations (Shapiro et al. 2002, Johnson et al. 2010, Murphy et al. 2011, Austin et al. 2013, Besnard et al. 2016). Moreover, the relevance of these collections-based approaches to conservation is now being explored (e.g., Li et al. 2014). Given the rapid growth of these new technologies, most of which were developed in the last 10 to 20 years, it is clear that the diversity of additional questions that can be addressed with traditional museum specimens will only grow. Just as collectors in the early 20th century could not possibly have foreseen isotopic, genomic, and microscanning uses for the specimens that they collected, we cannot foresee all of the ways in which specimens collected now will be used for analyses in the future.

The aforementioned advances have been made primarily through new methods that make use of traditional museum specimens, in particular study skins, skeletons, and fluid-preserved whole specimens. Equally exciting and promising are less traditional specimen and preparation types, such as stomach content and egg collections, spread wings, endo- or ectoparasites, and even plucked feathers, all of which hold great potential for research into a diversity of questions (Remsen et al. 1986, Barker and Vestjens 1991, Chesser 1995, Smith et al. 2003, Block et al. 2015; Chapter 8, this volume). Similarly, as the "genetic revolution" unfolded in the latter part of the 20th century, more than 30 ornithological collections began to develop curated frozen tissue collections to serve as sources of high-quality, relatively undegraded DNA (Stoeckle and Winker 2009). These tissue collections have proven indispensable to providing material for thousands of comparative phylogenetic and phylogeographic studies, and are even more valuable now following the advent of powerful modern genomic approaches (Chapter 9, this volume). Simultaneously, new methods for extracting genetic material have also increased the value of more traditional specimen types as sources of material for genetic/genomic studies (e.g., Cheviron et al. 2011, Hykin et al. 2015). Yet the development of advanced genomic methodologies has also increased the need for new types of tissue preservation, such as in buffers that allow for molecular studies of gene expression and transcriptomics as well as "standard" genomics (e.g., Edwards et al. 2005, Cheviron et al. 2011, Gayral et al. 2011, Buerki and Baker 2016).

Finally, as ornithologists delve deeper into interdisciplinary studies, they now collect data and specimens that traditionally were the purview of other disciplines. One excellent and timely example, which stretches across taxonomic disciplines, is given by Lutz et al. (Chapter 10, this volume). As those authors discuss, the recognition that an individual bird is actually an ecosystem populated by a diversity of microorganisms and, in some cases, not so "micro" internal and external parasites, opens whole new areas of inquiry that can be facilitated by ornithological research collections (e.g., Bain and Mawson 1981, Price and Johnson 2007, Price et al. 2008, Parker et al. 2011, Johnson et al. 2012, Block et al. 2015). Research in this area requires that the invertebrate specimens be properly preserved and curated, and that they be linked to the bird specimen from which they were collected (see later). Databases summarizing these interconnections are proving to be important research tools, not just for ornithologists and biodiversity scientists, but also for medical professionals and conservation biologists. Similarly, modern research on gene expression, brain anatomy, and animal behavior can be facilitated by properly preserved specimens, but all require specimen or data types that differ from those traditionally handled by ornithological collections. Accordingly, interdisciplinary research that will prevail in 21st century biology will dramatically expand the types of specimens and data that ornithological collections preserve and make available, either directly or through key partnerships.

BLURRING THE LINES BETWEEN "SPECIMEN" AND "DATA"

A biological specimen is an exemplar, or a representative individual (i.e., a sample), taken from a particular population at a particular place and time. The metadata associated with the specimen describe aspects of it, in particular when and where it was collected, but also other biological information, such as the state of the reproductive tract, gonad size, condition of oviduct, amount and location of subcutaneous fat, and development

or coloration of eyes, and bill. In many of the analyses described earlier, the specimen is examined, data are collected from it, and the associated metadata help place those specimen data in proper context for analyses, for example, in comparing genetic or phenotypic traits of individuals from different regions. Increasingly, though, researchers are making use of the metadata themselves for their analyses and use the specimen as a voucher for those data. Peterson and Navarro-Sigüenza (Chapter 7, this volume) illustrate how specimen metadata, of the sort typically found on the specimen label, can be used for powerful analyses of species distributions and the processes underlying those distributions. Similarly, metadata from specimen labels can be used to better understand avian demography across space and time (e.g., Beissinger and Peery 2007, Green 2008, Gilby et al. 2009, Ricklefs et al. 2011), including migration (Marantz and Remsen 1991, Joseph and Stockwell 2000). Although these uses of ornithological specimens are not particularly new (e.g., Foster and Cannell 1990), the number of such studies, as well as their power and sophistication, has increased dramatically in recent years as computational methods have been developed and refined (see Chapter 7, this volume). As scientists uncover more detailed information about past landscapes for analyses of historical patterns, this also leads to an understanding of how birds and other organisms have interacted with and responded to landscape changes (e.g., Byrne et al. 2008, Carnaval et al. 2009, Kearns et al. 2014, Mason and Taylor 2015).

The analysis of metadata associated with specimens has led to an expansion in the types of data and materials that ornithological collections preserve, and this expansion has been occurring for decades. Today it is common for ornithologists to collect a wide diversity of metadata, and also other "ancillary" materials, associated with the specimens that they collect and prepare. For example, there may be audio or video recordings of the animal in life, photos of the habitat in which it was collected, or tables of data (e.g., a list of birds seen/heard at the site but not necessarily collected), all associated with a specimen or collecting event. Some of these materials are "specimens" in the true sense of the word (e.g., the physical study skins collected), whereas others are not (e.g., photographs and lists of birds seen). Yet they are all mutually enriching, making possible analyses that could not be done with one data or specimen type alone.

This is well illustrated by *biodiversity media*: audio recordings, video recordings, and photographs collected from animals in the wild, as pioneered by workers such as Lanyon (1978). Such recordings and photos capture the behavioral phenotype of an individual at a particular time, a particular place, and a particular context (see Chapter 1, this volume). Accordingly, and as illustrated nicely by Mason et al. (Chapter 4, this volume) and Bostwick et al. (Chapter 5, this volume), biodiversity media such as these are extremely valuable for ornithological research, and have become increasingly so over time as accessibility and analytical techniques have improved (Figure 13.1). These chapters also illustrate that biodiversity media are most valuable when they are associated with physical specimens, as these associations allow the behavioral phenotype to be linked to the morphological phenotype, whether plumage, skeletal, or syringeal characters, for example, as well as to the genotype (for an early example, see Groth 1993). Accordingly, clear and transparent links between specimen types reinforce modern collection databases as valuable research tools rather than mere summaries of specimen types.

Because biodiversity media capture important aspects of the individual phenotype, they are sometimes referred to as "media specimens" (Budney et al. 2014, Webster and Budney 2016; see also Chapter 5, this volume). Some have gone further to suggest that such media can substitute for the collection of physical specimens (e.g., Minteer et al. 2014). Our position is that this is not the case. Arguing whether media or other ancillary items are "specimens" distracts from the key point: *Each specimen and data type captures a unique set of information about the individual organism at a particular time and a particular place, and, as such, they are best viewed as complementary rather than as alternatives to each other.* In fact, data complementary to the physical specimen extend its utility and enhance its value. Biodiversity media capture aspects of the behavioral phenotype in ways that physical specimens simply cannot (Chapter 1, this volume), and, at the same time, detailed or internal anatomical measurements, chemical isotopes, and DNA cannot be extracted from an audiovisual recording or photo (e.g., Bates and Voelker 2015). Similarly, tissue samples yield DNA, RNA, and other biomolecules of quality and quantity that cannot typically be obtained from standard physical specimens. Different types of preparation and data types reveal—and preserve—varied aspects of the individual phenotype and genotype. All are valuable and mutually reinforcing biodiversity data that can and should be handled by ornithological and other biological

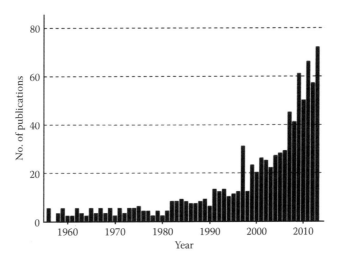

Figure 13.1. The number of publications using biodiversity media has increased dramatically over the years. Data show the number of publications using media from a single natural history media archive, the Macaulay Library at the Cornell Lab of Ornithology. To date, more than 900 publications have used media assets from this collection. The anomalous peak in 1997 is due to publication of a monograph honoring Theodore Parker (Remsen 1997). The strong upward step in the mid- to late 2000s is likely due in part to digitization of the Macaulay collection, which dramatically improved accessibility.

collections, which have the history and expertise in curating specimens and other biodiversity data.

THE ACCESSIBILITY CHALLENGE

The advances in specimen-based research described earlier have led to a dizzying array of possible specimen preparation and data types. Today, several different preparation types may be made from a single individual, such as a tissue sample, study skin, spread wing, and/or skeletal preparation. Likewise, there may be an array of ancillary materials also associated with that same specimen, such as audiovisual recordings of its behavior; stomach and gastrointestinal tract contents for diet or microbiome study; cloacal swabs and blood smears for endoparasite, bacterial, and viral study; or entomological samples of ectoparasites. Complicating matters further, some of these ancillary materials (e.g., audio recordings of birds singing, photographs) may be gathered for individuals that are not collected to yield a physical specimen, but are nonetheless from the same collecting event (place and time) tied to collected specimens. This diverse suite of specimen and data types creates a rich and mutually reinforcing database for ornithological research. Yet at the same time the number of specimen and data types handled by ornithological collections creates major challenges for both the curator in the collection and the collector in the field (Figure 13.2).

For the curator and collection manager, the diversity of specimen and data types creates two major challenges. The first of these is preservation, as each new specimen type creates needs for storage space and associated infrastructure as well as expertise. This is true for spread wings, tissue samples (whether frozen or preserved in a stabilizing reagent such as RNA*later*), whole birds in fluid preparations, and biodiversity media. Addressing this challenge will benefit from community-wide discussion about the most broadly valuable specimen and tissue types to preserve, taking into account both curation constraints and pragmatic trade-offs for collectors in the field (see later). These discussions need to take place consistently and collaboratively through time as we strive to maximize the value of specimens in an ever-changing research landscape. Another solution will likely come through increased specialization and partnerships among ornithological collections. Although some larger institutions may

be able to handle multiple specimen types (e.g., through specialized and focused departments within the institution), most institutions will find it difficult to be a jack-of-all-trades and instead will form key partnerships with other institutions focused on different specimen types (e.g., frozen tissue or media collections).

This brings us to the second major challenge for curation and management of multiple specimen types: accessibility. For the full research potential of the various associated specimens and data types to be realized, it is critical that they be strongly and transparently linked to each other. Within a collection or database, this is relatively easy to do by creating appropriate database fields and populating them with the relevant specimen numbers and collecting events (Figure 13.2). Across collections, however, such linkages are challenging because the institutions involved may manage data using very different systems, and catalog numbers are typically not unique across institutions. For example, a bird specimen housed in an ornithological collection may be associated with a frozen tissue sample in a different museum, with a sample of collected parasites in an entomological collection, and with audio recordings of songs in a natural history media archive. Moreover, it is also possible and important to link individual specimens with data and materials associated with the broader collecting event, such as field notes, photographs, bird lists, and recordings of birds that were not collected with a physical specimen. Finally, specimens can and should be linked to the data that have been generated from them (e.g., DNA sequences in GenBank and associated genetic datasets, or morphological data on MorphBank or DigiMorph), as well as to any scientific publications in which they were included.

Clearly, the issue of data and specimen accessibility is a big one, but it also is one that the collections community is striving to meet: the same technologies and advances used by data aggregators such as VertNet, iDigBio, and the Global Genome Biodiversity Network (GGBN) can be modified to address the additional growing complication of linkages across specimen parts and data types housed in different collections. Institutions also can use Globally Unique Identifiers, or GUIDs, to provide each specimen and data type with an identifier that is stable as well as unique across institutions (Guralnick et al. 2015, p. 2008). In short, the accessibility

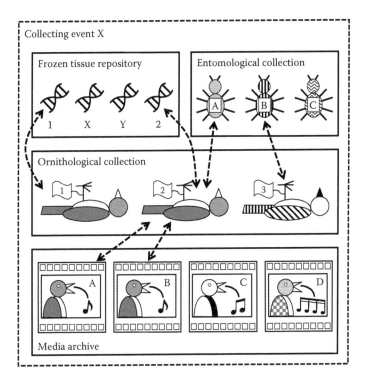

Figure 13.2. Specimens and data gathered during scientific collecting events often include physical specimens and other assets that are deposited in one or more ornithological research collections, as well as specimens and assets that typically are not handled by ornithological collections. This hypothetical example shows three physical specimens (1–3) that were collected during a single collecting event and deposited in an ornithological collection, along with other samples and data that were collected at the same place/time (i.e., the same collecting event) and deposited in other collections. All specimens and samples/data collected during a single collecting event can be connected to each other across databases (dashed arrows), for example through database exchange of specimen identifiers (preferably GUIDs) and collecting event identifiers. In the example illustrated here, tissue/DNA samples were collected from individuals 1 and 2 during specimen preparation and are deposited in a frozen tissue collection, ecto- and endoparasite samples were collected from specimens 2 and 3 and are deposited in an entomological collection, and two audiovisual recordings (i.e., biodiversity media) were recorded from individual 2 before it was collected as a specimen and are deposited at a biodiversity media archive. Accordingly, specimen 2 serves as a physical voucher specimen for a DNA sample (2), a parasite sample (A), and two media recordings (A and B), and other specimens serve as vouchers for some of the other samples/data. Note also that several samples were collected during the same collecting event (i.e., expedition), but from individuals that were not collected as specimens: DNA samples X and Y, parasite sample C, and audiovisual recordings C and D all came from uncollected individuals. Although these samples lack physical voucher specimens, they are nonetheless valuable, and in particular are tied to the other specimens (which can serve as a "population voucher") by being collected at the same time and place.

challenge is not really one of technology, but rather one of implementation and, perhaps, sociology. By working with participating institutions, data aggregators can provide guidelines and best practices, and other community-based mechanisms can be developed to facilitate interinstitutional linkages and relationships. A related issue is that of data quality, as it will be important for different institutions to converge on a common language and set of data values (see Chapter 10, this volume). Thus, more than ever before there is a growing need within the collections community

for mechanisms and expertise associated with data curation (aka "biocuration"; see Howe et al. 2008). This presents both an opportunity and a challenge for the institutions that house ornithological collections, as the only way these databases can be built, maintained, and augmented is with sufficient staffing and relevant expertise. Accordingly, individual institutions as well as the entire collections community will have to invest significant time and resources (and thus fundraising) aimed at enhanced data management. Fundraising for such efforts is challenging, as data management is inherently less enticing to donors than, say, expeditions, but the short- and long-term payoff will be the ability to more effectively and rigorously monitor and study the ecology and evolution of birds into the future.

THE EVOLVING COLLECTING EXPEDITION

For the collector in the field, the expanding list of preparation and data types to be gathered greatly increases the handling time needed per individual collected, the number of hands needed to do the handling, and the overall expense and complexity of an expedition. Additional field time needed for activities outside of specimen preparation (e.g., for audiovisual recording) leads to further substantial and conflicting demands on personnel time. Thus, the exciting array of specimen and data types that can be collected today amplifies the fundamental challenge that has always faced biological collections: so much biodiversity, so little time (and so little funding!). Nonetheless, carefully planned collecting expeditions can bring back a wide diversity of specimens and data, especially if they are conducted as partnerships of fieldworkers having diverse skill sets. For example, collectors with expertise in other taxa (e.g., parasitologists and entomologists) or data collection techniques (e.g., field recordists) could join ornithologists as part of an expedition team. Given the high likelihood of cross-institution partnerships in the future, and notwithstanding logistic difficulties in running such interdisciplinary field trips, collaborative expeditions with experienced field workers who are each collecting and processing the specimens and material that they are best equipped to handle will have the highest productivity, and likely also the highest probability of attracting external funding.

Equally important, the broader collections community as well as individual collectors will need to make decisions regarding the types of specimen and data types that can and should be collected. It is simply not possible to collect all of the specimen and sample/data types that could be collected, because the required field processing time and storage and curation constraints make doing so prohibitive. Tissues for RNA analysis illustrate this: whereas a cryopreserved tissue sample today has a high probability of being able to meet many if not virtually all future research needs requiring extraction of DNA, the same cannot be said

for RNA studies, because gene expression is often tissue-specific. It may or, more likely, may not be feasible to collect and preserve in RNA*later* or other similar buffers (see Gayral et al. 2011) a large number of tissues from every individual collected on an expedition. But it may be feasible for a subset of tissues to be collected and preserved from some or even many of the individuals collected. Thus, for maximum research impact, decisions must be made about which limited set of tissues (e.g., gonads and possibly brains) would be most valuable to collect and preserve and efficiently curate and database. Similar decisions must be made regarding collection of other specimen types (e.g., fluid versus round skin specimens versus spread wings) and from how many individuals representing a population to balance the research value with collection constraints.

Another solution to the challenge of adequately building our ornithological collections, particularly in the face of limited funding and personnel, is to rely more heavily on others to do some (or even much) of the collecting. One strategy that has been used by ornithological collections for decades is to rely on researchers from outside of the museum itself to collect material. For example, it is standard practice for some researchers to collect physical specimens (e.g., study skins and/or skeletons) and deposit those in an appropriate collection. However, this is not the norm in many areas of ornithological research, such as behavior, physiology, and population and community ecology (e.g., Jones et al. 2006). Yet specimens and other materials (e.g., photos, audiovisual recordings, and field notes) from such studies are scientifically valuable, and already are used by other researchers whenever they are archived and accessible. A central communications task before the collections community is to make the broader research community aware of the scientific value of these materials and their associated data, as well as the researcher's obligation to deposit the materials in an appropriate collection and, where feasible, help support curating those collections by including appropriate funds in their grant budgets (see Gardner et al. 2014). The time is ripe for such an effort, as there is growing recognition of the need, and now even mandates from funding agencies and scientific journals (Whitlock 2011), to archive data that are used for published and publicly funded research. What is needed now is a push to extend this recognition and mandate beyond simple data tables and spreadsheets (see Huang and Qiao 2011) to include specimens and biodiversity media. Just as important is the development of tools and workflows that facilitate the ability of researchers to collect and deposit their materials (e.g., uploading of data and/or media directly to collection databases). If the collections community would like to involve other researchers or even the general public (see later) in collecting and depositing materials, then the process needs to be made as simple as possible.

A related and similar strategy is the use of students (Chapter 11, this volume), volunteers, and citizen scientists to collect those materials that require only moderate training. The Internet revolution has created rich opportunities for collection and deposition of data in ornithological and other biodiversity collections, just as it has for accessibility of those data. Internet tools that allow researchers to quickly and easily record metadata associated with their specimens (e.g., GPS coordinates for precise locality and elevation) are being implemented, thereby saving museum staff the time of finding and keystroking those data themselves, the need for checking the data notwithstanding. Such tools are only the tip of the iceberg, however. Equally powerful and exciting are tools that allow citizen scientists with only moderate training to input natural history observations and other biodiversity data that are then incorporated into accessible databases. One prime ornithological example is eBird.org, which is currently receiving and vetting data at a rate of more than 2 million observations per month, which in turn are being used for a multitude of biodiversity studies (Wood et al. 2011, Hochachka et al. 2012, Sullivan et al. 2014). Recently, eBird expanded to allow users to upload digital media (photos and audio files, with video to be added in the near future) along with their observational data, and these media are being deposited directly into the Macaulay Library, a biodiversity collection with expertise in digital media. The media serve as a type of "voucher" for the observational data, and, in turn, are enriched by being associated with metadata that include bird lists describing the avian community where the media were collected. Similar efforts are currently underway to develop biodiversity databases and "virtual museums" with media for numerous taxa, for example, iNaturalist, Xeno-Canto, Wikiaves, and the Encyclopedia of Life. The potential impact of these data on research and conservation is enormous (see Pimm et al. 2014), particularly if properly curated and associated with biological specimens and other data. In some cases, these biodiversity databases are housed at, or closely associated with, more traditional specimen-based collections (e.g., iNaturalist is now part of the California Academy of Sciences, and the Macaulay Library is housed along with the Cornell Museum of Vertebrates at the Cornell Lab of Ornithology). This makes perfect sense, as these citizen-collected data describe patterns of biodiversity and complement specimen collections, and biological collections have the expertise for curating large volumes of data (see Howe et al. 2008).

CONCLUSIONS: BACK TO THE FUTURE

Specimens document biodiversity at a particular place and time. In collections, specimens become part of a time series that allows researchers to study and understand how biodiversity is evolving, both within and across species, and how biodiversity responds to environments that are increasingly affected by human activities. Thus, ornithological collections document the past and the present, but they enable valuable studies far into the future—they are today's data for tomorrow's research.

As our chapter demonstrates, specimen-based ornithological research is undergoing a renaissance, and its future is bright. Emerging technologies have created and will always create exciting new opportunities for the use of both traditional and new specimen types, and developments in biodiversity informatics have made those specimens and their metadata more accessible than ever before. The challenge is to expand the necessary resources and support to create and curate collections broadly defined. This will mean continued development and support for the infrastructure and expertise needed to handle diverse data and specimen types, increased attention to biocuration, and broader collaborative relationships among institutions.

The future of collections rests in the hands of the institutions that house them, but this biological heritage is important for all of humanity. We have focused on the amazing and ever-changing science made possible by ornithological collections, but their value for public engagement (e.g., Macnamara et al. 2008, 2013; Hauber 2014) should not be ignored and cannot be underestimated. The technological developments so important to the changing nature and scope of collections also let us communicate about collections and natural history science. They allow us to reach broad audiences in ways that were not possible before. For example, Cornell's All About Birds website (AllAboutBirds.org), which aims to engage the public and teach about the biology of birds, is visited by more than 14 million unique visitors each year, and the Atlas of Living Australia similarly reaches vast numbers with downloads numbering in the millions. Public support for ornithological collections is essential to ensuring the funding needed for them to evolve (Krishtalka and Humphrey 2000, Bates 2007). There are enormous benefits to be gained by enhancing our ornithological collections, in particular by embracing a broader concept of the "extended specimen" that brings together and preserves a diversity of specimen and data types for the broadest possible uses.

ACKNOWLEDGMENTS

Our views on the future of ornithological collections have been shaped and refined through conversations with colleagues too numerous to list, though not all of them would agree with the opinions expressed here! Thanks in particular to K. Bostwick, R. Bowie, E. Scholes, and D. Winkler. We express the community's debt to the permit-granting agencies that have enabled the collection of specimens and whose support into the future will be essential.

LITERATURE CITED

Austin, J., L. Joseph, L. Pedler, and A. Black. 2013. Uncovering cryptic evolutionary diversity in extant and extinct populations of the southern Australian arid zone Western and Thick-billed Grasswrens (Passeriformes: Maluridae: *Amytornis*). Conservation Genetics 14:1173–1184.

Bain, O., and P. M. Mawson. 1981. On some oviparous nematodes mainly from Australian birds. Records of the South Australian Museum 18:265–284.

Barker, R. D., and W. J. M. Vestjens. 1991. The food of Australian birds. 2. Passeriformes. CSIRO Press, Canberra, ACT, Australia.

Bates, J. M. 2007. Natural history museums: world centers of biodiversity knowledge, now and in the future. Systematist 29:3–6.

Bates, J. M., R. C. K. Bowie, D. E. Willard, G. Voelker, and C. Kahindo. 2004. A need for continued collecting of avian voucher specimens in Africa: why blood is not enough. Ostrich 75:187–191.

Bates, J. M., and G. Voelker. 2015. Scientific collecting in Malawi, a response to Dowsett-Lemaire et al. Bird Conservation International 25:270–279.

Becker, B., and S. Beissinger. 2006. Centennial decline in the trophic level of an endangered seabird after fisheries decline. Conservation Biology 20:470–479.

Beissinger, S. R., and M. Z. Peery. 2007. Reconstructing the historic demography of an endangered seabird. Ecology 88:296–305.

Besnard, G., J. A. M. Bertrand, B. Delahaie, Y. X. C. Bourgeois, E. Huillier, and C. Thebaud. 2016. Valuing museum specimens: high-throughput DNA sequencing on historical collections of New Guinea crowned pigeons (*Goura*). Biological Journal of the Linnean Society 117:71–82.

Block, N. L., S. M. Goodman, S. J. Hackett, J. M. Bates, and M. J. Raherilalao. 2015. Potential merger of ancient lineages in a passerine bird discovered on evidence from host-specific ectoparasites. Ecology and Evolution 5:3743–3755.

Brulez, K., I. Mikšík, C. R. Cooney, M. E. Hauber, P. G. Lovell, G. Maurer, S. J. Portugal, D. Russell, S. J. Reynolds, and P. Cassey. 2016. Eggshell pigment composition covaries with phylogeny but not with life history or with nesting ecology traits of British passerines. Ecology and Evolution 6:1637–1645.

Budney, G. F., W. McQuay, and M. S. Webster. 2014. Transitioning the largest archive of animal sounds from analogue to digital. Journal of Digital Media Management 2:212–220.

Buerki, S., and W. J. Baker. 2016. Collections-based research in the genomic era. Biological Journal of the Linnean Society 117:5–10.

Byrne, M., D. Yeates, L. Joseph, M. Kearney, J. Bowler, M. A. J. Williams, S. Cooper, S. C. Donnellan, J. S. Keogh, R. Leys, J. Melville, D. J. Murphy, N. Porch, and K. H. Wyrwoll. 2008. Birth of a biome: insights into the assembly and maintenance of the Australian arid zone biota. Molecular Ecology 17:4398–4417.

Carnaval, A. C., M. J. Hickerson, C. F. Haddad, M. T. Rodrigues, and C. Moritz. 2009. Stability predicts genetic diversity in the Brazilian Atlantic forest hotspot. Science 323:785–789.

Chesser, R. T. 1995. Comparative diets of obligate ant-following birds at a site in northern Bolivia. Biotropica 27:382–390.

Cheviron, Z. A., M. D. Carling, and R. T. Brumfield. 2011. Effects of postmortem interval and preservation method on RNA isolated from field-preserved avian tissues. Condor 113:483–489.

Clemann, N., K. M. C. Rowe, K. C. Rowe, T. Raadik, M. Gomon, P. Menkhorst, J. Sumner, D. Bray, M. Norman, and J. Melville. 2014. Value and impacts of collecting vertebrate voucher specimens, with guidelines for ethical collection. Memoirs of Museum Victoria 72:141–151.

Collar, N., C. Fisher, and C. Feare (editors). 2003. Why museums matter: avian archives in an age of extinction. Bulletin of the British Ornithologists' Club (Suppl.) 123A:1–360.

DiEuliisa, D., K. R. Johnson, S. S. Morsec, and D. E. Schindel. 2016. Specimen collections should have a much bigger role in infectious disease research and response. Proceedings of the National Academy of Sciences USA 113:1–7.

Edwards, S. V., S. Birks, R. T. Brumfield, and R. Hanner. 2005. Future of avian genetic resources collections: archives of evolutionary and environmental history. Auk 122:979–984.

Foster, M. S., and P. F. Cannell. 1990. Bird specimens and documentation: critical data for a critical resource. Condor 92:277–283.

Gardner, J., T. Amano, W. J. Sutherland, L. Joseph, and A. Peters. 2014. Natural history collections: the end of time series? Frontiers in Ecology and Environment 12:436–438.

Gayral, P., L. Weinert, Y., Chiari, G. Tsagkogeorga, M. Ballenghien, and N. Galtier. 2011. Next-generation sequencing of transcriptomes: a guide to RNA isolation in nonmodel animals. Molecular Ecology Resources 11:650–661.

Gilby, A. J., S. R. Pryke, and S. C. Griffith. 2009. The historical frequency of head-colour morphs in the Gouldian Finch (Erythrura gouldiae). Emu 109:222–229.

Graham, C. H., S. Ferrier, F. Huettman, C. Moritz, and A. T. Peterson. 2004. New developments in museum-based informatics and applications in biodiversity analysis. Trends in Ecology and Evolution 19:497–503.

Green, R. E. 2008. Demographic mechanism of a historical bird population collapse reconstructed using museum specimens. Proceedings of the Royal Society B 275:2381–2387.

Grellet-Tinner, G., N. A. Spooner, and T. H. Worthy. 2016. Is the "Genyornis" egg of a mihirung or another extinct bird from the Australian dreamtime? Quaternary Science Reviews 133:147–164.

Groth, J. G. 1993. Evolutionary differentiation in morphology, vocalizations and allozymes among nomadic sibling species in the North American red crossbill (Loxia curvirostra) complex. University of California Publications in Zoology 127:1–143.

Guralnick, R. P., N. Cellinese, J. Deck, R. Pyle, J. Kunze, L. Penev, R. Walls, G. Hagedorn, D. Agosti, J. Wieczorek, T. Capatano, and R. D. M. Page. 2015. Community next steps for making globally unique identifiers work for biocollections data. Zookeys 494:133–154.

Hauber, M. 2014. The book of eggs. B. Becker and J. M. Bates (editors). University of Chicago Press, Chicago, IL.

Hickey, J. J., and D. W. Anderson. 1968. Chlorinated hydrocarbons and eggshell changes in raptorial and fish-eating birds. Science 162:271–273.

Hochachka, W. H., D. Fink, R. A. Hutchinson, D. Sheldon, W.-K. Wong, and S. Kelling. 2012. Data-intensive science applied to broad-scale citizen science. Trends in Ecology and Evolution 27:130–137.

Holmes, M. W., T. T. Hammond, G. O. U. Wogan, R. E. Walsh, K. LaBarbera, E. A. Wommack, F. M. Martins, J. C. Crawford, K. L. Mack, L. M. Bloch, and M. W. Nachman. 2016. Natural history collections as windows on evolutionary processes. Molecular Ecology 25:864–881.

Howe, D., M. Costanzo, P. Fey, T. Gojobori, L. Hannick, W. Hide, D. P. Hill, R. Kania, M. Schaeffer, S. St Pierre, S. Twigger, O. White, and S. Y. Rhee. 2008. Big data: the future of biocuration. Nature 455:47–50.

Huang, X., and G. Qiao. 2011. Biodiversity databases should gain support from journals. Trends in Ecology and Evolution 26:377–378.

Hykin, S. M., K. Bi, and J. A. McGuire. 2015. Fixing formalin: a method to recover genomic-scale DNA sequence data from formalin-fixed museum specimens using high-throughput sequencing. PLoS One 10:e0141579.

Jetz, W., G. H. Thomas, J. B. Joy, K. Hartmann, and A. O. Mooers. 2012. The global diversity of birds in space and time. Nature 491:444–448.

Johnson, K., D. Clayton, J. Dumbacher, and R. Fleischer. 2010. The flight of the Passenger Pigeon: phylogenetics and biogeographic history of an extinct species. Molecular Phylogenetics and Evolution 57:455–458.

Johnson, K. P., S. M. Shreve, and V. S. Smith. 2012. Repeated adaptive divergence of microhabitat specialization in avian feather lice. BMC Biology 10:52.

Jones, M. B., M. P. Schildhauer, O. J. Reichman, and S. Bowers. 2006. The new bioinformatics: integrating ecological data from the gene to the biosphere. Annual Review of Ecology, Evolution, and Systematics 37:519–544.

Joseph, L. 2011. Museum collections in ornithology: today's record of avian biodiversity for tomorrow's world. Emu 111:i–vii.

Joseph, L., and D. Stockwell. 2000. Temperature-based models of the migration of Swainson's Flycatcher *Myiarchus swainsoni* across South America: a new use for museum specimens of migratory birds. Proceedings of the Academy of Natural Sciences 150:293–300.

Kaplan, S., and J. W. Moyer. [online]. 2015. Moustached Kingfisher: Why a scientist killed a bird that hadn't been seen in half a century. The Independent, 12 October 2015. <http://www.independent.co.uk /environment/nature/moustached-kingfis...illed -a-bird-that-hadnt-been-seen-in-half-a-century -a6690846.html>.

Kearns, A., L. Joseph, A. Toon, and L. Cook. 2014. Australia's arid-adapted butcherbirds experienced range expansions during Pleistocene glacial maxima. Nature Communications 5:3994.

Kearns, A., L. Joseph, L. C. White, J. J. Austin, C. Baker, A. Driskell, J. Malloy, and K. E. Omland. 2016. Norfolk Island Robins are a distinct endangered species: ancient DNA unlocks surprising relationships and phenotypic discordance within the Australo-Pacific Robins. Conservation Genetics 17:321.

Kiff, L. F. 2005. History, present status, and future prospects of avian eggshell collections in North America. Auk 122:994–999.

Krishtalka, L., and P. S. Humphrey. 2000. Can natural history museums capture the future? BioScience 50:611–617.

Lanyon, W. E. 1978. Revision of the *Myiarchus* flycatchers of South America. Bulletin of the American Museum of Natural History 161:429–627.

Lee, J., S. D. Sarre, L. Joseph, and J. Robertson. 2016. Microscopic characteristics of the plumulaceous feathers of Australian birds: a preliminary analysis of taxonomic discrimination for forensic purposes. Australian Journal of Forensic Sciences 48:421–444.

Li, S., B. Li, C. Cheng, Z. Xiong, Q. Liu, J. Lai, H. V. Carey, Q. Zhang, H. Zheng, S. Wei, H. Zhang, L. Chang, S. Liu, S. Zhang, B. Yu, X. Zeng, Y. Hou, W. Nie, Y. Guo, T. Chen, J. Han, J. Wang, J. Wang, C. Chen, J. Liu, P. J. Stambrook, M. Xu, G. Zhang, M. T. Gilbert, H. Yang, E. D. Jarvis, J. Yu, J. Yan. 2014.

Genomic signatures of near-extinction and rebirth of the Crested Ibis and other endangered bird species. Genome Biology 15:557.

Lujan, N. K., and L. M. Page. 2015. Libraries of life. New York Times, 27 February 2015, p. A25.

Macnamara, P., J. M. Bates, and J. H. Boone. 2008. Architecture by birds and insects: a natural art. University of Chicago Press, Chicago, IL.

Macnamara, P., J. Bates, and J. Boone. 2013. The art of migration: birds, insects, and changing seasons in Chicagoland. University of Chicago Press, Chicago, IL.

Marantz, C. A., and J. V. Remsen. 1991. Seasonal distribution of the Slaty Elaenia, a little-known austral migrant of South America. Journal of Field Ornithology 62:162–172.

Mason, N. A., and S. A. Taylor. 2015. Differentially expressed genes match morphology and plumage despite largely homogeneous genomes in a Holarctic songbird. Molecular Ecology 24:3009–3025.

Michener, W. K., and M. B. Jones. 2012. Ecoinformatics: supporting ecology as a data-intensive science. Trends in Ecology and Evolution 27:85–93.

Minteer, B. A., J. P. Collins, K. E. Love, and R. Puschendorf. 2014. Avoiding (re)extinction. Science 344:260–261.

Murphy, S., L. Joseph, A. Burbidge, and J. Austin. 2011. A cryptic and critically endangered species revealed by mitochondrial DNA analyses—the Western Ground Parrot. Conservation Genetics 12:595–600.

Olsen, E. 2015. [online]. Museum specimens find new life online. New York Times, 19 October 2015. <http://www.nytimes.com/2015/10/20/science /putting-museums-samples-of-life-on-the-internet .html?_r=0>.

Page, R. D. 2008. Biodiversity informatics: the challenge of linking data and the role of shared identifiers. Briefings in Bioinformatics 9:345–354.

Parker, P. G., E. L. Buckles, H. Farrington, K. Petren, N. K. Whiteman, R. E. Ricklefs, J. L. Bollmer, and G. Jiménez-Uzcátegui. 2011. 110 Years of Avipoxvirus in the Galapagos Islands. PLoS One 6(1): e15989.

Parr, C. S., R. Guralnick, N. Cellinese, and R. D. Page. 2012. Evolutionary informatics: unifying knowledge about the diversity of life. Trends in Ecology and Evolution 27:94–103.

Pimm, S. L., C. N. Jenkins, R. Abell, T. M. Brooks, J. L. Gittleman, L. N. Joppa, P. H. Raven, C. M. Roberts, and J. O. Sexton. 2014. The biodiversity of species and their rates of extinction, distribution, and protection. Science 344:1246752.

Price, R. D., and K. P. Johnson. 2007. Three new species of chewing lice (Phthiraptera: Ischnocera: Philopteridae) from Australian parrots (Psittaciformes: Psittacidae). Proceedings of the Entomological Society of Washington 109:523–521.

Price, R. D., K. P. Johnson, and R. L. Palma. 2008. A review of the genus Forficuloecus Conci (Phthiraptera: Philopteridae) from parrots (Psittaciformes: Psittacidae), with descriptions of four new species. Zootaxa 1859:49–62.

Prum, R. O. 2006. Anatomy, physics, and evolution of avian structural colors. Pp. 295–353 in G. E. Hill and K. J. McGraw (editors), Bird coloration, volume 1. Harvard University Press, Cambridge, MA.

Pyle, P. 1997. Identification guide to North American birds, part 1. Slate Creek Press, Bolinas, CA.

Pyle, P. 2008. Identification guide to North American birds, part 2. Slate Creek Press, Bolinas, CA.

Remsen, J. V. 1995. The importance of continued collecting of bird specimens to ornithology and bird conservation. Bird Conservation International 5:146–180.

Remsen, J. V. (editor). 1997. Studies in Neotropical ornithology honoring Ted Parker. Ornithological Monographs 48:1–918.

Remsen, J. V., F. G. Stiles, and P. E. Scott. 1986. Frequency of arthropods in stomachs of tropical hummingbirds. Auk 103:436–441.

Ricklefs, R. E., T. Tsunekage, and R. E. Shea. 2011. Annual adult survival in several new world passerine birds based on age ratios in museum collections. Journal of Ornithology 152:481–495.

Rocha, L. A., A. Aleixo, G. Allen, F. Almeda, C. C. Baldwin, M. V. Barclay, J. M. Bates, A. M. Bauer, F. Benzoni, C. M. Berns, M. L. Berumen, D. C. Blackburn, S. Blum, F. Bolaños, R. C. Bowie, R. Britz, R. M. Brown, C. D. Cadena, K. Carpenter, L. M. Ceríaco, P. Chakrabarty, G. Chaves, J. H. Choat, K. D. Clements, B. B. Collette, A. Collins, J. Coyne, J. Cracraft, T. Daniel, M. R. de Carvalho, K. de Queiroz, F. Di Dario, R. Drewes, J. P. Dumbacher, A. Engilis, Jr., M. V. Erdmann, W. Eschmeyer, C. R. Feldman, B. L. Fisher, J. Fjeldså, P. W. Fritsch, J. Fuchs, A. Getahun, A. Gill, M. Gomon, T. Gosliner, G. R. Graves, C. E. Griswold, R. Guralnick, K. Hartel, K. M. Helgen, H. Ho, D. T. Iskandar, T. Iwamoto, Z. Jaafar, H. F. James, D. Johnson, D. Kavanaugh, N. Knowlton, E. Lacey, H. K. Larson, P. Last, J. M. Leis, H. Lessios, J. Liebherr, M. Lowman, D. L. Mahler, V. Mamonekene, K. Matsuura, G. C. Mayer, H. Mays, Jr., J. McCosker, R. W. McDiarmid, J. McGuire, M. J. Miller, R. Mooi, R. D. Mooi, C. Moritz, P. Myers, M. W. Nachman, R. A. Nussbaum, D. Ó. Foighil, L. R. Parenti, J. F. Parham, E. Paul, G. Paulay, J. Pérez-Emán, A. Pérez-Matus, S. Poe, J. Pogonoski, D. L. Rabosky, J. E. Randall, J. D. Reimer, D. R. Robertson, M. O. Rödel, M. T. Rodrigues, P. Roopnarine, L. Rüber, M. J. Ryan, F. Sheldon, G. Shinohara, A. Shor, W. B. Simison, W. F. Smith-Vaniz, V. G. Springer, M. Stiassny, J. G. Tello, C. W. Thompson, T. Trnski, P. Tucker, T. Valqui, M. Vecchione, E. Verheyen, P. C. Wainwright, T. A. Wheeler, W. T. White, K. Will, J. T. Williams, G. Williams, E. O. Wilson, K. Winker, R. Winterbottom, C. C. Witt. 2014. Specimen collection: an essential tool. Science 344:814–815.

Shapiro, B., D. Sibthorpe, A. Rambaut, J. Austin, G. M. Wragg, O. R. P. Bininda-Emonds, P. L. M. Lee, and A. Cooper. 2002. Flight of the Dodo. Science 295:1683.

Smith, T. B., P. P. Marra, M. S. Webster, I. J. Lovette, H. L. Gibbs, R. T. Holmes, K. A. Hobson, and S. Rohwer. 2003. A call for feather sampling. Auk 120:218–221.

Starling, M., R. Heinsohn, A. R. Cockburn, and N. Langmore. 2006. Cryptic gentes revealed in Pallid Cuckoos *Cuculus pallidus* using reflectance spectrophotometry. Proceedings of the Royal Society B 273:1929–1934.

Stavenga, D. G., H. L. Leertouwer, N. J. Marshall, and D. Osorio. 2011. Dramatic colour changes in a bird of paradise caused by uniquely structured breast feather barbules. Proceedings of the Royal Society B 278:2098–2104.

Stoeckle, M., and K. Winker. 2009. A global snapshot of avian tissue collections: state of the enterprise. Auk 126:684–687.

Suarez, A. V., and N. D. Tsutsui. 2004. The value of museum collections for research and society. BioScience 54:66–74.

Sullivan, B. L., J. L. Aycrigg, J. H. Barry, R. E. Bonney, N. Bruns, C. B. Cooper, T. Damoulas, A. A. Dhondt, T. Dietterich, A. Farnsworth, D. Fink, J. W. Fitzpatrick, T. Fredericks, J. Gerbracht, C. Gomes, W. M. Hochachka, M. J. Iliff, C. Lagoze, F. A. La Sorte, M. Merrifield, W. Morris, T. B. Phillips, M. Reynolds, A. D. Rodewald, K. V. Rosenberg, N. M. Trautmann, A. Wiggins, D. W. Winkler, W.-K. Wong, C. L. Wood, J. Yu, S. Kelling 2014. The eBird enterprise: an integrated approach to development and application of citizen science. Biological Conservation 169:31–40.

Vinther J., D. E. G. Briggs, R. O. Prum, and V. Saranathan. 2008. The color of fossil feathers. Biology Letters 4:522–525.

Wandeler, P., P. E. Hoeck, and L. F. Keller. 2007. Back to the future: museum specimens in population genetics. Trends in Ecology and Evolution 22:634–642.

Webster, M. S., and G. F. Budney. 2016. Sound archives and media specimens in the 21st century. Pp. 479–503 in C. H. Brown and T. Riede (editors), Comparative Bioacoustic Methods eBook. Bentham Science Publishers, Oak Park, IL.

Whitlock, M. C. 2011. Data archiving in ecology and evolution: best practices. Trends in Ecology and Evolution 26:61–65.

Winker, K. 2004. Natural history museums in a post-biodiversity era. BioScience 54:455–459.

Winker, K., J. M. Reed, P. Escalante, R. A. Askins, C. Cicero, G. E. Hough, and J. Bates. 2010. The importance, effects, and ethics of bird collecting. Auk 127:690–695.

Wood, C., B. Sullivan, M. Iliff, D. Fink, and S. Kelling. 2011. eBird: engaging birders in science and conservation. PLoS Biology 9:e1001220.

INDEX

Broadbill, Dusky (*Corydon sumatranus*), 188, 189 (table)
Broadbill, Green (*Calyptomena viridis*), 189 (table)
Broadbill, Whitehead's (*Calyptomena whiteheadi*), 189 (table)
Budgerigar (*Melopsittacus undulatus*), 38
Bunting, Indigo (*Passerina cyanea*), 60

C

Calidris alpina, see Dunlin
Calidris melanotos, see Sandpipers, Pectoral
Calyptomena viridis, see Broadbill, Green
Calyptomena whiteheadi, see Broadbill, Whitehead's
Campylorhamphus trochilirostris, 67 (figure)
Capito fitzpatricki, see barbet
carbon isotope, 101
Carduelis carduelis, see Goldfinch, European
Catamblyrhynchus diadema, 63 (figure)
Catharus frantzii, see Nightingale-thrush, Ruddy-capped
Centurus (*Melanerpes*) woodpecker, 112
Ceratophyllus altus ex Campephilus magellanicus, see flea
Ceratopipra, 77 (figure)
Ceratopipra mentalis, 78
Certhiasomus stictolaemus, 67 (figure)
Charitospiza eucosma, 63 (figure)
Cloacotaenia megalops, 172 (figure)
coloration, see pigments and coloration, advanced methods
    for studying
Complete Specimen Package (CSP), 187, 194, 197
computed tomography (CT), 11, 12
computer graphics, 84
Condor, California (*Gymnogyps californianus*), 94
conservation planning, 213
Coot, Eurasian (*Fulica atra*), 172 (figure)
Cornell Expeditions in Field Ornithology (CEFO), 188
Corvus brachyrhynchos, see Crow, American
Corvus corax, see Raven, Common
Corydon sumatranus, see Broadbill, Dusky
Cotingacola lutzae ex Laniocera hyppopyrra, see Louse, Chewing
Coturnix japonica, see Quail, Japanese
courtship phenotype, 82, 85, 86
crickets, North American brown sword-tailed (*Anaxipha* spp.),
    69
Crossbills (*Loxia* spp.), 15
Crow, American (*Corvus brachyrhynchos*), 94 (table)
Crytopipo, 77 (figure)
Cymbirhynchus macrorhynchos, see Broadbill, Black-and-red
CyVerse, 205

D

Darwin Core, 202, 203, 205, 210 (table)
data duality, see biodiversity informatics and data duality
    (global)
DataONE, 205
Digital Accessible Knowledge (DAK), 114, 115 (figure)
digital archiving initiatives, 58
digital media, 78, 186, 187, 197, 227
digital photography, 29–30, 43
Diglossa brunneiventris, 63 (figure)
Diomedea exulans, 131 (figure), 132 (figure)
Distributed Generic Information Retrieval (DiGIR) protocol,
    203
Distributed Information Network for Avian Biodiversity
    Data (NSF), 203
Dixiphia pipra, 77 (figure)

DNA sequencing, 143, 144
Dodo (*Raphus cucullatus*), 150
double-digest RADseq (ddRADseq), 147
Dunlin (*Calidris alpina*), 94 (table)

E

ecological niche, 113, 114, 117, 119
Ectopistes migratorius, see Pigeon, Passenger
egg pigmentation, 40, 41, 42
Egret, Little (*Egretta garzetta*), 172 (figure)
Emberiza citrinella, see Yellowhammer
Empidonax alnorum, see Flycatcher, Alder
energy-dispersive x-ray spectroscopy (EDS), 42
Equus quagga quagga, see quagga
Erythropitta arquata, see Pitta, Blue-banded
Erythropitta ussheri, see Pitta, Black-crowned
Eudocimus ruber, see Ibis, Scarlet
Eupetes macrocerus, see Rail-babbler, Malaysian
Eurylaimus javanicus, see Broadbill, Banded
Eurylaimus ochromalus, see Broadbill, Black-and-yellow
expeditions, 188, see also student-led expeditions
extended specimen, 1–9
    biodiversity media, 5
    definition of, 6
    description, 6–7
    extended phenotype, 2
    ornithology, 1, 2
    philosophy example, 12
    research collections, 1, 4, 6
    specimen concept, 4–6
    "specimen-free" data, 6
    specimens, 1, 2, 6
    traditional specimen, 4

F

field workflow, 161, 168
Finch, Small Ground (*Geospiza fuliginosa*), 27 (figure), 63 (figure)
Finch, Zebra (*Taeniopygia guttata*), 18
FishNet, 203
Flamingo, American (*Phoenicopterus ruber*), 12 (figure), 31 (figure)
flea (*Ceratophyllus altus ex Campephilus magellanicus*), 169 (figure)
flight muscles, 129, 137
fly, hippoboscid (*Icosta Americana ex Accipiter cooperi*), 169 (figure)
Flycatcher, Alder (*Empidonax alnorum*), 59–60
Fulica atra, see Coot, Eurasian

G

gas chromatograph (GC), 92
GenBank, 102, 186
genomic resources, transforming museum specimens into,
    143–156
    ancient DNA, problems with sequence data specific to,
        148–149
    birds, 144, 147
    case studies, 149–150
    DNA extraction, 150–151
    genomes, 144, 150
    natural history collections, 149
    next-generation sequencing, 143, 145–148
    phylogenetics, 148, 151
Geospiza fuliginosa, see Finch, Small Ground
Geothlypis trichas, see Yellowthroat, Common

S

T

U

V

W

wing shape, 128, 129, 139
Wren-Babbler, Striped (*Kenopia striata*), 3 (figure)

X

Xiphocolaptes promeropirhynchus, 67 (figure)
Xiphorhychus triangularis, 134 (figure)

Y

Yellowhammer (*Emberiza citronella*), 60
Yellowthroat, Common (*Geothlypis trichas*), 59

Z

Zosterops senegalensis, see White-eye, African Yellow

STUDIES IN AVIAN BIOLOGY
Series Editor: Kathryn P. Huyvaert
http://americanornithology.org

31. *The Northern Goshawk: A Technical Assessment of its Status, Ecology, and Management.* Morrison, M. L., editor. 2006.

32. *Terrestrial Vertebrates of Tidal Marshes: Evolution, Ecology, and Conservation.* Greenberg, R., J. E. Maldonado, S. Droege, and M. V. McDonald, editors. 2006.

33. *At-Sea Distribution and Abundance of Seabirds off Southern California: A 20-Year Comparison.* Mason, J. W., G. J. McChesney, W. R. McIver, H. R. Carter, J. Y. Takekawa, R. T. Golightly, J. T. Ackerman, D. L. Orthmeyer, W. M. Perry, J. L. Yee, M. O. Pierson, and M. D. McCrary. 2007.

34. *Beyond Mayfield: Measurements of Nest-Survival Data.* Jones, S. L., and G. R. Geupel, editors. 2007.

35. *Foraging Dynamics of Seabirds in the Eastern Tropical Pacific Ocean.* Spear, L. B., D. G. Ainley, and W. A. Walker. 2007.

36. *Status of the Red Knot (Calidris canutus rufa) in the Western Hemisphere.* Niles, L. J., H. P. Sitters, A. D. Dey, P. W. Atkinson, A. J. Baker, K. A. Bennett, R. Carmona, K. E. Clark, N. A. Clark, C. Espoz, P. M. González, B. A. Harrington, D. E. Hernández, K. S. Kalasz, R. G. Lathrop, R. N. Matus, C. D. T. Minton, R. I. G. Morrison, M. K. Peck, W. Pitts, R. A. Robinson, and I. L. Serrano. 2008.

37. *Birds of the US–Mexico Borderland: Distribution, Ecology, and Conservation.* Ruth, J. M., T. Brush, and D. J. Krueper, editors. 2008.

38. *Greater Sage-Grouse: Ecology and Conservation of a Landscape Species and Its Habitats.* Knick, S. T., and J. W. Connelly, editors. 2011.

39. *Ecology, Conservation, and Management of Grouse.* Sandercock, B. K., K. Martin, and G. Segelbacher, editors. 2011.

40. *Population Demography of Northern Spotted Owls.* Forsman, E. D., et al. 2011.

41. *Boreal Birds of North America: A Hemispheric View of Their Conservation Links and Significance.* Wells, J. V., editor. 2011.

42. *Emerging Avian Disease.* Paul, E., editor. 2012.

43. *Video Surveillance of Nesting Birds.* Ribic, C. A., F. R. Thompson, III, and P. J. Pietz, editors. 2012.

44. *Arctic Shorebirds in North America: A Decade of Monitoring.* Bart, J. R., and V. H. Johnston, editors. 2012.

45. *Urban Bird Ecology and Conservation.* Lepczyk, C. A., and P. S. Warren, editors. 2012.

46. *Ecology and Conservation of North American Sea Ducks.* Savard, J.-P. L., D. V. Derksen, D. Esler and J. M. Eadie, editors. 2014.

47. *Phenological Synchrony and Bird Migration: Changing Climate and Seasonal Resources in North America.* Wood E.M. and J.L. Kellermann, editors. 2015.

48. *Ecology and Conservation of Lesser Prairie-Chickens.* Haukos, D.A. and C. Boal, editors. 2016.

49. *Golden-winged Warbler Ecology, Conservation, and Habitat Management.* Streby, H. M., D. E. Andersen, and D. Buehler, editors. 2016.

Milton Keynes UK
Ingram Content Group UK Ltd.
UKHW050007161223
434481UK00014B/137